Triangulation Short-Cut Layouts

THIRD EDITION
Joseph J. Kaberlein

Glencoe Publishing Co, Inc.
17337 Ventura Boulevard
Encino, CA 91316

KABERLEIN SHEET METAL SERIES
- Short Cuts for Round Layouts
- Triangulation Short-Cut Layouts
- Air Conditioning Metal Layout

Copyright © 1973 by Glencoe Publishing Co., Inc.

Printed in the United States of America

All rights reserved. No part of this book may be reproduced or transmitted in any form or by any means, electronic or mechanical, including photocopying, recording, or by any storage and retrieval system without permission in writing from the Publisher.

Glencoe Publishing Co., Inc.
17337 Ventura Boulevard
Encino, California 91316
Collier Macmillan Canada, Ltd.

Library of Congress Catalog Card Number 73–79332
ISBN 0-02-819410-1
6 7 8 9   84 83 82 81

PREFACE

TRIANGULATION SHORT-CUT LAYOUTS is intended for use as a text in vocational, trade, and technical high schools and in teacher training schools where pattern layout is being taught. Each problem is practical and adaptable, drawn to scale with dimensions that are of ample size for metal construction. This book has also been designed to serve as a text and guide for men in the sheet metal field.

A special aid for the reader has been incorporated into this book. Each projection line in each diagram is marked by an arrow to show the direction in which the line should be drawn. This feature will enable anyone—students as well as more experienced men—to determine at a glance the means to obtain the various profile developments without studying the text at any great length.

A course in beginning drawing that embraces geometrical construction and isometric views is a prerequisite for understanding and making use of this text. The student with little practical experience in sheet metal layout will find it best to study the first plate in each section before attempting the more complex layouts that follow.

CONTENTS

Preface iii

Plate

1	Dividing Quarter Circles	2
2	Simple Method for Dividing a Line into Equal Spaces . . .	2
3	Dividing into Four, Eight, or Sixteen Equal Spaces.	4
4	Dividing into Six, Twelve, or Twenty-Four Equal Spaces . . .	4
5	Ellipse Drawn by Compass	6
6	Ellipse Drawn by Use of String	6
7	Illustrating True-Length Triangles for a Square-To-Round . . .	8
8	Square-To-Round.	10
9	Modern Method for Laying Out Patterns for Square-To-Round . . .	12
10	Rectangular-To-Round	14
11	Rectangular-To-Round	16
12	Square-To-Round.	18
13	Welding Offset Seams, Metal 18 Gauge and Lighter	18
14	Four-Piece Square-To-Round	20
15	Rectangular-To-Round with Gable-Pitched Roof Flange	22
16	Rectangular-To-Round, One Side Straight	24
17	Rectangular-To-Round Offset.	26
18	Rectangular-To-Round Double Offset	28
19	Rectangular-To-Round with Pitch at Base	30
20	Square-To-Round with Fan Bracket	32
21	Round Equal-Taper Joint	34
22	Small- and Large-End Pipe Patterns	36
23	Pattern for Taper Joint.	36
24	Round Taper Joint with One Side Straight	38
25	Round Taper Joint with Two Sides Straight	40
26	Elliptical or Oval-to-Round Equal Tapering.	42
27	Elliptical-To-Round with One Side Straight	44
28	Welding Offset Seams, Metals 18 Gauge and Lighter	44
29	Oblong-To-Oblong Transformer on Center	46
30	Oblong-To-Oblong Off Center	48
31	Elbow-Boot Center Taper	50
32	Elbow Boot, One Side Straight	52
33	Offset Elbow Boot.	54
34	Elbow Boot On Center with Slant Throat	56
35	Elbow Boot On Center with Slant Throat	58
36	Elbow Boot, Straight on One Side with Slant Throat	60
37	Rectangular-To-Round with Pitch at the Top	62
38	Rectangular-To-Round, Straight on One Side, with Pitch at the Top . .	64
39	50-Deg. Double-Offset Rectangular-To-Round	66
40	50-Deg. Twisted Double-Offset Rectangular-To-Round	68
41	Square-To-Intersecting Hip on Roof	70
42	Two-Piece Elbow Boot.	72
43	Three-Piece Rectangular-To-Round Elbow	74
44	Three-Piece Rectangular-To-Round Elbow, One-Side Straight . . .	76
45	Three-Piece Oblong-To-Rectangular Elbow	78
46	Three-Piece Oblong-To-Rectangular Elbow	80
47	Rectangular-To-Round with Curved Back	82
48	Rectangular-To-Rectangular Transition Elbow	84
49	Rectangular-To-Rectangular Transition Equal Taper	86

CONTENTS

50	Transition Elbow with Slant Throat	88
51	Transition Elbow with Slant Throat	90
52	Transition Elbow with Slant Throat	92
53	Double-Flare Transition Elbow	94
54	Irregular-Shaped Elbow	96
55	Transition Elbow Irregular-To-Rectangular	98
56	Transition Elbow with Curved Heel and Throat Patterns	100
57	Rectangular-To-Irregular Shaped Transition	102
58	Rectangular-To-Triangular Transition	104
59	Rectangular-To-Irregular-Shaped Transition	106
60	Transformer in One Piece	108
61	Equal-Tapering Transformer in One Piece	110
62	Double-Offset Transformer in One Piece	112
63	Double-Offset Transformer in One Piece	114
64	Double-Offset Transformer in One Piece	116
65	Twisted Transformer	118
66	Twisted Rectangular Offset	120
67	Transformer Quarter Round-To-Rectangular	122
68	Round Taper with Base Mitered 30 Deg.	124
69	Round Taper with Base Mitered 30 Deg.	126
70	Round Tapering Offset	128
71	Oblong-To-Round Offset	130
72	Oblong-To-Round Offset	132
73	Tapering Double Offset	134
74	Rectangular-To-Round Double Offset	136
75	Bullhead Y Branch	138
76	Y Branch with Equal Spread	140
77	Short Method: Y Branch Equal Spread	142
78	Y Branch with Different Spreads	144
79	Y Branch of Different Diameters with Equal Spread	146
80	Short Method: Y Branch of Different Diameters with Equal Spread	148
81	Y Branch with Round Main and Oblong Secondary Branches	150
82	Short Method: Round-To-Oblong Y Branch	152
83	Y Branch with Round Main and Oblong Secondary Branches	154
84	Short Method: Round-To-Oblong Y Branch	156
85	Y Branch with Secondary Branches at 30 Deg.	158
86	Y Branch with Equal Spread	160
87	Y Branch with Horizontal Branches	162
88	Y Branch with Horizontal and Vertical Branches	164
89	Y Branch with Horizontal and Vertical Branches of Different Diameters	166
90	Y Branch with Round Main and Rectangular Branches	168
91	Short Method: Y Branch with Round Main and Rectangular Branches	170
92	Y Branch with Rectangular Main and Round Branches	172
93	Short Method: Y Branch with Rectangular Main and Round Branches	174
94	Y Branch with a Flat Crotch	176
95	Round Y Branch with Center Splitter	178
96	Round Y Branch Flat on One Side	180
97	30-Deg. Y Branch Flat on One Side	182
98	Y Branch, One Side Flat with Horizontal Branches	184
99	Y Branch, One Side Flat with Rectangular Main and Round Branches	186
100	Y Branch, One Side Flat with Round Main and Rectangular Branches	188

CONTENTS

101	Y Branch, One Side Straight with Flat Crotch	190
102	Three-Pronged, Flat Crotch Branch Fitting	192
103	Three-Pronged Branch, 30 Deg. and 45 Deg.	194
104	30-Deg. Tapering T on a Taper Joint	196
105	40-Deg. Tapering T on a Taper Joint, Flat on One Side	198
106	Three-Pronged Branch, Two at 35 Deg.	200
107	Three-Way Y Branch	202
108	Four-Pronged Branch Fitting	204
109	Streamlined Two-Way Branch	206
110	Streamlined Branch Fitting	208
111	Streamlined Three-Way Branch Fitting	210
112	Rectangular 45 Deg. T, Intersecting a Cylinder on a Tangent	212
113	Tapering T Straight on One Side Intersecting Cylinder on a Tangent	214
114	Tapering Offset T Intersecting Cylinder on a Tangent	216
115	Tapering Offset T Straight on One Side, Intersecting Cylinder on a Tangent	218
116	45-Deg. T on a Taper Straight on One Side	220
117	Tapering T Centered on Cylinder, at 45 Deg.	222
118	Tapering T Off Center, at 45 Deg.	224
119	Tapering T Flat on One Side, at 45 Deg.	226
120	Short Method: Round T Intersecting a Taper at 45 Deg.	228
121	Round T at 45 Deg. Intersecting a Taper, One Side Straight	230
122	Two Elbows on a Tapering Base	232
123	Quarter Pattern for Round Elbows	234
124	Tapering-End Elbow	236
125	Center-Tapered Elbow	238
126	Welding Elbow Gore Seams, Metal 18 Gauge and Lighter	238
127	Equal-Tapering Elbow with Center Radius	240
128	Tapering Elbow, with Throat Radius	244
129	Tapering Elbow, with Heel Radius	246
130	Ship's Ventilator with Round Mouth	248
131	Rectangular-To-Round Tapering Elbow	250
132	Round-To-Oblong Tapering Elbow	254
133	Two-Quart Measure	256
134	360-Deg. Spiral Conveyor Chute	258
135	Toolbox with Pitched Cover	260
136	Flat-Cover, Suitcase-Style Toolbox	262
137	Aligning Round Pipes with Draw Band when Arc Welding	264
138	Forming Cone in Roller with the Aid of an Angle Iron	264
139	Conical-Cap Cutout	264
140	Finding the Degree of Cutout for Stack Cap	266
141	Finding Cutout for Stack Cap	266
142	Obtaining Right Angles	268
143	Obtaining the Diameter of the Main Branch on Two- or Three-Pronged Fittings	268
144	Finding the Length of Pipe Between Two Angles	270
145	Bend Allowance	270
146	Metal Lost in Rolling	272
147	Small and Large Ends	272
148	Area of Ellipse—Circumference of Ellipse	274
149	Riveted Lap Seam	274

CONTENTS

150	Grooved Lock Seam	276
151	Hammer-Grooved Lock Seam	276
152	Grooved Seam on a Cylinder Off Center	278
153	Grooved Seam on a Cylinder On Center	278
154	Forming Wire Edge	280
	Thickness of Galvanized Iron	282
	Comparison of Thicknesses of Sheet Iron and Copper	283
	Nonferrous Metal Gauge	284
	Metal Gauge and Rivet Sizes	285
	Measures	286
	Decimals to Fractions	288
	Decimal Equivalents of One Inch	289
	Circumferences and Areas of Circles	290
	Index	293

TRIANGULATION SHORT-CUT LAYOUTS

PLATE 1 DIVIDING QUARTER CIRCLES

When making cylindrical objects, the circle must be divided into as many spaces as may be desired, with all spaces an equal distance apart; or it may be divided into the required number of spaces according to its diameter.

Figure 1 shows how a quarter circle may be divided into three equal spaces, by spanning the dividers to equal the radius of the circle. Use point 1 as a center to strike an arc crossing the large circle at point 3. Then use point 4 as a center, strike an arc at point 2, thus dividing the quarter circle into three equal spaces. Should six equal spaces be required in a quarter circle, then each one of the three spaces should be divided in half.

Figure 2 shows how a quarter circle may be divided into four equal spaces. Set the dividers to any radius desired, then use point 1 as a center, striking an arc toward point D. Keep the dividers set to the radius, use point 5 as a center, and strike an arc crossing the first arc drawn at point D. Draw a straight line from point D to point A, crossing the circle to obtain point 3. Again set the dividers to any radius desired. Draw arcs from points 1 and 3, 3 and 5, to cross each other at points E and F. Draw straight lines from points E and F toward point A, obtaining points 2 and 4 which will divide the circle into four equal spaces.

Should eight equal spaces be required in a quarter circle, then each one of the four spaces is divided in half.

PLATE 2 SIMPLE METHOD FOR DIVIDING A LINE INTO EQUAL SPACES

Line A to B may be any length desired. Line A to C may be any length that will represent a number of equal spaces that are divided by even inches on the ruler.

EXAMPLE: Line A to B is equal to $54\frac{7}{16}$ in. and should be divided into nine equal spaces. Then 7 is the nearest number that will allow nine equal spaces on the ruler without encountering fractions. Thus, $7 \times 9 = 63$, with one end of the ruler set on point A. The other end of the ruler then is moved up or down until number 63 on the ruler crosses line $B-C$ at point C. Now, with the aid of a pencil or scratch awl, place a mark from A to C at every 7 in. such as from 7 to 56 in., as shown. Draw a straight line down from each division point on line $A-C$ to intersect line $A-B$. This will give the required number of equal spaces on line $A-B$ without the use of the dividers. This method of dividing a line is fast and accurate.

PLATE 1

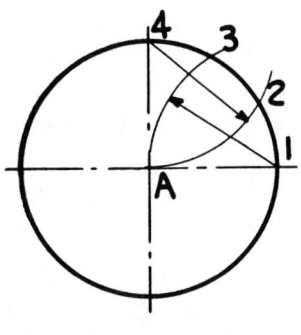

1 THREE SPACES

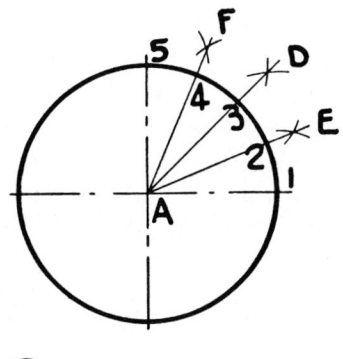

2 FOUR SPACES

PLATE 2

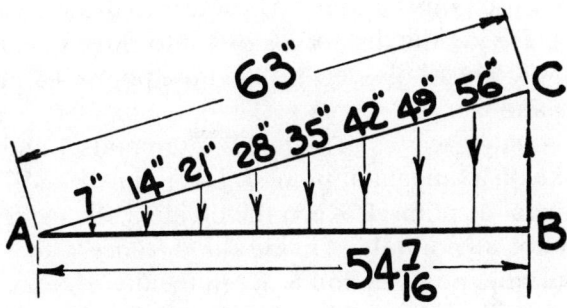

PLATE 3   DIVIDING INTO FOUR, EIGHT, OR SIXTEEN EQUAL SPACES

Linear measurements or circumferences are divided into four, eight, or sixteen, and six, twelve, or twenty-four equal spaces to represent rivet or bolt holes for joining to another object.

To divide a line into four equal spaces as in Figure 1, set the dividers by trial and error until they span half the distance 1 to 5. Then use points 1 and 5 as centers. Draw arcs to strike each other on a tangent at point 3. Span the dividers to equal half the distance 1 to 3. Use points 1 and 3 as centers. Draw arcs to strike each other on a tangent at point 2. Keep the dividers set and use point 3 as a center, striking an arc at point 4. The result will be four equal spaces.

To divide a line into eight equal spaces as in Figure 2, span the dividers to equal half the distance 1 to 9. Use points 1 and 9 as centers. Draw arcs to strike each other on a tangent at point 5. Span the dividers to equal half the distance 1 to 5. Use points 1 and 5 as centers to draw arcs striking each other on a tangent at point 3. Keep the dividers set, and using 5 as a center, strike an arc at point 7. Span the dividers to equal half the distance 3 to 5. Draw arcs from 3 and 5 to strike on a tangent at point 4. Keep the dividers set, and strike arcs at points 2, 6, and 8 from point 7.

PLATE 4   DIVIDING INTO SIX, TWELVE, OR TWENTY-FOUR EQUAL SPACES

To divide a line into six equal spaces as in Figure 1, span the dividers to equal half the distance 1 to 7, and use points 1 and 7 as centers to draw arcs to strike on a tangent at point 4. Divide the distance 1 to 4 into three equal spaces resulting in points 2 and 3. Keep the dividers set, and use points 4 and 7 as centers to strike an arc at 5 and 6.

To divide a line into twelve equal spaces as in Figure 2, use points 1 and 13 as centers to draw arcs to strike on a tangent at point 7. Use points 1 and 7 as centers to draw arcs on a tangent at point 4. Keep the dividers set, using point 7 as a center to strike an arc at point 10. Divide the distance 4 to 7 into three equal spaces, thus obtaining points 5 and 6. Keep the dividers set, using points 1 and 4 as centers, to draw arcs at points 2 and 3. Also use points 7, 10, and 13 as centers to draw arcs at points 8, 9, 11, and 12.

To divide a line into twenty-four equal spaces, use points 1 and 25 as centers to draw arcs to strike on a tangent at point 13, as in Figure 3. Use points 1 and 13 as centers to draw arcs on a tangent at point 7. Keep the dividers set, using point 13 as a center to strike an arc at 19. Use points 7 and 13 as centers to strike arcs on a tangent at point 10. Keep the dividers set, and use points 7, 13, and 19 as centers to strike arcs at points 4, 16, and 22. Divide the distance 10 to 13 into three equal spaces.

PLATE 3

To divide a line into sixteen equal spaces as in Figure 3, use the above procedure to obtain the division points. Use points 1 and 17 as centers to strike arcs on a tangent at point 9. Use 1 and 9 as centers to strike on a tangent at point 5. Keep the dividers set to strike an arc at 13. Use 5 and 9 as centers to strike on a tangent at point 7. Keep the dividers set to strike an arc at 11 and 3; also use point 13 as a center to strike an arc at point 15. Use 9 and 7 as centers to strike on a tangent at point 8. Keep the dividers set to strike an arc at 6; also use point 3 as a center to strike arcs at 4 and 2. Use points 11 and 15 as centers to strike arcs at 10 and 12, 14 and 16.

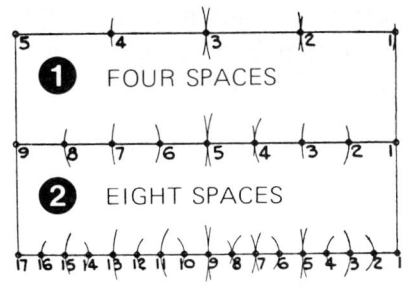

1 FOUR SPACES

2 EIGHT SPACES

3 SIXTEEN SPACES

PLATE 4

Keep the dividers set, and use points 10, 7, 4, and 1 as centers to strike arcs at points 9, 8, 6, 5, 3, and 2. Also use points 13, 16, 19, 22, and 25 as centers to strike arcs at points 14, 15, 17, 18, 20, 22, and 23.

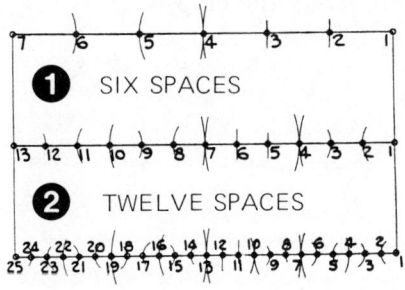

1 SIX SPACES

2 TWELVE SPACES

3 TWENTY-FOUR SPACES

PLATE 5 ELLIPSE DRAWN BY COMPASS

To draw an ellipse to any dimension, the center lines referred to as the major and minor axis must be drawn first.

To construct an ellipse as in Plate 5, draw the major axis A to B and the minor axis C to D equal to the dimensions shown. Draw the diagonal line D to B; then use point B as a center to draw an arc from point 1 intersecting the diagonal line at point 2. Use point 1 as a center to draw an arc from point D intersecting the diagonal line at point 3. Use the distance 2 to 3 as a radius, and points A and B as centers to draw arcs at E and F. Keep the dividers set, and use point E as a center to draw an arc from A to G crossing the arc drawn from E. Use point F as a center to draw an arc from B to H crossing the arc drawn from F. Bisect the distance $D-G$ and draw a straight line through the bisecting arcs to intersect the center line at point J. Use point J as a center to draw the arc G, D, H. The same radius length will be used to draw the arc G, C, H.

PLATE 6 ELLIPSE DRAWN BY USE OF STRING

To draw an ellipse by use of string to construct the curves, draw the major axis A to B, and the minor axis C to D, equal to the dimensions shown. Use one half of the major axis as a radius; then use point C as a center to draw an arc at E and F. Stick a sharp pin in points E and F, fasten a piece of string or fine wire to the pins, the length to equal the distance $E-C-F$. This string should be long enough, when pulled tight, to allow the center of the pencil point to rest on the center of point C. Swing the pencil (inside of the string), drawing an arc from C to A and C to B. Repeat the procedure to draw the arc from D to A and D to B completing the ellipse.

PLATE 5

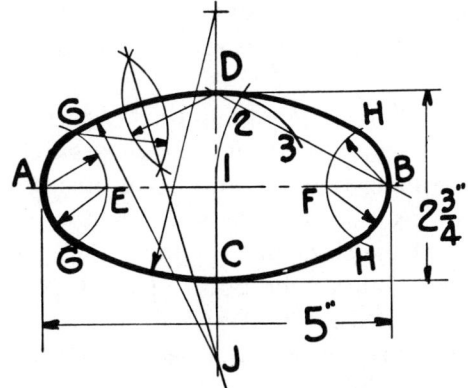

PLATE 6

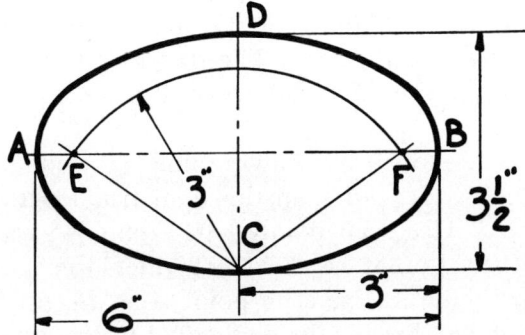

PLATE 7   ILLUSTRATING TRUE-LENGTH TRIANGLES FOR A SQUARE-TO-ROUND

Triangulation is the means of finding the length of the slant side of a right triangle. This slant line is called the true-length line whenever a pattern is developed by means of triangulation.

To illustrate a practical example as to what is meant by the term *true-length line* in the language of sheet-metal workers, take the 45-deg. triangle that you have in your drawing kit and set this triangle in an upright position so that one leg will rest flat on your drawing board, and the other leg will point up in a vertical position. Now look at the slant line (or the slant edge of your triangle) from the point at the base line to the point at the top of the vertical leg of the triangle. This slant line is referred to by sheet-metal workers as the true-length line whenever a pattern is developed by the use of triangulation. However, an engineer may refer to this slant line as the hypotenuse of a right triangle.

Lines *A* to 1*B*, *A* to 2*B*, *A* to 3*B*, *A* to 4*B*, and *A* to 5*B* in the top view, Figure 1, represent the distance that each point is from the corner point *A* when looking down into the object from the top. This is also illustrated in the isometric view in Figure 2. The distances 1 to *B*, 2 to *B*, and 3 to *B* represent the height, and *A* to *B* represents the base of the triangle. The slant lines *A* to 1, *A* to 2, and *A* to 3 represent the true slant length lines.

To obtain the true-length lines to lay out a pattern as in Figure 4, erect a triangle as in Figure 3. Draw the base line *A* to *A'* to any length desired. The height is drawn to equal the desired height of the fitting as shown in the isometric view.

The distances *A* to 1*B*, *A* to 2*B*, *A* to 3*B*, and *C* to 1*D* are transferred from the top view to the base line *A–A'* in Figure 3. The true-length lines for the pattern are the slant lines from points *E* to 1*B'*, *E* to 4'–2', *E* to 3', and *E* to 1*D'* in Figure 3.

The pattern in Figure 4 may be developed by drawing line *A–A* equal to the dimensions as shown. Set the dividers to span the slant true-length line from point *E* to 5*B'* in Figure 3. Use each point *A* in Figure 4 as a center to strike arcs to cross each other at point 5*B'*. Set the dividers to span the slant true-length line *E* to 4' in Figure 3. Use each point *A* as a center to strike an arc on each side of point 5*B'*. Set the dividers equal to any one of the spaces on the circle in Figure 1. Use point 5*B'* in Figure 4 as a center to draw an arc to cross the arcs drawn on each side of point 5*B'* establishing points 4'. Keep the dividers set and use points 4' as centers to draw an arc at one side of point 4'. Set the dividers to span the slant true-length line *E* to 3' in Figure 3. Use each point *A* as a center to strike an arc crossing the arc drawn at one side of point 4' establishing point 3'.

This procedure is continued until point 1*D'* has been obtained. Use the slant length *E* to 1*D'* in Figure 3 as a radius, then use point 1*D'* in Figure 4 as a center to strike an arc near point *C*. Use the distance *A* to *C* in Figure 1 as a radius, using point *A* in Figure 4 as a center to strike an arc crossing the arc to establish point *C*.

No seams or allowances are shown; this will be illustrated in the plates to follow.

PLATE 7

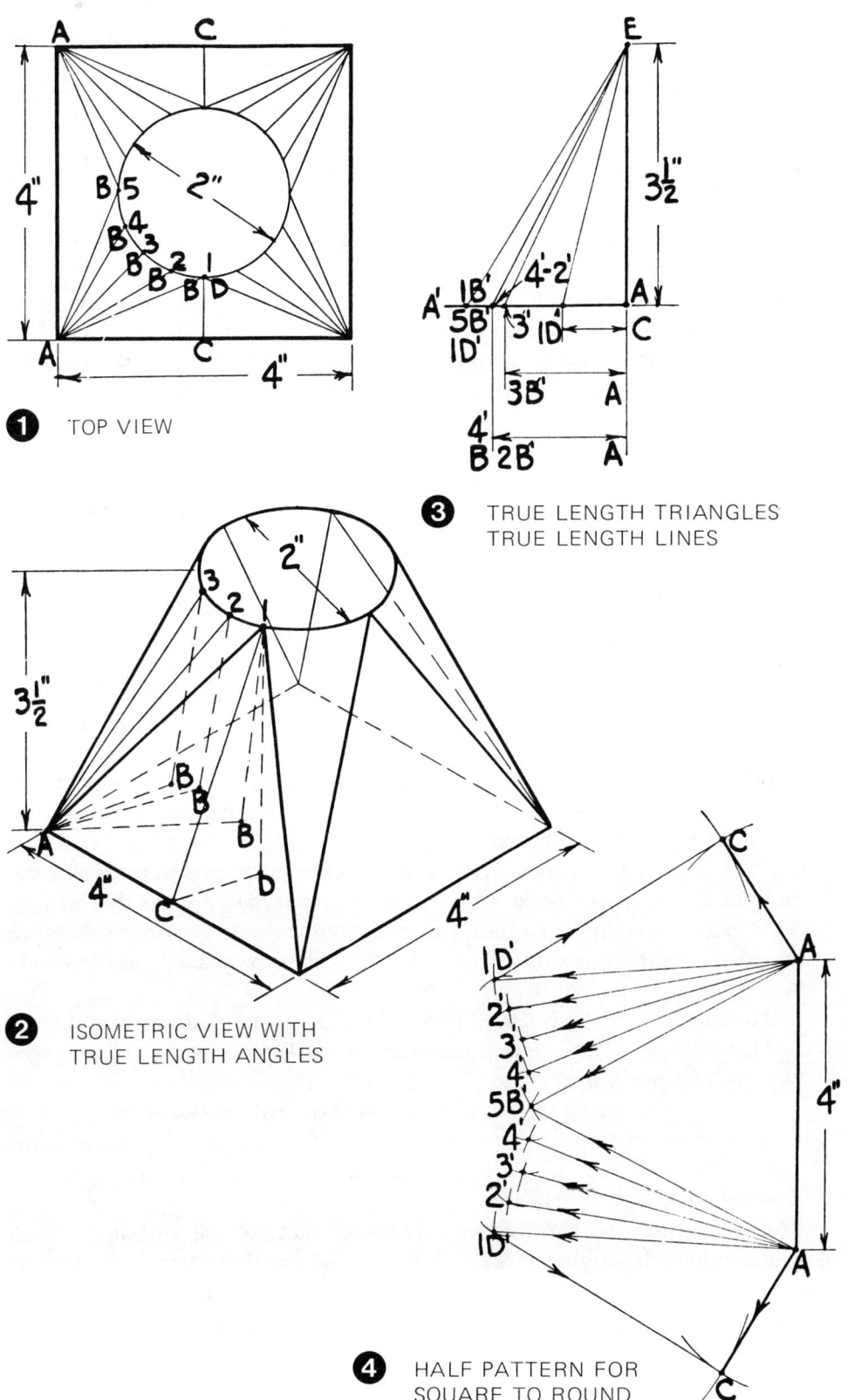

PLATE 8   SQUARE-TO-ROUND

This plate illustrates a square-to-round with all the patterns complete with allowances for riveting.

The half pattern in Figure 4 may be laid out using the same method as in Plate 7. The rivet holes on the top should be placed at every second space as shown at 1′, 3′, and 5′.

The length of the collar in Figure 5 is obtained by multiplying the diameter of the round opening in Figure 1 by 3.1416. Thus $2\frac{1}{4}$ or $2.25 \times 3.1416 = 7.068$ or $7\frac{1}{16}$ in. This length is then divided in half by striking an arc, using each end point 1 as a center to intersect each other on a tangent at the center point 1. Use the center point 1 and the end point 1 as centers to strike arcs on a tangent at point 5. Use points 5 and 1 as centers to strike arcs on a tangent at point 3, thereby spacing the rivet holes equally apart to match the holes on the square-to-round patterns when assembling.

The collar should be flanged to meet the taper of the square-to-round. This flange should begin at a height above the riveting holes at least equal to the riveting-edge allowance on the square-to-round. This will allow the collar to fit over the square-to-round with the rivet holes aligning.

NOTE: It is not practical or necessary in industry to set all the various lengths of the base line of the true-length triangle, and number each point as shown in Figure 3. But it is advisable for the beginning student to follow this procedure and number each point until he is proficient enough to avoid any errors. Therefore you will notice that throughout this book this numbering procedure is followed to illustrate to the student the method for obtaining the true-length lines.

In practical work for industry, as soon as a length is set on the base line of the triangle, the true slant length is obtained and immediately transferred to its respective place on the pattern.

On large fittings it may be advisable to number the points to recheck true-length lines when the proper contour of the pattern may appear doubtful.

NOTE: Any markings on a pattern with crayon, chalk, or other substance for purposes of identification, size, and so on, shall be placed on the inside of the pattern for the following reasons:
1. The exterior of the finished job will have a neat, clean appearance.
2. Many job sites require that the exterior of all pipes and fittings be free from any markings.
3. All of the better shops require this as standard procedure.
4. Most important, doing this habitually will readily permit you to identify the inside of the pattern and, therefore, eliminate many errors when forming the patterns.

You will notice that throughout this book, many of the kink lines, bend lines, and other identifying markings are placed on the outside of the patterns. This was done purposely to match the formation of the pattern with the top and bottom views, or with the isometric view to illustrate the manner in which the patterns are formed and assembled.

PLATE 8

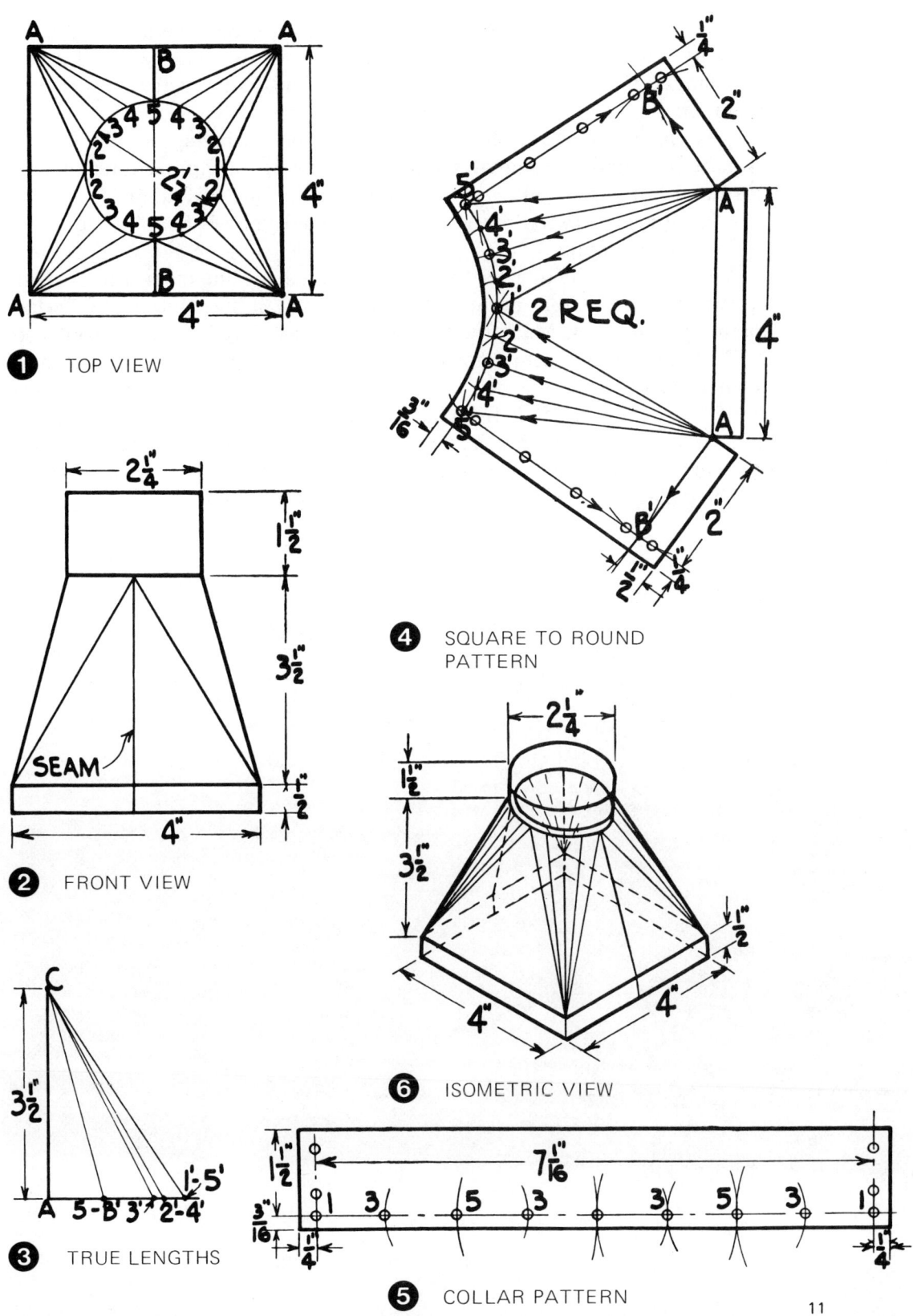

PLATE 9 MODERN METHOD FOR LAYING OUT PATTERNS FOR SQUARE-TO-ROUND

The two previous plates illustrated the practical procedure that may be followed by novice sheet-metal workers who are not thoroughly familiar with the triangulation procedures for developing the patterns for a square-to-round, etc.

This plate illustrates a short, modern method for laying out patterns for a square-to-round. It eliminates some of the elementary procedures that are illustrated in the previous plates, thereby saving time and labor.

It is not necessary to draw a full profile as in Figure 1, but only a quarter-top view as in Figure 3.

After drawing the quarter-top view, draw a slant line from A to C crossing the quarter circle at point 3. Divide the remaining circle 1 to 3 in half obtaining point 2–4. Mark the height of the true-length triangle from A to D. Use point A as a center to draw the arcs from points 5–1, 2–4, and 3 to intersect the base line A–A obtaining points 1'–5', 2'–4', and 3'; also transfer the distance 5–B from line C–B to the base line A–A obtaining point 5'–B'. The slant lengths from these points on the base line to point D will represent the true-length lines to lay out the pattern as in Figure 4.

The collar in Figure 5 is equal to the diameter times 3.1416.

PLATE 9

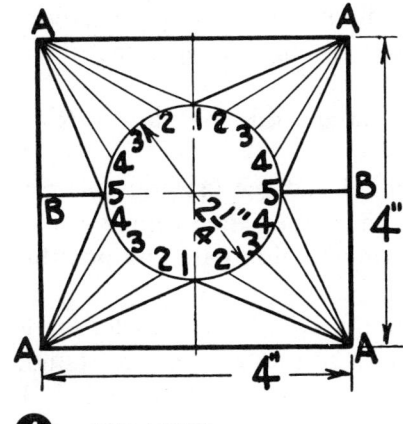

① TOP VIEW

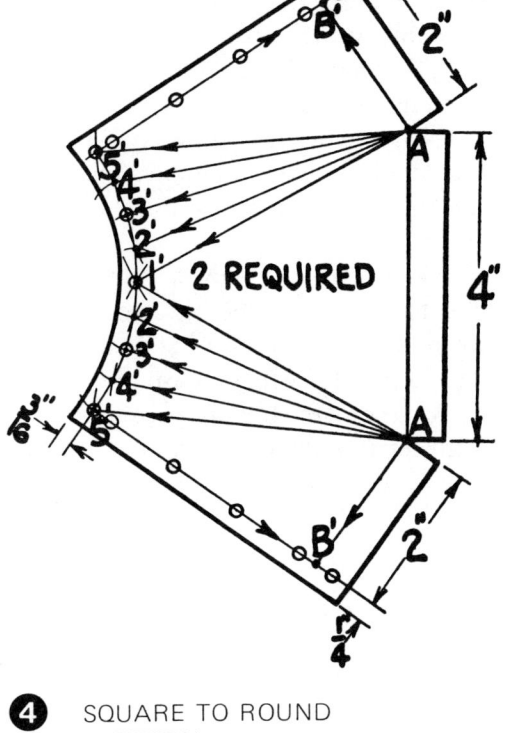

④ SQUARE TO ROUND PATTERN

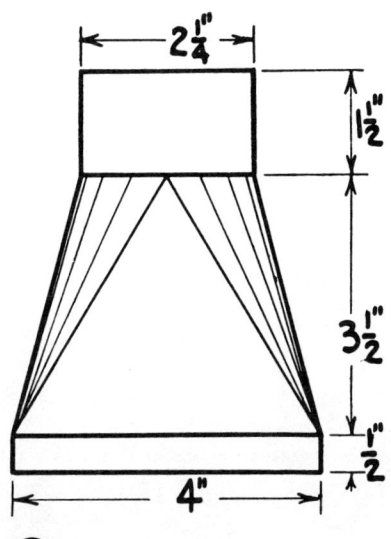

② FRONT VIEW

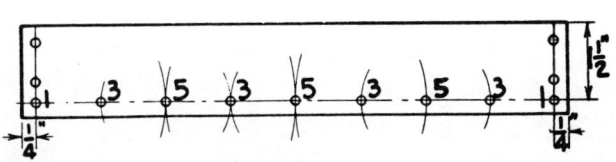

⑤ COLLAR PATTERN

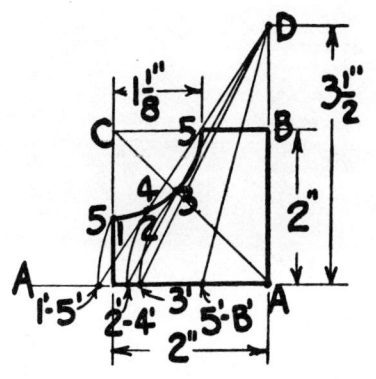

③ 1/4 TOP VIEW

13

PLATE 10 RECTANGULAR-TO-ROUND

The previous plate illustrated a short, modern method that may be followed in developing the pattern for a square-to-round by eliminating some of the elementary procedures.

This plate illustrates a similar method for laying out the pattern for a rectangular-to-round.

NOTE: The only difference between this plate and the previous plate is that the base is rectangular. The procedure is the same. The average novice should follow this procedure with a little caution to eliminate any possible errors.

Draw only the quarter-top view as in Figure 3. Divide the quarter circle into four equal spaces as 1 to 5. Mark the height of the true-length triangle from A to C. Use point A as a center to draw an arc from points 1, 2, 3, 4, and 5, to intersect the base line $A-A$, thus obtaining points $1'$, $2'$, $3'$, $4'$, and $5'$. Also transfer the distance $5-B$ to the base line obtaining point $5'-B'$. The slant lengths from these points on the base line to point C will represent the true-length lines to lay out the pattern in Figure 4.

PLATE 10

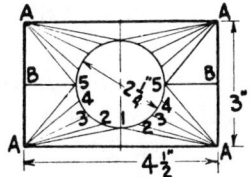

① TOP VIEW

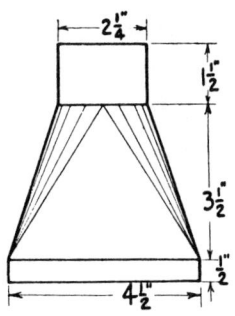

② FRONT VIEW

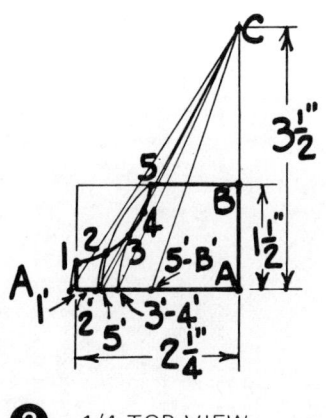

③ 1/4 TOP VIEW

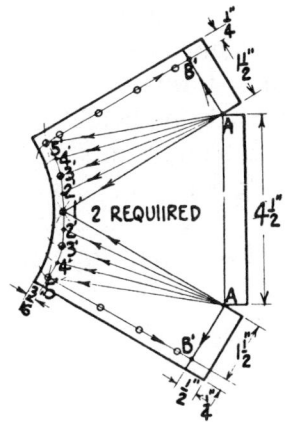

④ RECTANGULAR TO ROUND PATTERN

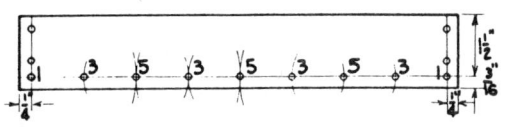

⑤ COLLAR PATTERN

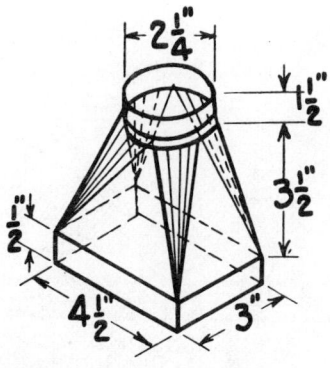

⑥ ISOMETRIC VIEW

PLATE 11   RECTANGULAR-TO-ROUND

This plate illustrates how the short method may be used on a rectangular-to-round with the diameter larger than the rectangle.

Draw the quarter-top view as in Figure 3. Divide the quarter circle into equal spaces and mark the height of the triangle A to C. Use point A as a center to draw arcs from points 1, 2, 3, 4, and 5 to intersect the base line $A-A$, thus obtaining points $1'$, $2'$, $3'$, $4'$, and $5'$. Also transfer $5-B$ to the base line obtaining point $5'-B'$. The slant lengths from these points on the base line to point C will represent the true-length lines to lay out the pattern in Figure 4.

PLATE 11

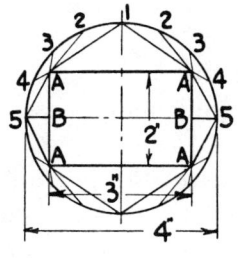

1 TOP VIEW

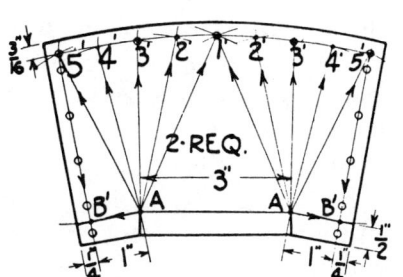

4 RECTANGULAR TO ROUND

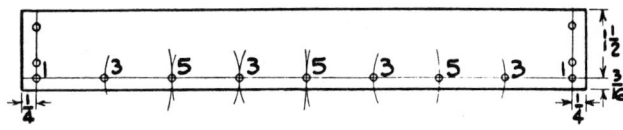

5 COLLAR PATTERN

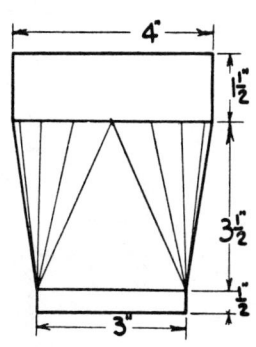

2 FRONT VIEW

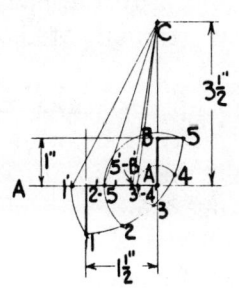

3 1/4 TOP VIEW

PLATE 12   SQUARE-TO-ROUND

This plate illustrates the half pattern with the seams on the corners. This method has its advantages in many cases. When assembling, the rivet at the bottom corner should be inserted and flattened; then the patterns should be pulled into position so that the remaining holes will align. This procedure must be followed for each side.

To lay out the pattern, follow the same procedure as in the previous plates until point 1 has been established; then use point 1 as a center and the distance 1 to B in Figure 4 as a radius to strike an arc to cross the arc at A'. Points 2 and 3 are obtained by arcs drawn with A' as a center.

PLATE 13   WELDING OFFSET SEAMS,
METAL 18 GAUGE AND LIGHTER

This plate illustrates the welding of center seams on the square to round, and the round collar to the square to round. This will allow overlapping of metal at the seams with a smooth surface and greater strength at the seams.

When the patterns are laid out, allow $\frac{1}{8}$ inch at the top edge, and at each end for center seam on square to round, also allow $\frac{1}{8}$ inch at the bottom edge and end of collar. When duct continues from collar, then also allow $\frac{1}{8}$ inch at top of collar. Use burring machine and set gauge back $\frac{5}{16}$ inch to offset one edge at each center seam of square to round and collar at about depth of bottom roll on machine, form patterns, then weld center seams. Next, use the burring machine with the gauge set back $\frac{5}{16}$ inch to offset or shrink the top edge of square to round, and also to flange out bottom of collar. This will allow the offset or shrunken end to slide into flanged collar.

PLATE 12

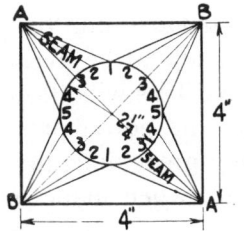

1 TOP VIEW

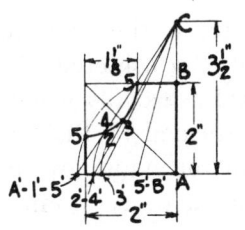

3 1/4 TOP VIEW

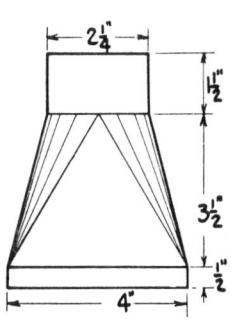

2 FRONT VIEW

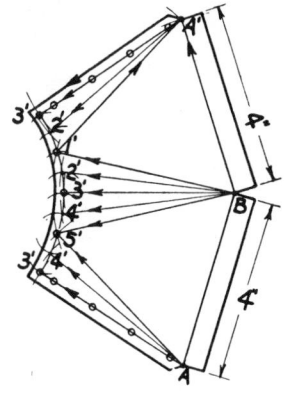

4 HALF PATTERN

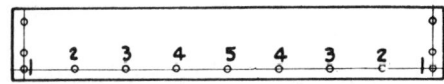

5 COLLAR PATTERN

PLATE 13

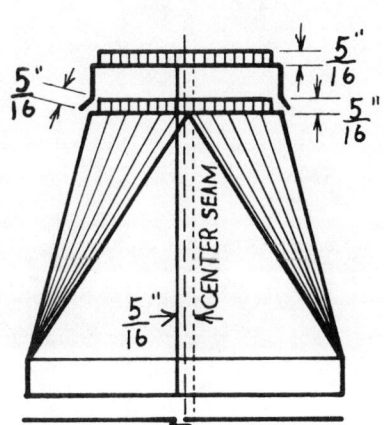

PLATE 14 FOUR-PIECE SQUARE-TO-ROUND

The method illustrated in this plate for laying out the square-to-round patterns in four pieces is considered ideal when the patterns are to be made with heavy metal and especially when they are to be welded. This will allow the patterns to be formed to the desired shape either in the power-press forming brake or in the hand forming brake without any great difficulty. This will also prove to be economical in the use of metal when made in quantities.

The method for laying out the patterns is the same as for previous plates and with the use of the quarter-top view.

PLATE 14

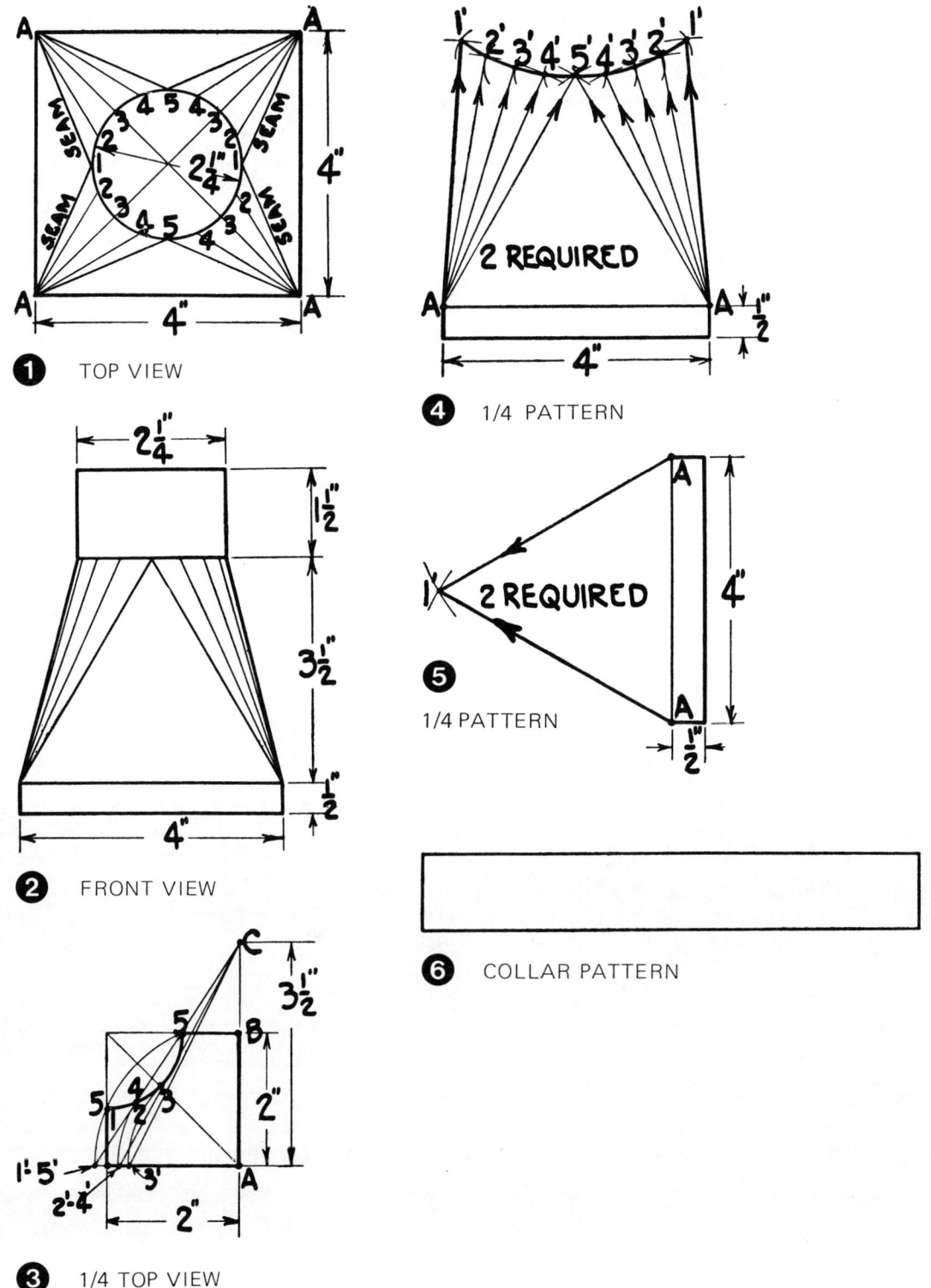

❶ TOP VIEW

❷ FRONT VIEW

❸ 1/4 TOP VIEW

❹ 1/4 PATTERN

❺ 1/4 PATTERN

❻ COLLAR PATTERN

PLATE 15 RECTANGULAR-TO-ROUND WITH GABLE-PITCHED ROOF FLANGE

The procedure for laying out the pattern in Figure 4 is identical to the procedure in the previous plate, except for obtaining the base point C' which must conform to the pitch of the roof.

Draw the quarter-top view as in Figure 3 and establish the points for the true-length lines at the base line $A-5'$. The distance $5-B$ in Figure 1 represents one half of the difference of the 2-in. base width, and the $3\frac{1}{2}$-in. diameter. (In this problem the difference is 1 in.) Transfer this distance to the base line $A-5'$ in Figure 3, establishing point $5-B'$; draw line A to E, representing the height of the true-length triangle, equal to the height in Figure 2.

Lay out the half pattern in Figure 4 until points $5'$ and B' have been obtained. NOTE: To obtain point C' in Figure 4, transfer the height $E-D$ from Figure 2 to line $E-A$ in Figure 3, establishing point D. Now draw a horizontal line across from point D to intersect with the slant true-length line drawn from the base-line point $5'-B'$ to point E, thereby establishing point C'. To complete the pattern in Figure 4, transfer the true length from point E to C' in Figure 3 to line $5'-B'$ to represent the distance $5'$ to point C' in Figure 4.

To lay out the roof flange as in Figure 5, draw a horizontal center line equal to the length of the slant pitch lines $A-D-A$ in Figure 2. Then draw a vertical center line across the horizontal line equal to the $3\frac{1}{2}$-in. diameter. Transfer the distance $C'-D$ from Figure 3 to each side of the vertical center line in Figure 5, as represented by $C'-D$, establishing point C' and obtaining the width C' to C', the cutout for the roof flange. The distance A to A is equal to 2 in., the width of the end of the rectangular base in Figure 1. Allow a lap on the inside of the roof flange opening to be turned up to fasten the flange to the rectangular base.

NOTE: The distance A to C' on the roof flange in Figure 5 is equal to A to C' in Figure 4.

PLATE 15

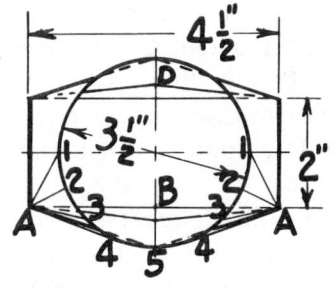

① TOP VIEW

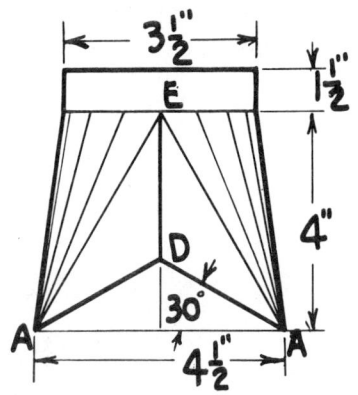

② FRONT VIEW

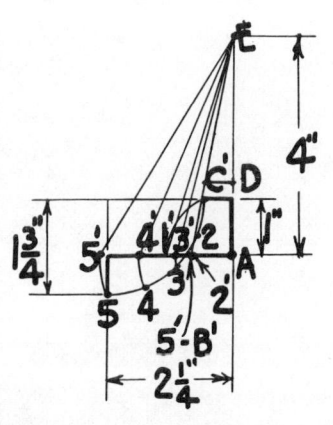

③ 1/4 TOP VIEW

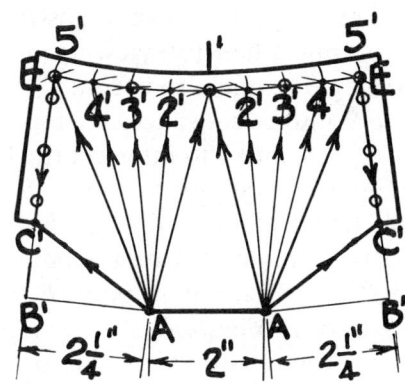

④ HALF PATTERN

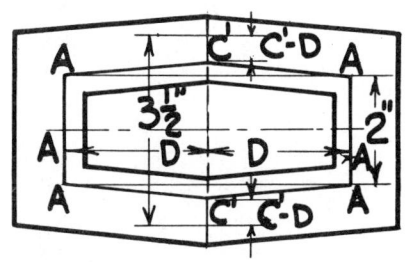

⑤ ROOF FLANGE

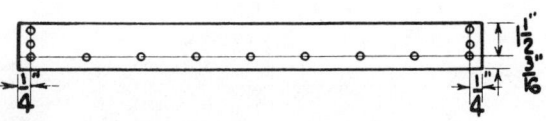

⑥ COLLAR PATTERN

PLATE 16 RECTANGULAR-TO-ROUND, ONE SIDE STRAIGHT

Since this rectangular-to-round is straight on one side, it may be more convenient to draw a half-top view as in Figure 3. Mark the height of the true-length triangle at each end from C to E and B to E, and divide the half circle 1 to 9 into equal spaces. Use point B as a center to draw an arc from points 1, 2, 3, 4, and 5 to intersect the base line $C-B$. Use point C as a center to draw an arc from points 5, 6, 7, 8, and 9 to intersect the base line $C-B$. The slant lengths from points 1', 2', 3', 4', and 5' to point E represent the true lengths from point B on the pattern in Figure 4. The slant lengths from points 5', 6', 7', 8', and 9' to point E represent the true lengths from point C on the pattern in Figure 4.

NOTE: A quarter-top view may be drawn if desired and only if the reader is thoroughly familiar with this short method.

To obtain point B, draw a line from point 9 parallel to line $C-E$ to intersect the base line $C-B$ and equal to the height of line $C-E$. Point B will then be used as a center to draw arcs from points 5, 6, 7, 8, and 9 to represent the true-length lines from point B in Figure 4.

PLATE 16

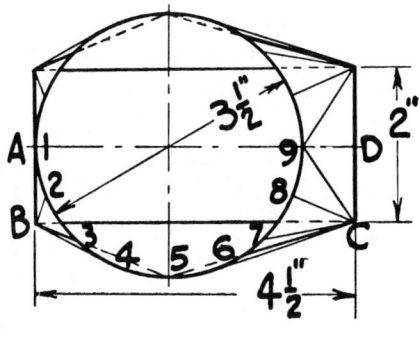

1 TOP VIEW

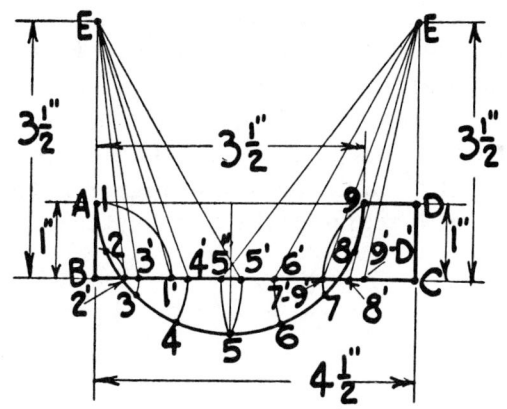

3 HALF TOP VIEW

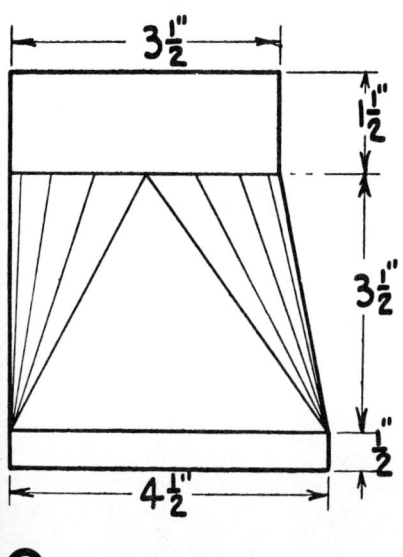

2 FRONT VIEW

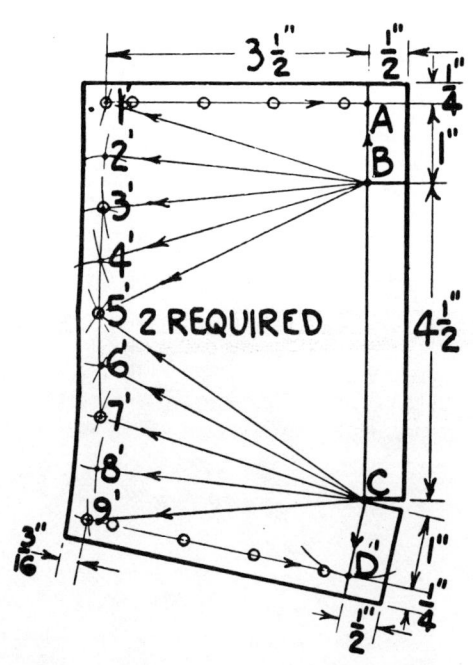

4 RECTANGULAR TO ROUND PATTERN

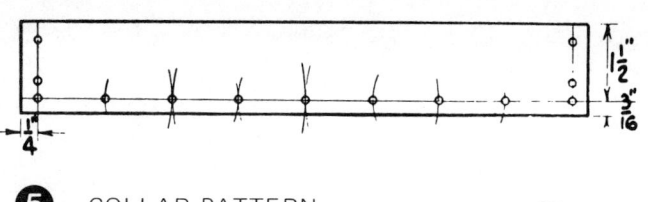

5 COLLAR PATTERN

PLATE 17 RECTANGULAR-TO-ROUND OFFSET

Draw a half-top view as in Figure 3. Mark the height of the true-length triangles at each end from *B* to *E* and *C* to *E*. Use point *C* as a center to draw arcs from points 1, 2, 3, 4, and 5. Use point *B* as a center to draw arcs from points 5, 6, 7, 8, and 9 to intersect the base line *C–B*. The slant lengths from points 1', 2', 3', 4', and 5' to point *E* represent the true lengths from point *C* on the pattern in Figure 4. The slant lengths 5", 6', 7', 8', and 9' to point *E* represent the true lengths from point *B* on the pattern in Figure 4.

PLATE 17

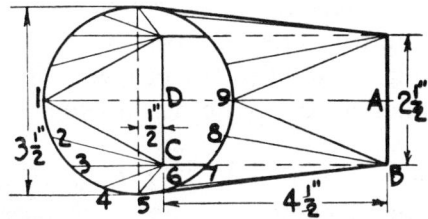

① TOP VIEW

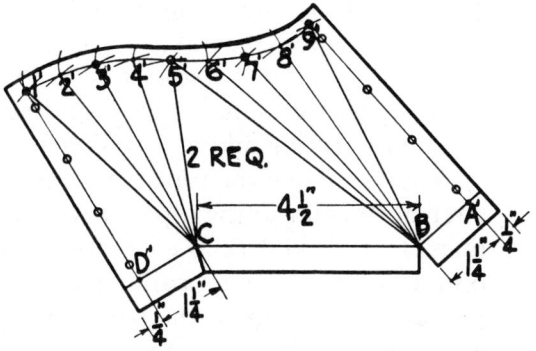

④ RECTANGULAR TO ROUND PATTERN

② FRONT VIEW

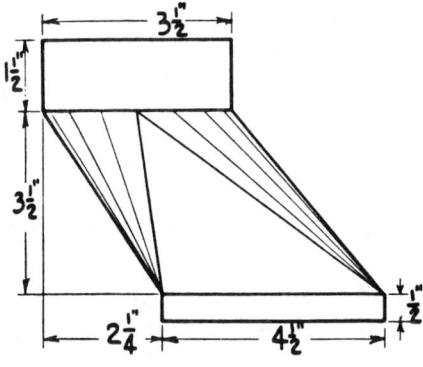

③ HALF TOP VIEW

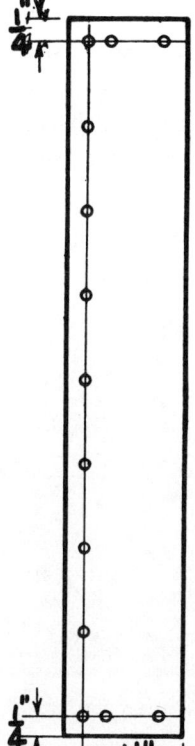

⑤ COLLAR PATTERN

PLATE 18 RECTANGULAR-TO-ROUND DOUBLE OFFSET

When drawing the top view of a square- or rectangular-to-round fitting where the round opening at the top is radically offset in two directions from the opening at the base, it is important to establish the proper points for locating the kink or bend and seam lines on the patterns, in order to facilitate forming and assembling.

After drawing the top view—such as the base points A, B, D, E, and the circular opening at the top—draw a square or a rectangle parallel to the base lines A, B, C, and E, to encompass and strike on a tangent the curved lines on the circular or elliptical opening (depending on the shape obtained in the process of developing the top view, such as may be obtained in Plates 39 and 40). Now those four tangent points, or the points that are the nearest to each tangent-line striking point will represent the points where the kink or bend and seam lines will be placed on the patterns, as represented by points 1, 5, 9, and 13.

For this type of fitting a full top view must be drawn as in Figure 1. It is not necessary to draw two true-length triangles as in Figure 3, providing the reader will not get confused with the many points on the base line of one true-length triangle.

To lay out the patterns as in Figures 4 and 5, use the slant length lines from the true-length triangle $A-B$ for the pattern in Figure 4. Use the slant length lines from the true-length triangle $D-E$ for the pattern in Figure 5. The spaces 1 to 9 on the patterns are equal to the spaces on the circle in Figure 1.

PLATE 18

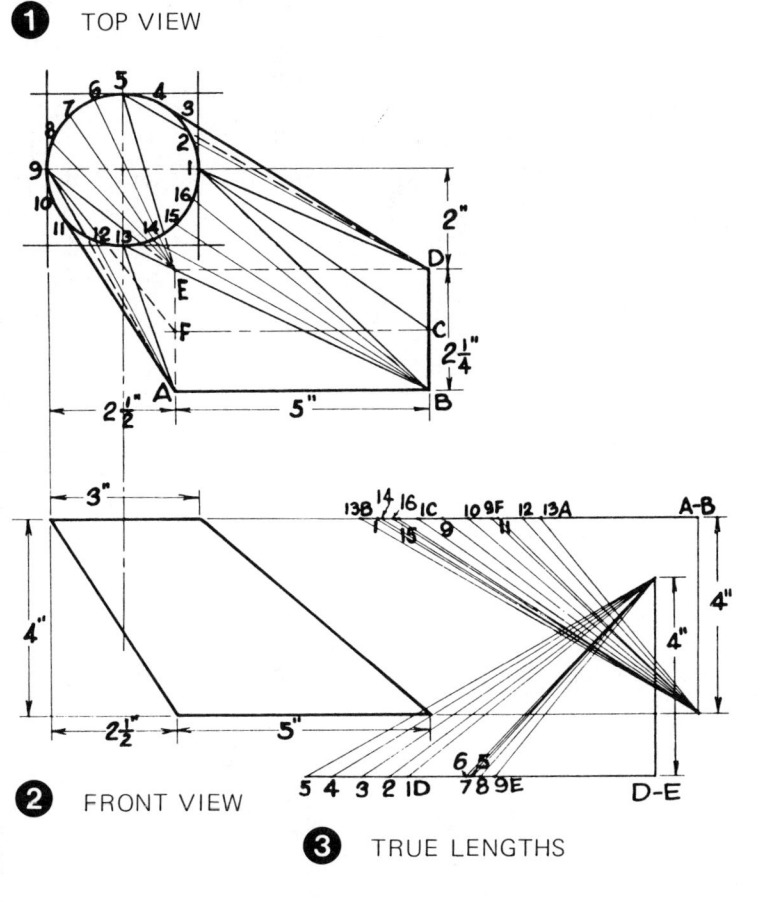

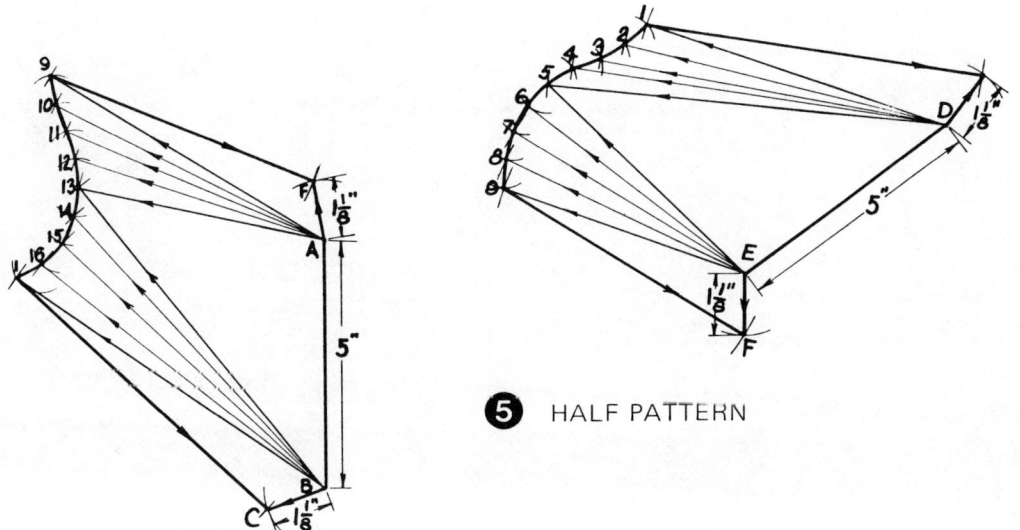

PLATE 19   RECTANGULAR-TO-ROUND WITH PITCH AT BASE

The procedure for laying out the patterns for a rectangular-to-round on a pitch is the same as for previous plates, except that the length for the half-top view must be obtained by drawing the pitch as in Figure 3.

The slant lengths from point F to points $1'$, $2'$, $3'$, $4'$, and $5'$ in Figure 4 represent the true lengths from point C in Figure 5. The slant lengths from point E to points $5'$, $6'$, $7'$, $8'$, and $9'$ in Figure 4 represent the true lengths from point B in Figure 5.

PLATE 19

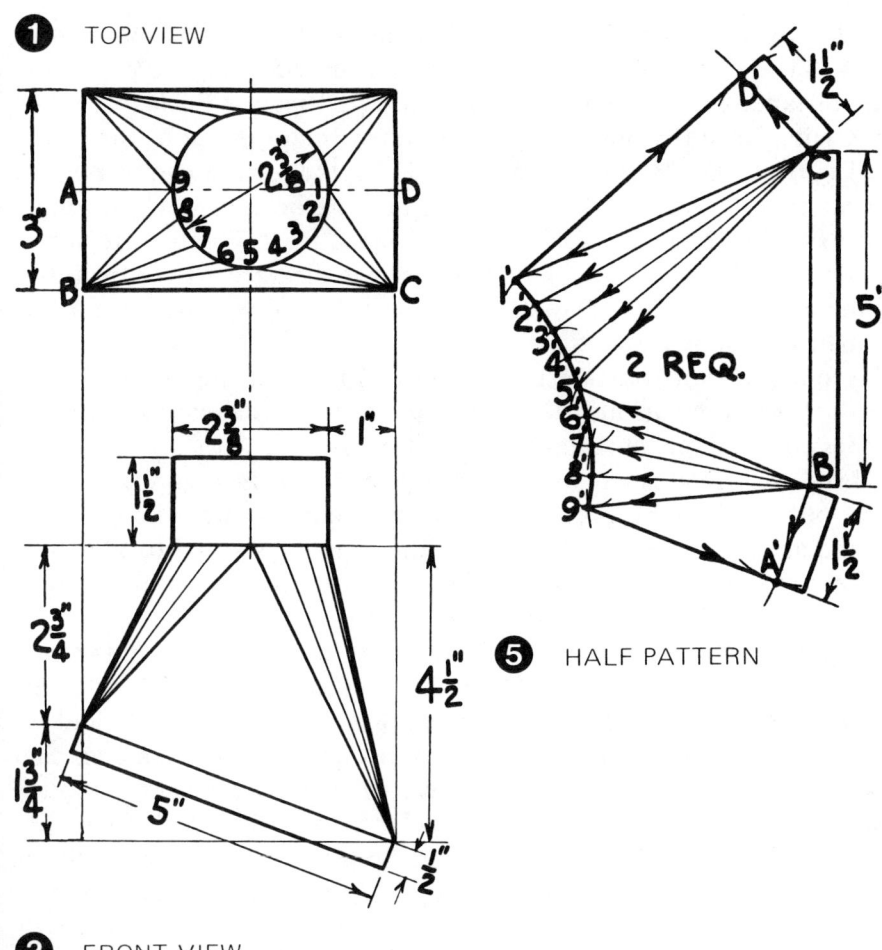

1. TOP VIEW
2. FRONT VIEW
5. HALF PATTERN — 2 REQ.

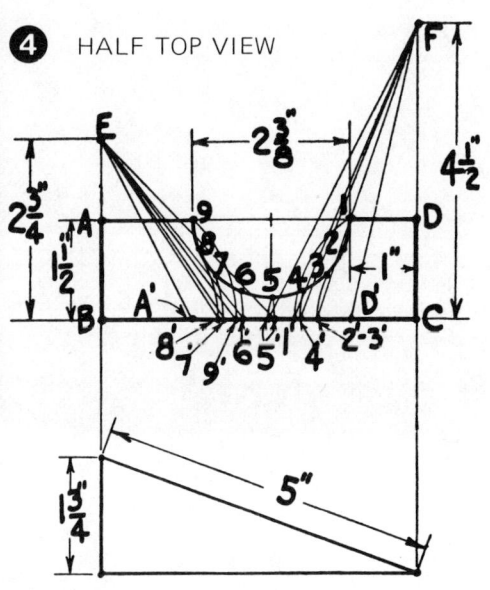

4. HALF TOP VIEW
3. TRUE LENGTHS FOR TOP VIEW

31

PLATE 20 SQUARE-TO-ROUND WITH FAN BRACKET

To draw the top view as in Figure 1, divide line E–1 in half as represented by point 4. Since one half of the height in the side view, Figure 2, is 3 in., divide the distance 4 to 1, Figure 1, into three equal spaces. Each space then represents 1 in. Since the bracket is $2\frac{1}{2}$ in. from the top edge of the round opening, Figure 2, take $2\frac{1}{2}$ spaces on line 1–E, Figure 1, as shown by point A to represent the top edge of the bracket. Since the bracket is $1\frac{1}{2}$ in. high, as shown in Figure 2, set $1\frac{1}{2}$ spaces from point A to B, Figure 1, to represent the bottom edge of the bracket. Transfer the distance 1 to A from line 1–E to line F–G from point F to D. Draw line C–C in position according to the dimensions shown.

To lay out the short-bracket pattern as in Figure 3, draw line C to C equal to the given dimensions. Note that line C to A is equal to $2\frac{1}{2}$ times the thickness of the bracket metal less than the distance C to A in Figure 1, and the distance A to B is equal to the distance A to B in Figure 1.

To lay out the long-bracket pattern as in Figure 4, draw line C to C equal to the given dimensions. Note that line C to D is equal to $2\frac{1}{2}$ times the thickness of the bracket metal less than the distance C to D in Figure 1. The distance A to B is equal to the distance A to B in Figure 1.

The heights E to B' and E to A' in Figure 1 are equal to the heights representing the bottom and top edges of the bracket in Figure 2. Transfer the distances E to B and E to A to the base line thus obtaining points E' and E''.

Lay out the half square-to-round pattern in Figure 5 in the same manner as in Plate 9. The distance E to B' on line 1–E in Figure 5 is equal to the slant line E' to B' in Figure 1. The distance E to A' in Figure 5 is equal to the slant line E'' to A' in Figure 1. Lay out the rivet or bolt holes to fit the holes on the bracket patterns. Note that slight kinking across from point C' to D on the long pattern in Figure 4 will facilitate aligning the holes with those on the square-to-round when assembling.

PLATE 20

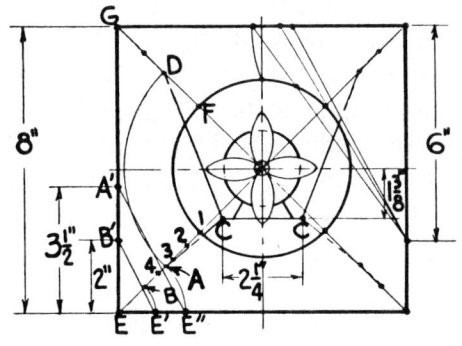

1 TOP VIEW

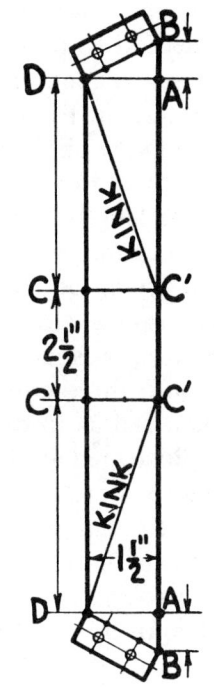

4 LONG BRACKET PATTERN

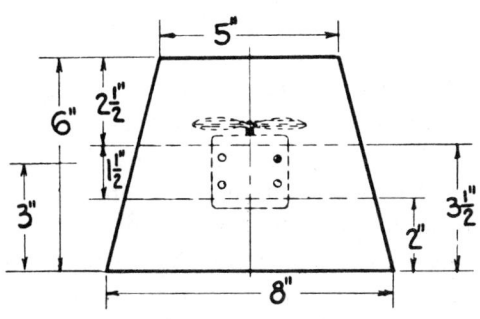

2 SIDE VIEW

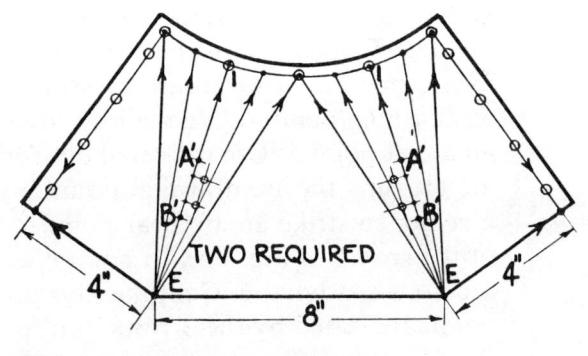

5 HALF PATTERN FOR SQUARE TO ROUND WITH BRACKET HOLES

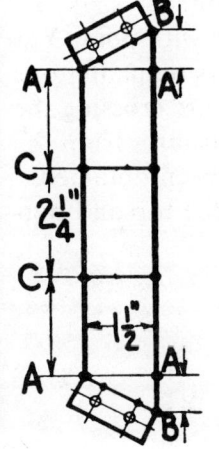

3 SHORT BRACKET PATTERN

PLATE 21 ROUND EQUAL-TAPER JOINT

This method for laying out the pattern for a round equal-taper joint is an easy and simple method to follow and will save time and effort.

First establish the true length of the center line as represented by points 1 and 2 on the pattern in Figure 5, which is equal to the distance 1–2 to point B in Figure 3. Now observe how the remaining true-length lines from point 2 to 3, and 1 to 4, and 4 to 5 crisscross each other. Notice that these lines are all the same length and equal to the distance 2–3 in Figure 3.

The student who cannot visualize this procedure may follow the remaining instructions for laying out the full pattern.

Draw a quarter-top view as in Figure 4. Divide the quarter outside circle 2 to 8 into as many equal spaces as desired. To obtain point 4 by dividing the quarter circle into three equal spaces, use point 8 as a center and the distance 8 to A as a radius to draw an arc crossing the larger circle in the same manner as illustrated in Plate 1, dividing a quarter circle. Draw a straight line from 4 to A crossing the inside circle, thus establishing point 3.

Erect a true-length triangle as in Figure 3 by transferring the distances 1 to 2 and 2 to 3 to the base line.

To lay out the pattern as in Figure 5, draw the center line 1 to 2 equal to the slant true-length line 1–2 to B in Figure 3. Set the dividers to span the length of the slant true-length line B to 2–3, 1–4, in Figure 3. (The dividers will remain set to equal this span throughout the development of the entire pattern.) Use point 2 in Figure 5 as a center to strike an arc near point 3. Set a small divider to equal one of the spaces in Figure 6. Then use point 1 in Figure 5 as a center to strike an arc on each side of point 1 crossing the arcs establishing point 3. Use point 1 in Figure 5 as a center, and with the large dividers strike an arc near point 4. Set another small divider to equal one of the spaces in Figure 7. Then use point 2 in Figure 5 as a center to strike an arc crossing the arcs on each side of point 2 thus establishing point 4. Use the large dividers with point 4 as a center to strike an arc at point 5. Use the small dividers with point 3 as a center to strike an arc crossing the arc drawn at point 5. Use the large dividers with point 3 as a center to strike an arc near point 6. Use the small dividers (equaling one of the spaces in Fig. 7) with point 4 as a center to strike an arc crossing the arc drawn at point 6. Continue this procedure by drawing the arcs from the points indicated by the arrows until points 13 and 14 have been obtained.

Allowances for seams and for connecting the collars at the top and bottom may be made as desired.

PLATE 21

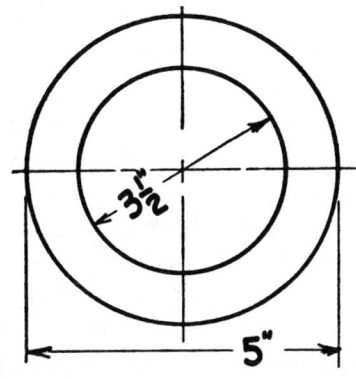

1 TOP VIEW

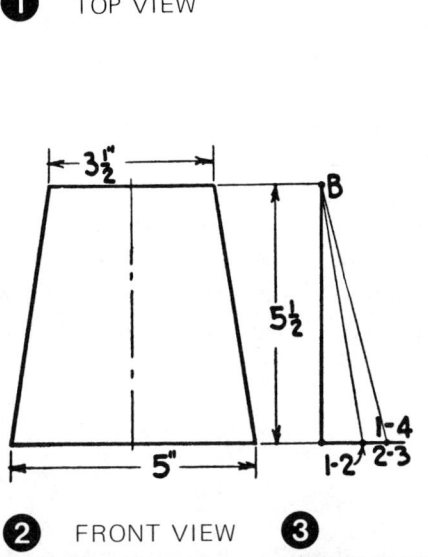

2 FRONT VIEW

3 TRUE LENGTH TRIANGLE

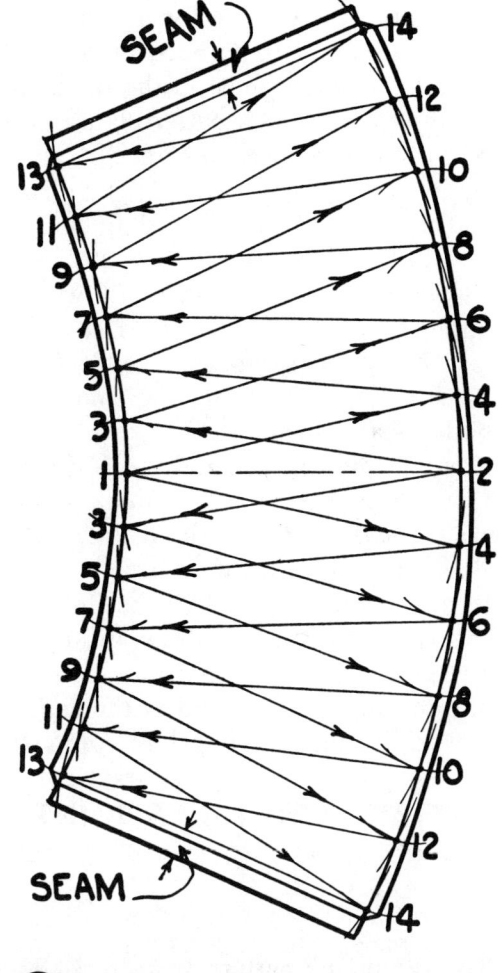

5 FULL PATTERN

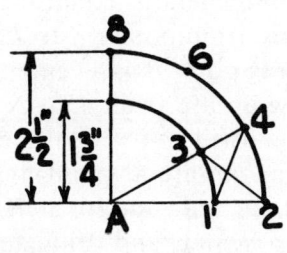

4 1/4 TOP VIEW

1 3 5 7 9 11 13

6 HALF CIRCUMFERENCE OF 3½ INCH DIAMETER

2 4 6 8 10 12 14

7 HALF CIRCUMFERENCE OF 5 INCH DIAMETER

PLATE 22 SMALL- AND LARGE-END PIPE PATTERNS

To lay out the pattern for a joint of pipe with a small end and a large end, the pattern will be tapered at the ends and curved slightly at the top and bottom. This depends on the thickness of the metal.

Figure 1 shows a pattern laid out with a small end and a large end made of $\frac{1}{4}$-in. metal. The distance A to B, which represents the small end, is equal to the given diameter. The distance C to D, which represents the large end, is equal to the given diameter plus $\frac{1}{3}$ of 7, times the thickness of the metal ($20 + \frac{1}{3} \times 7 \times .25 = 20.58$). If the pattern is laid out in the manner shown in Figure 1, this will allow the small end to be inserted inside of the large end of the next joint far enough to make a riveting connection.

To lay out the pattern as in Figure 1, use the above method to draw the large end C–D and the small end A–B. To complete the pattern, use point A as a center and A to C as a radius to draw an arc from C to E. Keep the dividers set, and use point B as a center to draw an arc from D to F. Use C as a center to draw an arc from A to G. Use D as a center to draw an arc from B to H. Use the distance A to B (the small end) as a radius, with A as a center, to draw an arc crossing the large arc at H. Use B as a center to draw an arc at G. Use the distance C to D (the large end) as a radius, with D as a center, to draw an arc at E. Use C as a center to draw an arc at F. Use any object that may be flexible enough (such as a ruler or a strip of light metal) to be held at the ends and pulled to form a light arch. Place this arch so that it will rest on points H, A, B, and G. Use a pencil or a scratch

PLATE 23 PATTERN FOR TAPER JOINT

To lay out the pattern for a taper joint as in Plate 23, draw the line C–D or small diameter, the line A–B or large diameter. Draw a line from D to G and another from B to F so that each is on a 90-deg. angle to the slant line D–B. Divide the distance G–H in half, thus obtaining point J. Divide the distance E–F in half, thus obtaining point K. Set the dividers to equal the distance D to K. Using point D as a center, draw an arc from points K to L. Keep the dividers set for all the remaining lengths. Using point B as a center, draw an arc from J to M. Using A as a center, draw an arc from J to N. Using C as a center, draw an arc from K to O. Set the dividers to equal the distance D to J. Then, using point D as a center, draw an arc at point M. Keep the dividers set, and use point M as a center, then draw an arc at R. Use C as a center, and draw an arc at N. Use N as a center, and draw an arc at Q. Set the dividers to equal the distance B to K. Then, using B as a center, draw an arc at L. Use L as a center, and draw an arc at S. Use A as a center, and draw an arc at O. Use O as a center, and draw an arc at P. Again set the dividers to equal the distance B to J, and, using point L as a center, draw an arc from D to R. Use M as a center, and draw an arc from

PLATE 22

awl to draw the arc at the top. Repeat the above procedure to draw an arc crossing points *E, D, C,* and *F*. Draw a straight line from *F* to *G* and *H* to *E*.

The circumference is equal to the diameter times 3.1416. The distance *H* to *G*, however, is equal to only three times the diameter. We must, therefore, multiply the diameter by .1416 (20 × .1416 = 2.832 ÷ 2 = 1.41) and add one half of the product to each side of the small end at points *G* and *H*. The same procedure must be repeated for the large end by multiplying the large diameter by .1416 (20.58 × .1416 = 2.914 ÷ 2 = 1.45 in.) and adding one half of the product to each side of the large end at points *E* and *F*.

The inside diameter of both ends will be the thickness of the metal less than the dimensions shown when rolled into a cylinder.

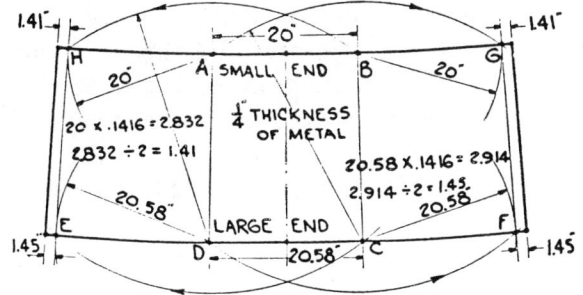

PLATE 23

B to *S*. Use *N* as a center, and draw an arc from *A* to *P*. Use *O* as a center, and draw an arc from *C* to *Q*.

Use any object that may be flexible enough to form an arch and rest on each point; then draw the arc from *Q* to *R*. Repeat the same procedure to draw the arc from *P* to *S*.

To obtain the full circumferences, multiply the small diameter by .1416 (20 × .1416 = 2.832 ÷ 2 = 1.41) and add one half the product to each side at *Q* and *R*. Repeat this procedure by multiplying the large diameter by .1416 (24 × .1416 = 3.398 ÷ 2 = 1.699) and add one half of the product to each side at *P* and *S*. This will allow the inside diameter of both ends to be the thickness of the metal less than the dimensions shown.

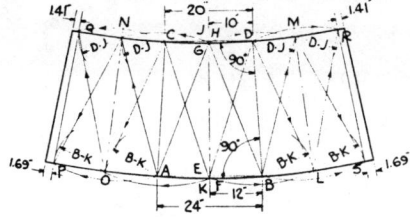

PLATE 24 ROUND TAPER JOINT WITH ONE SIDE STRAIGHT

Draw only one half of the top view in Figure 1. Divide each half circle into any number of equal spaces as desired.

Erect a true-length triangle as in Figure 3, by transferring, from the top view, the lengths of the broken lines to the base line on one side of the triangle and the solid lines to the opposite side. Number each as shown.

To lay out the pattern as in Figure 4, draw the center line 1 to 2 equal to the length of the slant true-length line 2 to A in Figure 3. Use the slant length line 3 to A in Figure 3 as a radius, and point 2 in Figure 4 as a center to strike an arc near point 3. Set a small divider to equal one of the spaces in Figure 6. Use point 1 in Figure 4 as a center to strike an arc crossing the arc at point 3. Use point 3 as a center, and the slant true-length 4 to A in Figure 3 as a radius to strike an arc at point 4. Set another small divider to equal one of the spaces in Figure 5, and use point 2 as a center to strike an arc crossing the arc at point 4. Use point 4 as a center and the slant true length 5 to A in Figure 3 as a radius to strike an arc at point 5. Use point 3 as a center, and with the small dividers strike an arc crossing the arc at point 5.

Follow this procedure by taking all the true lengths from the triangle in Figure 3 until point 13 has been obtained. The distance 13 to 14 is equal to the given height.

All allowances may be made as desired.

PLATE 24

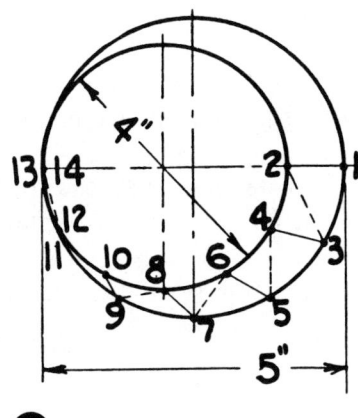

1 TOP VIEW

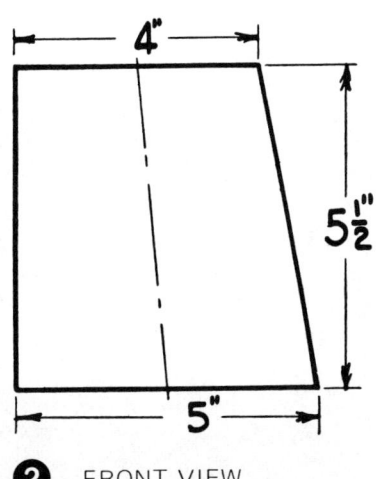

2 FRONT VIEW

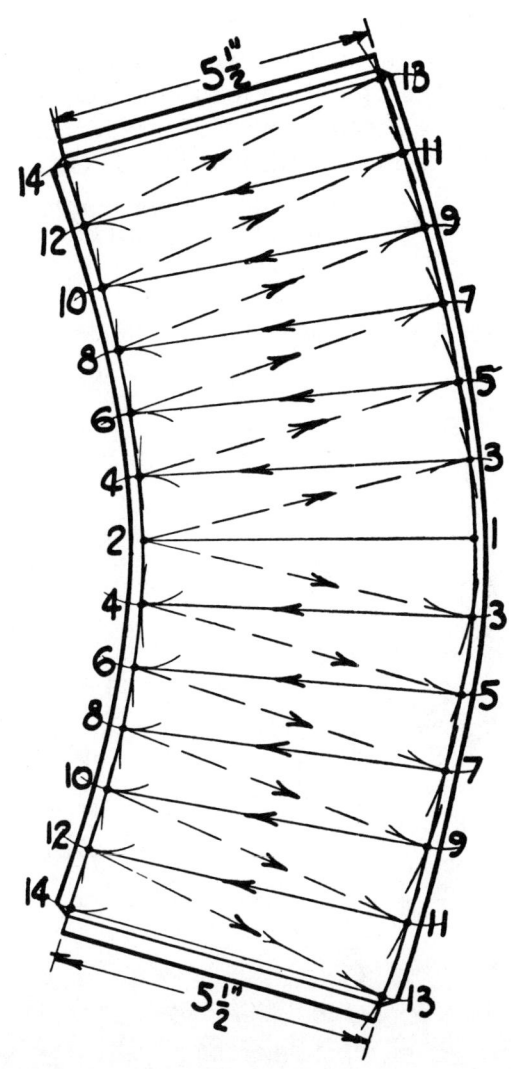

4 FULL PATTERN

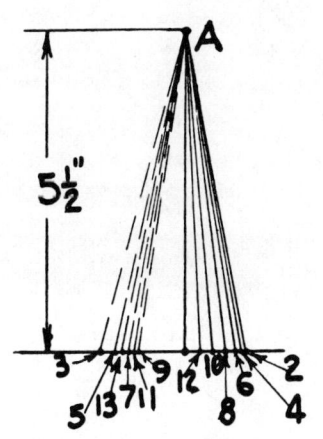

3 TRUE LENGTHS

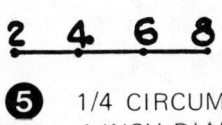

5 1/4 CIRCUMFERENCE OF 4 INCH DIAMETER

6 1/4 CIRCUMFERENCE OF 5 INCH DIAMETER

39

PLATE 25   ROUND TAPER JOINT WITH TWO SIDES STRAIGHT

This type of taper joint will allow a run of round pipes and tapers to hug the ceiling and wall with a minimum of clearance.

The procedure for this plate is identical with Plate 24. Draw only one half of the top view in Figure 1. Draw a line to cross through the two center points dividing the two circles in halves.

Erect a true-length triangle as in Figure 3, by transferring, from the top view, the lengths of the broken lines to one side of the base line and the solid lines to the opposite side.

To lay out the pattern as in Figure 4, follow the same procedure as in Plate 21.

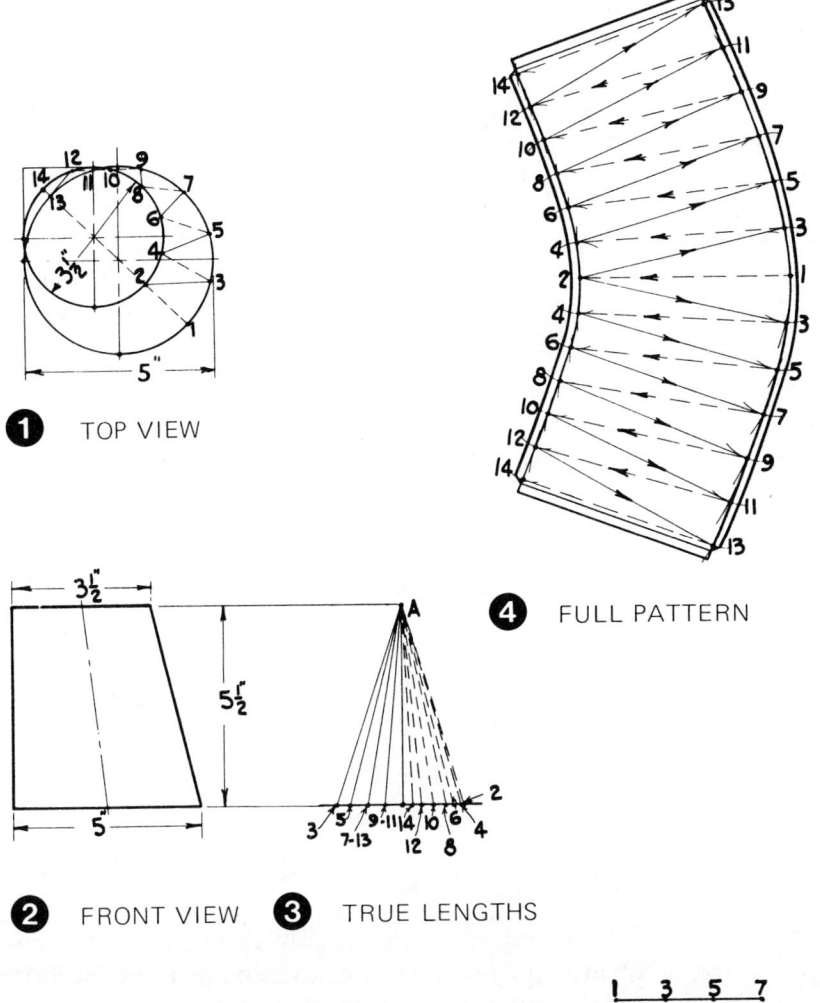

PLATE 25

① TOP VIEW
② FRONT VIEW
③ TRUE LENGTHS
④ FULL PATTERN
⑤ ¼ CIRCUMFERENCE OF 4 INCH DIAMETER
⑥ ¼ CIRCUMFERENCE OF 3½ INCH DIAMETER

PLATE 26 ELLIPTICAL OR OVAL-TO-ROUND EQUAL TAPERING

To draw the ellipse, follow the same procedure as for Plate 6. Draw one quarter of the top view in Figure 1, and divide each into the same number of equal spaces.

Draw line 2' to 10' in Figure 5 equal to one quarter of the circumference of the $3\frac{1}{4}$-in.-diameter circle.

Draw line 1' to 9' in Figure 6 equal to one-quarter circumference of the $2\frac{1}{2}$ by 5-in. ellipse.

The circumference of an ellipse is equal to two times the sum of one half of the major axis squared plus one half of the minor axis squared; then extract the square root of the product and multiply by 3.1416 (pi).

$2\frac{1}{2}$ by 5-in. ellipse. $1\frac{1}{4}$ in. = $\frac{1}{2}$ of minor axis; $2\frac{1}{2}$ in. = $\frac{1}{2}$ of major axis. Then the circumference is equal to $3.1416 \sqrt{2(\frac{1}{2}\text{ major}^2 + \frac{1}{2}\text{ minor}^2)}$. Thus $3.1416 \sqrt{2(2.5^2 + 1.25^2)} = 3.1416 \sqrt{2(6.25 + 1.5625)} = 3.1416 \sqrt{2(7.8125)} = 3.1416 \sqrt{15.625}$.

Extracting the square root of 15.625:

```
        3. 9  5
      √15.6250
        9
       ──────
    69   662
         621
       ──────
    785  4150
         3925
        ──────
          225
```

Then $3.1416 \times 3.95 = 12.409320$. $12.409 = 12\ 13/32$ in., circumference of ellipse.

This method is accurate for obtaining the circumference of any ellipse regardless of the size. (No attempt is made here to explain the procedure for extracting the square root because it is too lengthy.)

To lay out the pattern as in Figure 4, erect a true-length triangle as in Figure 3, following the same procedure as in Plate 21.

PLATE 26

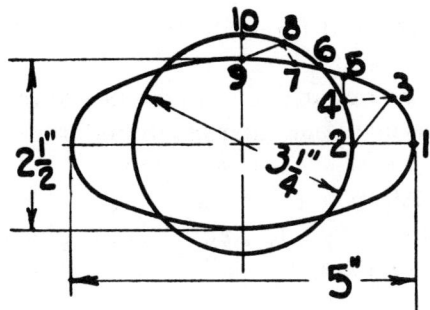

① TOP VIEW

⑤ 1/4 CIRCUMFERENCE OF 3¼ INCH DIAMETER

⑥ 1/4 CIRCUMFERENCE ELLIPSE

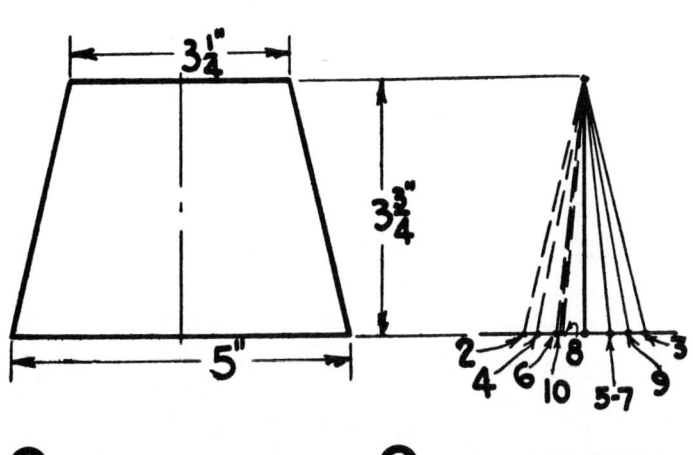

② FRONT VIEW

③ TRUE LENGTHS

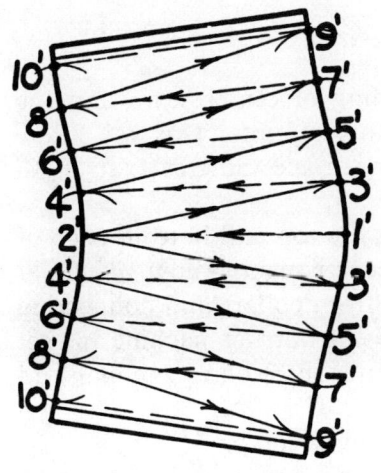

④ HALF PATTERN

PLATE 27 ELLIPTICAL-TO-ROUND WITH ONE SIDE STRAIGHT

The procedure for this plate is identical with that for Plate 26, but a half-top view must be drawn as in Figure 1.

Draw line 2' to 10' in Figure 5 equal to one-quarter circumference of the 3¼-in. circle. Draw line 1' to 9' in Figure 6 equal to one-quarter circumference of the 2½ by 5-in. ellipse.

Erect a true-length triangle as in Figure 3.

To lay out the pattern as in Figure 4, draw line 1 to 2 equal to the given height, proceeding in the same manner as for Plate 21.

PLATE 28 WELDING OFFSET SEAMS, METALS 18 GAUGE AND LIGHTER

This plate illustrates the preparation for welding of center seam and the collars at top and bottom for round or oval taper joints. This will allow overlapping of metal at the seams with a smooth surface and greater strength at the seams.

When patterns are laid out allow ⅛ inch at the top and bottom edge of the taper joint. Allow ⅛ inch at each end for center seam. Allow ⅛ inch at one edge of each collar. When duct continues from collars and connecting seams are welded, allow ⅛ inch at both edges. Using burring machine and set gauge back 5/16 inch, then offset one edge of center seam on each collar and taper joint about depth of bottom roll.

Form taper joint and collars. Weld connecting seams.

Use burring machine to offset or shrink, and flange each section as illustrated.

Procedure is illustrated and explained in Plate 12.

PLATE 27

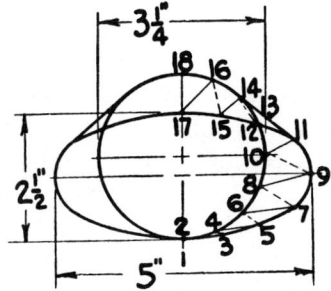

1 TOP VIEW

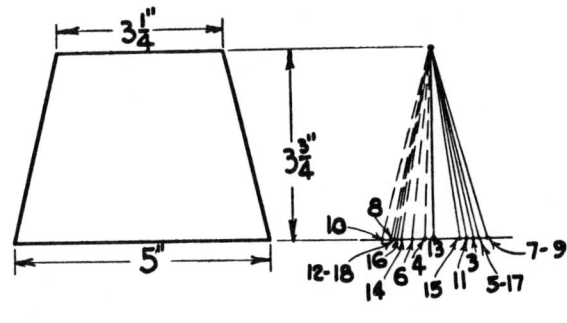

2 FRONT VIEW **3** TRUE LENGTHS

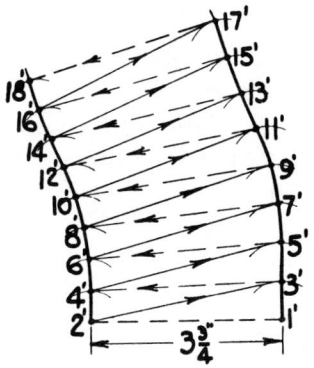

4 HALF PATTERN

5 1/4 CIRCUMFERENCE 3¼ INCH DIAMETER

6 1/4 CIRCUMFERENCE OF ELLIPSE

PLATE 28

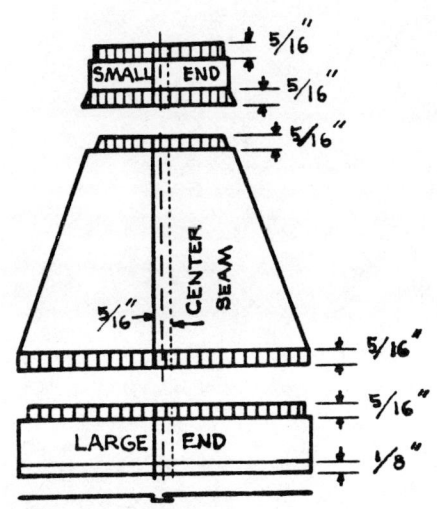

PLATE 29   OBLONG-TO-OBLONG TRANSFORMER ON CENTER

Draw only one quarter of the top view in Figure 1.

Erect a true-length triangle as in Figure 3.

To lay out the pattern as in Figure 5, draw the center line A to 1 equal to the slant true-length line $A-1$ to C in Figure 3. The distance A to 2 is equal to the distance A to 2 in Figure 1. The remaining pattern may be completed in the same manner as in Plate 21.

PLATE 29

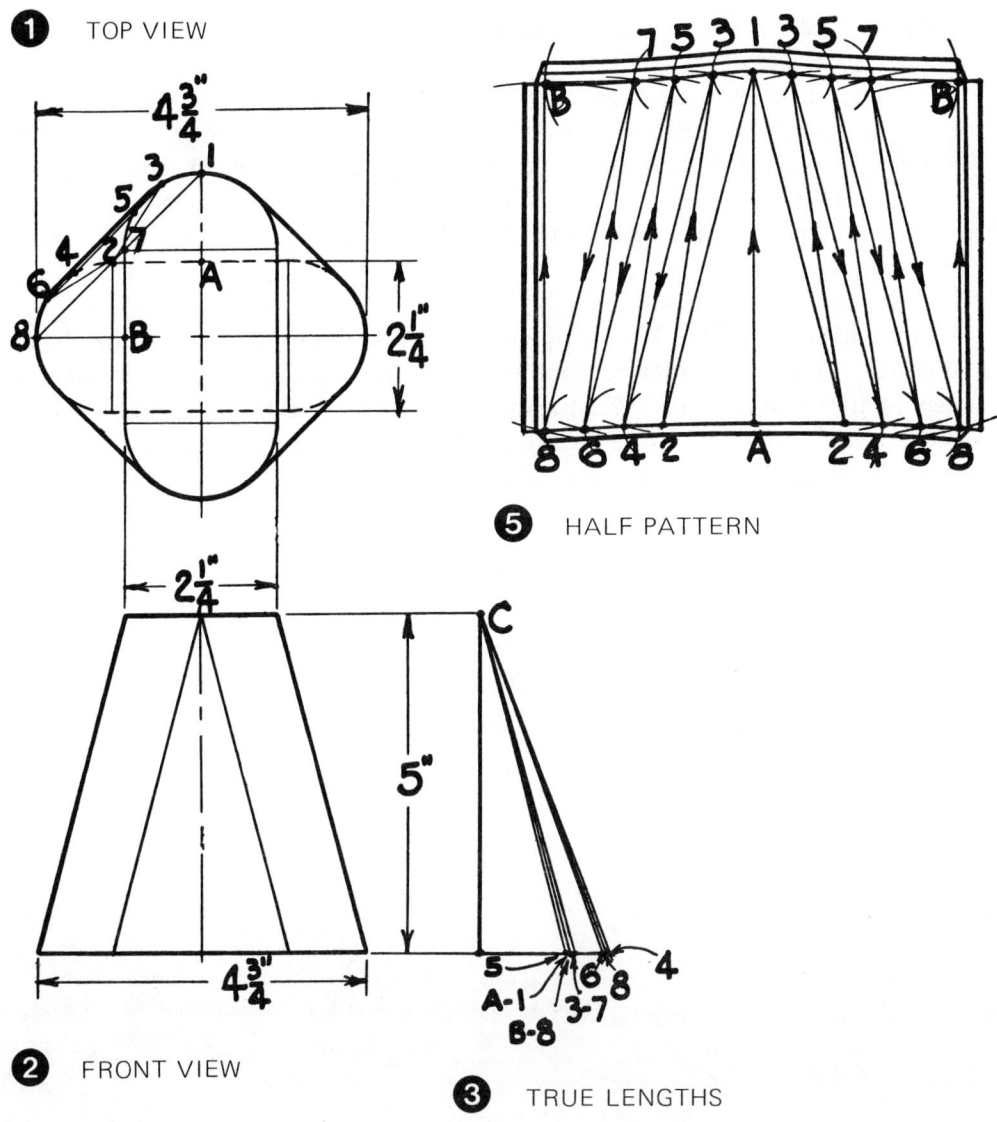

① TOP VIEW

② FRONT VIEW

③ TRUE LENGTHS

④ 1/4 CIRCUMFERENCE

⑤ HALF PATTERN

47

PLATE 30 OBLONG-TO-OBLONG OFF CENTER

The procedure for this plate is the same as for Plate 29 except that the full-top view must be drawn as in Figure 1.

It is not necessary to draw two true-length triangles as in Figure 3; this is only to avoid confusion in showing the various true-length lines for the pattern. These numbers may be eliminated by immediately transferring the true-length lines to the pattern as soon as they have been obtained, thus eliminating the necessity of numbering and returning to obtain the true-length lines.

The pattern in Figure 4 may be made in two half patterns, and the procedure will be identical. If a full pattern is to be made, then draw line 14 to 16 equal to the distance 14 to 16 in the top view. Obtain the two lengths for lines 14 to 15, and 16 to 15, using points 14 and 16 as centers to draw arcs to cross each other obtaining point 15. The remaining pattern may be completed by following the same procedure as for Plate 21.

PLATE 30

❶ TOP VIEW

❺ HALF CIRCUMFERENCE

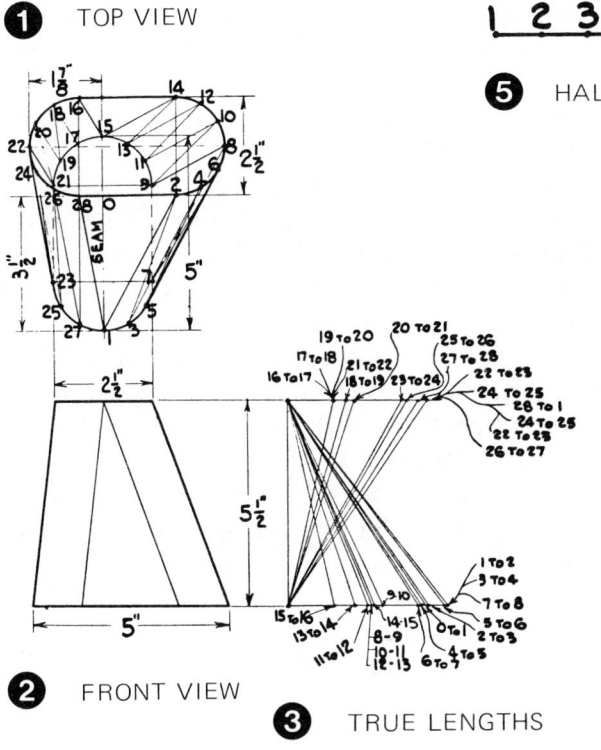

❷ FRONT VIEW

❸ TRUE LENGTHS

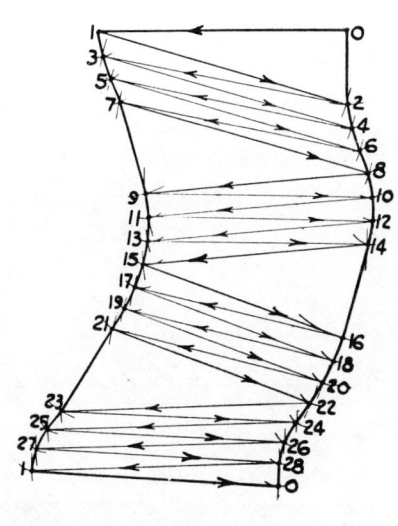

❹ FULL PATTERN

PLATE 31 ELBOW-BOOT CENTER TAPER

This type of fitting is generally used on residential heating and air conditioning where small, round pipes may be used very efficiently as basement runs.

The top view in Figure 1 and the side view in Figure 4 do not have to be drawn; they are only to show the size and shape of the fitting.

Draw the front view as in Figure 2 (only half of the view is necessary).

To erect the true-length triangle as in Figure 3, draw line A to B equal to the length of line $A-7$ in Figure 1. Use point $A7$ as a center to draw an arc from each point 1, 2, 3, 4, 5, and 6 on the circle to intersect the base line of the triangle, thus obtaining the true-length lines.

To lay out the pattern as in Figure 5, draw line A to A equal to the given dimension. Use the slant true-length line $1'$ to B in Figure 3 as a radius; then use each point A in Figure 5 as a center to strike arcs to cross each other at point $1'$. Use the slant line $2'$ to B in Figure 3 as a radius, then use point A in Figure 5 as a center to strike an arc at point 2. Set the dividers to span one of the spaces on the circle in Figure 2; then use the point $1'$ in Figure 5 as a center to strike an arc to cross the arc at point $2'$. Follow this same procedure to obtain the remaining points $3'$, $4'$, $5'$, $6'$, and 7, completing the pattern with the seam allowances as shown.

The front pattern and the snap collar are made as shown in Figure 6 and 7.

PLATE 31

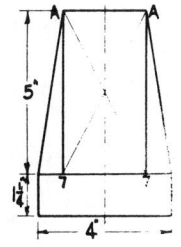

1 TOP VIEW

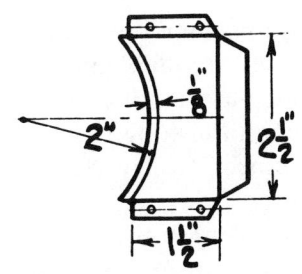

6 FRONT PATTERN

2 FRONT VIEW

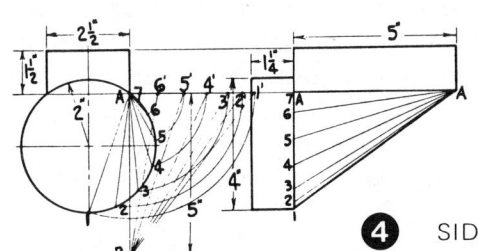

7 SNAP COLLAR PATTERN

4 SIDE VIEW

3 TRUE LENGTH TRIANGLE

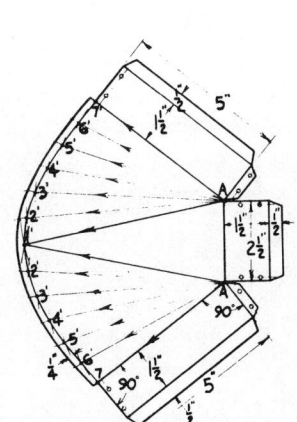

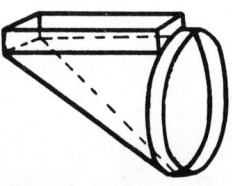

8 ISOMETRIC VIEW

5 SIDE & BACK PATTERN

51

PLATE 32 ELBOW BOOT, ONE SIDE STRAIGHT

Although the shape of this elbow boot differs from that shown in the previous plate, the procedures are identical.

Draw the front view as in Figure 2; then divide in half the distance A to B on the circle, establishing the center point 7.

Divide the distance 7 to B into equal spaces; the spaces 7 to A are identical and are not required.

To erect the true-length triangle in Figure 2, draw line B to $B'-A'$ equal to the distance A to A in Figure 3; then use point $B-1$ as a center point to draw arcs from points 2, 3, 4, 5, 6, and 7 from the circle to the base line of the true-length triangle as represented by points $2'$, $3'$, $4'$, $5'$, $6'$, and $7'$.

To lay out the pattern as in Figure 4, transfer the heights D to A and C to B from Figure 2 to establish lines D to A' and C to B' representing the back pattern in Figure 4. Use the slant true-length line $B'-A'$ to $7'$ in Figure 2 as a radius; then use points A' and B' in Figure 4 as centers to draw arcs to cross each other at point $7'$. Use the slant true-length line $B'-A'$ to $6'$ in Figure 2 as a radius; then use points B' and A' in Figure 4 as centers to strike an arc at each point $6'$. Set the dividers to span any one space on the circle in Figure 2; then use point $7'$ as a center to strike arcs crossing the arcs at each point $6'$ in Figure 4. Continue this procedure to complete the pattern with seam allowances as shown.

The front pattern and the snap-collar pattern are made as in Figures 5 and 6.

PLATE 32

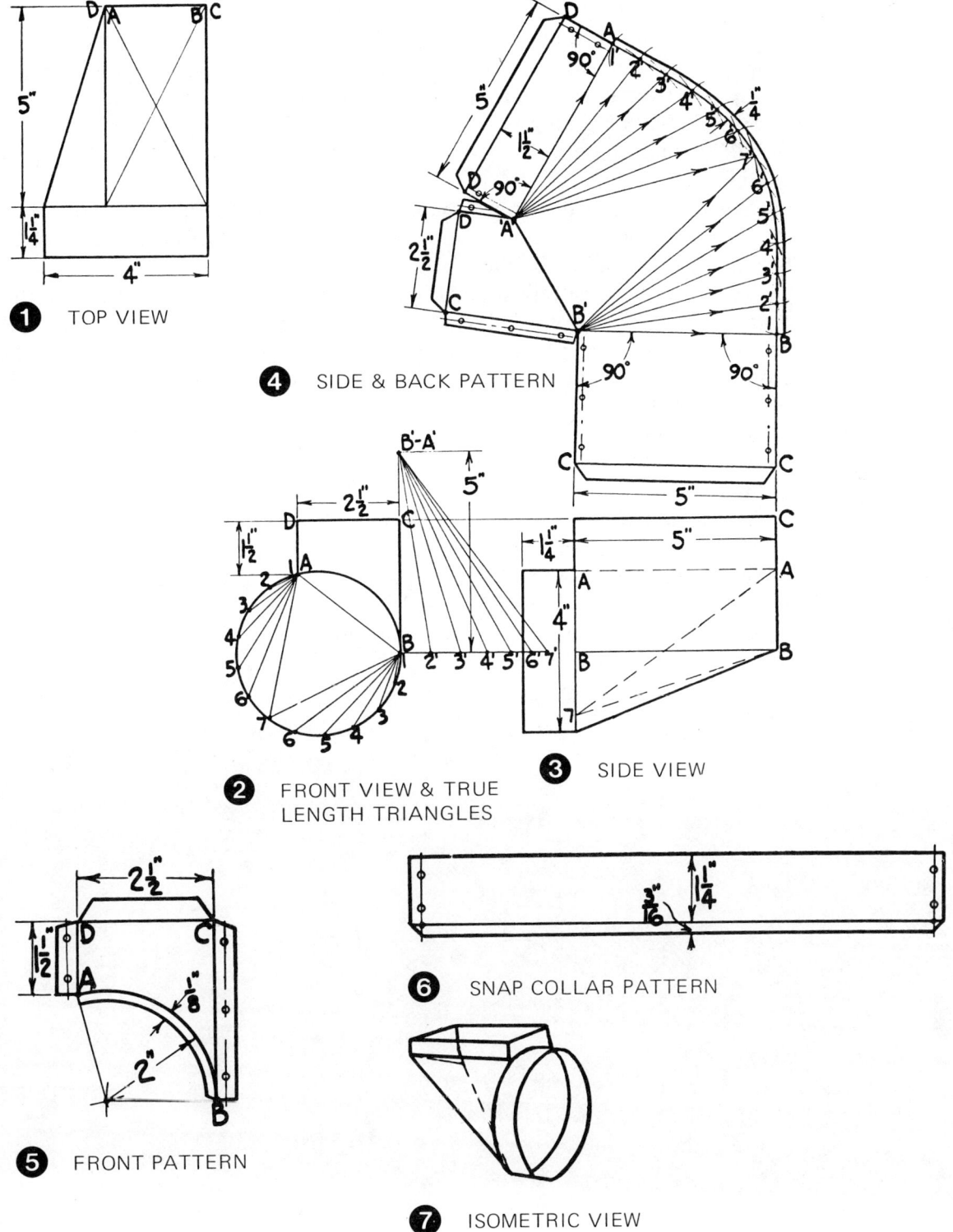

PLATE 33 OFFSET ELBOW BOOT

For the front view in Figure 2, first draw the circle and lines E to $A1$ and E to D, and then draw line $D-C$ to any length desired. Use point A as a center to draw an arc tangent to line $D-C$ toward the large circle. Draw a line tangent to the circle obtaining point $B1$, and tangent to the arc drawn to intersect line $D-C$, thus obtaining point C. Divide the distance A to B on the circle in half, obtaining point 7. Divide the distance A to 7 or B to 7 into equal spaces.

Erect a true-length triangle, the height $A1$ to $A'-B'$, Figure 2, is equal to the distance A to A in Figure 3. Transfer the distances from A to 2, 3, 4, 5, 6, and 7 to the base line on the triangle.

To lay out the pattern as in Figure 4, draw line A' to B' equal to the distance A to B in Figure 2. Use the true-length line $7'$ to $A'-B'$ in Figure 2 as a radius; then use point A' and B' in Figure 4 as centers to draw arcs to cross each other at point $7'$. Complete the pattern following the same procedure as in Plate 31.

To lay out the pattern as in Figure 5, transfer the lengths of lines E to A, D to C, C to B, and A to B from Figure 2 to obtain points A', C', and B' in Figure 5.

To lay out the pattern as in Figure 6, the same procedure is followed as for laying out that pattern in Figure 5 except that a center point is obtained to draw the arc from A to B.

A snap collar is shown in Figure 7.

Complete all patterns with seam allowances as shown.

PLATE 33

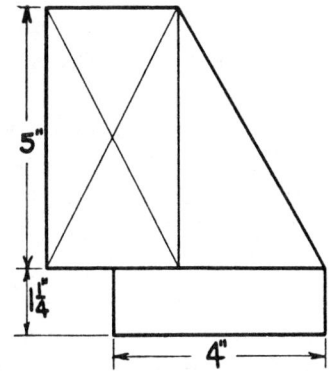

1 TOP VIEW

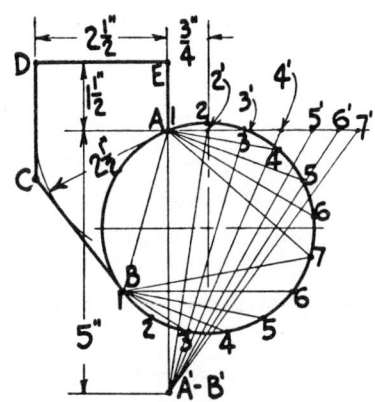

2 FRONT VIEW & TRUE LENGTH TRIANGLE

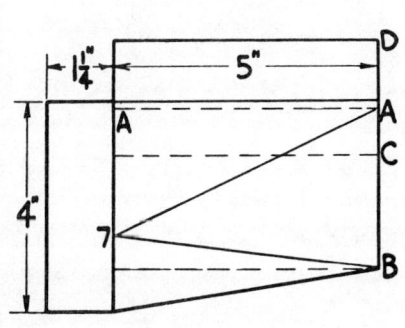

3 SIDE VIEW

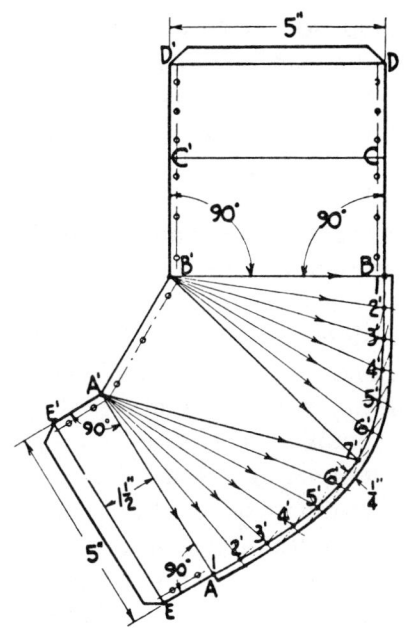

4 SIDE PATTERN

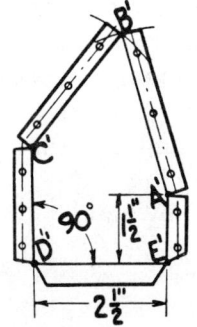

5 BACK PATTERN

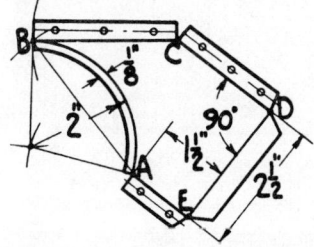

6 FRONT PATTERN

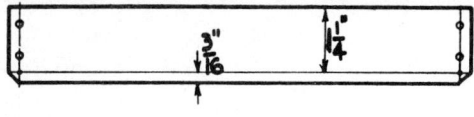

7 SNAP COLLAR

PLATE 34   ELBOW BOOT ON CENTER WITH SLANT THROAT

The top and side views are given only to show the position of the openings.

Draw the front view as in Figure 3 (only a half front view is necessary). Divide the half circle into equal spaces 1 to 7 as shown.

To erect a true-length triangle as in Figure 3, draw the base line $A-B$ to B' equal to the width of the base opening plus the width of the throat in Figure 2. Use point $A-B$ as a center to draw an arc from each point on the circle to intersect the triangle, obtaining points $1'$, $2'$, $3'$, $4'$, $5'$, $6'$, and $7'$ which represent the true-length triangle.

To lay out the pattern as in Figure 4, draw the base line $B'-B'$ equal to the dimensions shown. Use the slant true-length line $1'$ to B' in Figure 3 as a radius, and use each point B' in Figure 4 as centers to draw arcs to cross each other at point $1'$. Transfer the remaining slant true-length lines from B' to $2'$, $3'$, and $4'$ in Figure 3 to Figure 4, obtaining points $2'$, $3'$, and $4'$ in the same manner as in Plate 31.

Erect another true-length triangle as represented by $A-B$ to A', Figure 3, equal to the width of the throat.

Continue laying out the pattern by using the slant true-length line $4'$ to A' in Figure 3 as a radius and point $4'$ in Figure 4 as a center to strike an arc near point A'. Set the dividers to equal the length of the base ($4\frac{1}{2}$ in.) in Figure 2, and use point B' as a center to strike an arc to cross the arc at point A', Figure 4. Transfer the remaining true lengths from A' to $5'$, $6'$, and $7'$ in Figure 3 to Figure 4, thus obtaining points $5'$, $6'$, and $7'$. Use the slant length C' to A' in Figure 3 as a radius, and point $7'$ in Figure 4 as a center to strike an arc near point C'. Set the dividers to span half of the width (1 in.) in Figure 3, and use point A' as a center to strike an arc crossing the arc at C'.

Complete the pattern by making the required allowances as shown.

PLATE 34

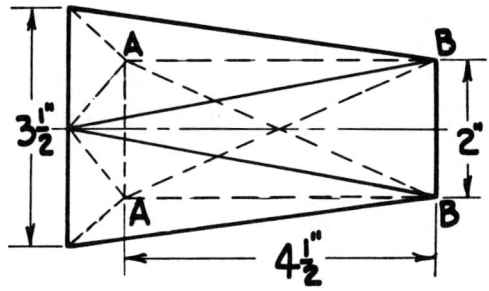

① TOP VIEW

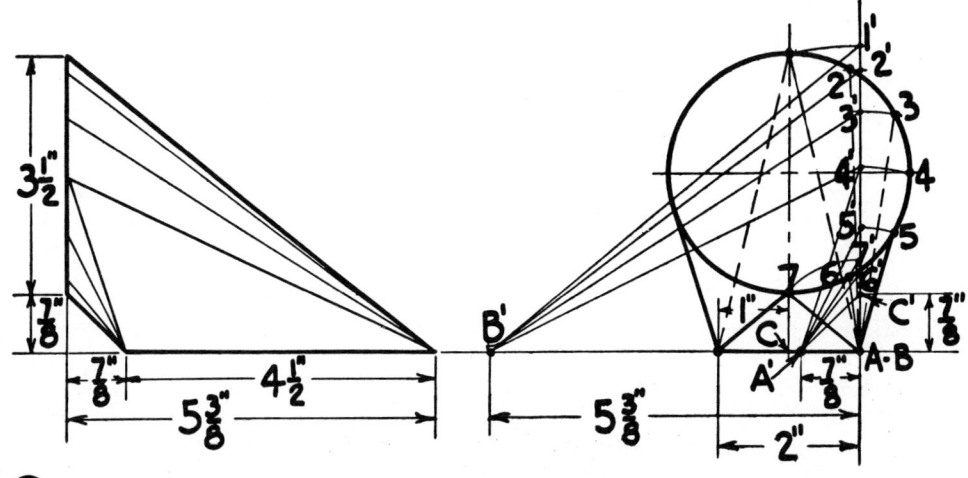

② SIDE VIEW   ③ FRONT VIEW & TRUE LENGTHS

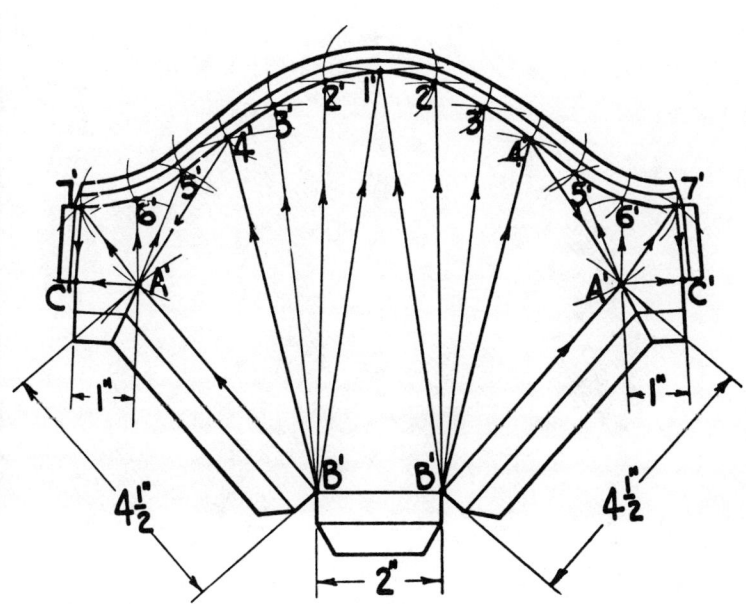

④ FULL PATTERN

PLATE 35 ELBOW BOOT ON CENTER WITH SLANT THROAT

Draw the front view as in Figure 3. (Only half of the front view is necessary.)

Erect two true-length triangles as $A-B$ to A' and $A-B$ to B'.

To lay out the pattern as in Figure 4, draw the base line $B'-B'$ equal to the dimensions given. Use the slant length line B' to $1'$ in Figure 3 as a radius, and points $B'-B'$ in Figure 4 as centers to strike arcs to cross each other at $1'$.

Complete the pattern by following the same procedure as in Plate 31.

PLATE 35

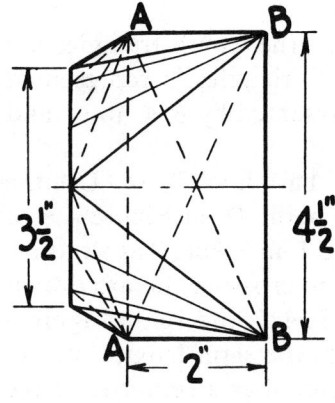

1 TOP VIEW

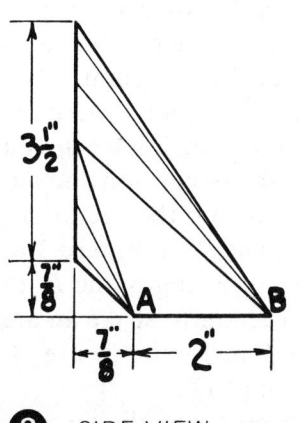

2 SIDE VIEW

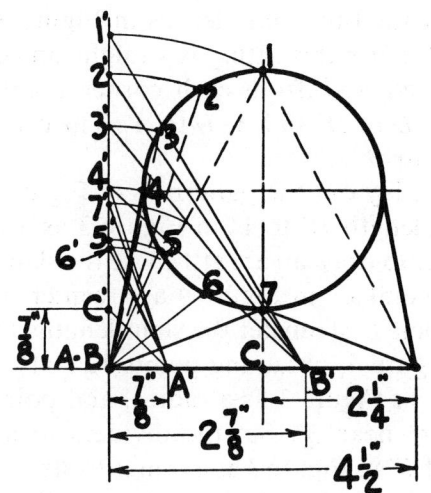

3 FRONT VIEW & TRUE LENGTHS

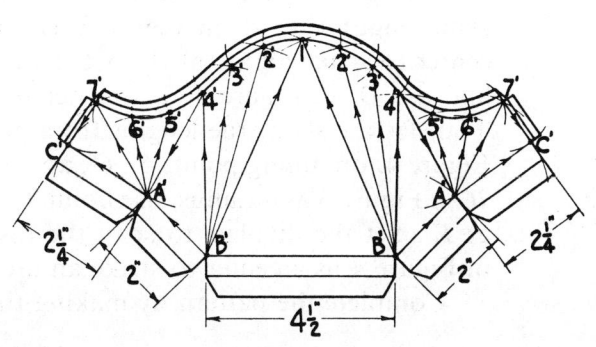

4 FULL PATTERN

PLATE 36 ELBOW BOOT, STRAIGHT ON ONE SIDE WITH SLANT THROAT

Draw the full front view as in Figure 3.

Because this fitting is straight on one side, four true-length triangles are required, two from each corner point. One set of triangles is represented by $A-D$ to D' and $A-D$ to A'. The other set is represented by $B-C$ to C' and $B-C$ to B'.

To lay out the pattern as in Figure 4, draw the base line $C'-D'$. Use the slant length D' to $1''$ in Figure 3 as a radius, and point D' in Figure 4 as a center to draw an arc at $1'$. Use the slant length C' to $1'$ in Figure 3 as a radius, and point C' in Figure 4 as a center to draw an arc crossing the arc drawn at point $1'$. Transfer the slant lengths from C' to $12'$, $11'$, and $10'$ in Figure 3 to Figure 4, obtaining points $12'$, $11'$, and $10'$. Use the slant length line $10'$ to B' in Figure 3 as a radius, and point $10'$ in Figure 4 as a center to strike an arc near B'. Set the dividers to span the dimensions shown, and use point C' in Figure 3 as a center to strike an arc crossing the arc at B', Figure 4. Transfer the lengths from B' to $9'$, $8'$, and $7'$ in Figure 3 to Figure 4 obtaining points $9'$, $8'$, and $7'$. Use the slant line E' to B' in Figure 3 as a radius, and point $7'$ in Figure 4 as a center to strike an arc at E'. Set the dividers to span the distance B to E in Figure 3, and use B' in Figure 4 as a center to strike an arc crossing the arc at E'. Transfer the slant true lengths from point D' to $2'$, $3'$, and $4'$ in Figure 3 to Figure 4, obtaining points $2'$, $3'$, and $4'$. Use the slant length $4'$ to A' in Figure 3 as a radius, and point $4'$ in Figure 4 as a center to strike an arc at A'. Set the dividers to span the dimension shown, and use D' in Figure 4 as a center to strike an arc crossing the arc at A'. Transfer the slant true lengths from point A' to $5'$, $6'$, and $7''$ in Figure 3 to Figure 4, obtaining points $5'$, $6'$, and $7''$. Use the slant true-length line E' to B' in Figure 3 as a radius, use point $7''$ in Figure 4 as a center to strike an arc at E''. Set the dividers to span the distance $A-D$ to E in Figure 3, use A' in Figure 4 as a center to strike an arc crossing the arc at E''.

Complete the pattern by making the required allowances as shown.

PLATE 36

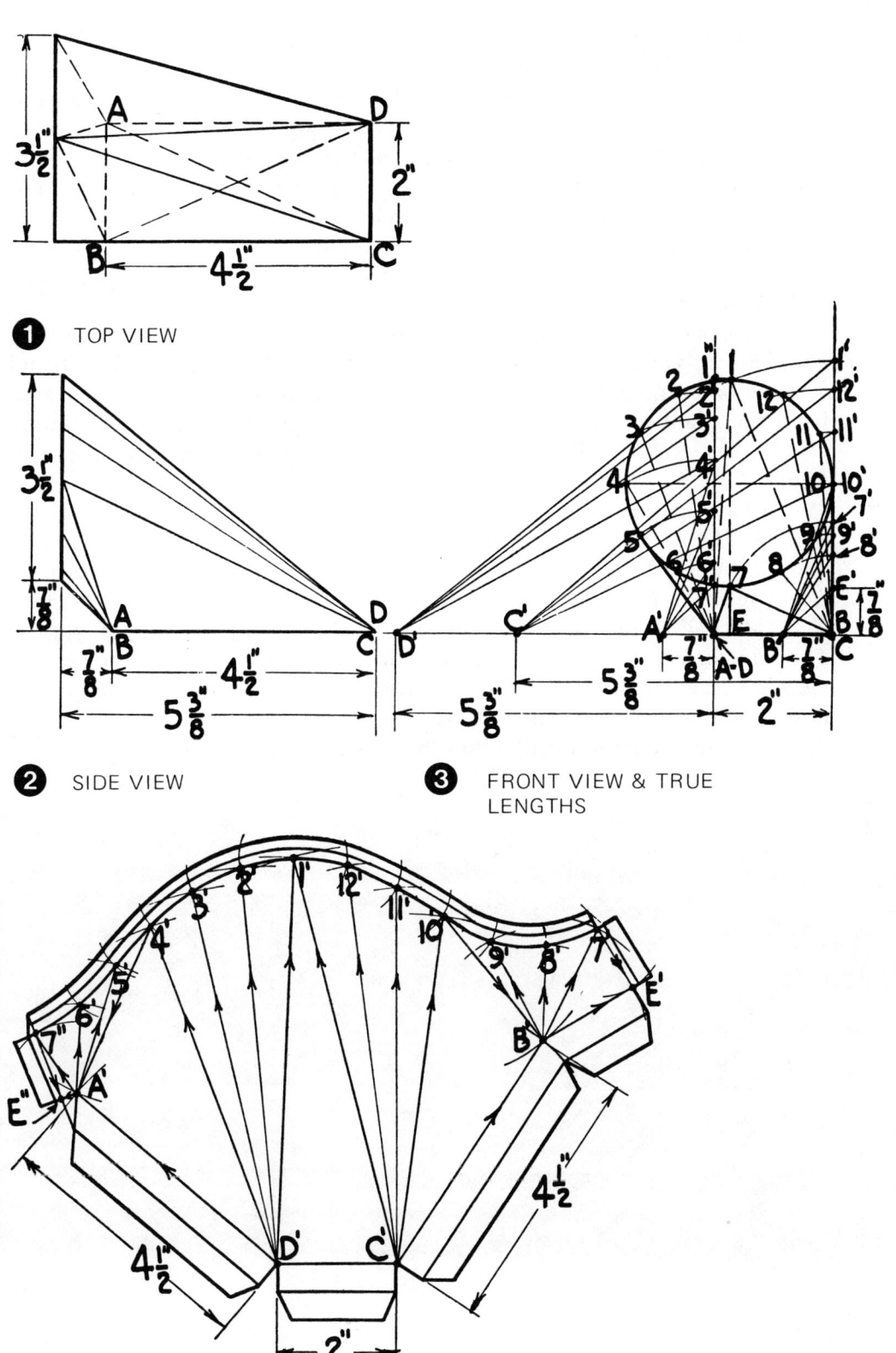

① TOP VIEW

② SIDE VIEW

③ FRONT VIEW & TRUE LENGTHS

④ FULL PATTERN

PLATE 37 RECTANGULAR-TO-ROUND WITH PITCH AT THE TOP

The procedure for laying out the pattern for a rectangular-to-round pitched at the top is quite different from the method used in previous plates.

To lay out the pattern by a practical short-cut method used in industry, draw only the side view as in Figure 2. The heights from B to $9A'$ and C to $1D'$ are equal to one half of the width of the top view $B-B$, Figure 1. Transfer the height of line 5–5 on the half circle to line B from point $9A'$ to $5'$, Figure 2. Transfer the height 6–6 on the half circle to line B from point $9A'$ to $6'$. Also transfer lines 7–7 and 8–8 to line B from point $9A'$ to $7'$ and $8'$. Transfer the heights of lines 2–2, 3–3, 4–4, and 5–5 to line $C-2D'$ from point $1D'$ to $2'$, $3'$, $4'$, and $5'$.

Transfer the slant lengths from point C to points 1, 2, 3, 4, and 5 to the base line; also transfer the slant lengths from point B to points 5, 6, 7, 8, and 9 to the base line in Figure 2, and number each as shown. The slant length from the numbers on the base line to their respective numbers on lines $C-1D'$ and $B-9A'$ will represent the true-length lines for the pattern in Figure 3.

To lay out the pattern in Figure 3, follow the same procedure as for Plate 9. The slant lines from points 5, 6, 7, 8, and 9 to points $5'$, $6'$, $7'$, $8'$, and $9A'$ on line $B-9A'$ in Figure 2 represent the true-length lines from point B in Figure 3. The slant lines from points 1, 2, 3, 4, and 5 to points $1D'$, $2'$, $3'$, $4'$, and $5'$ represent the true-length lines from point C in Figure 3. The slant lines from point 9 to $B-A$, and from point 1 to $C-D$ in the side view are true-length lines; therefore, transfer these two lines to the pattern in Figure 3, from $9'$ to A and $1'$ to D.

Complete the pattern with allowances as shown.

PLATE 37

1 TOP VIEW

2 SIDE VIEW & TRUE LENGTHS

3 HALF PATTERN

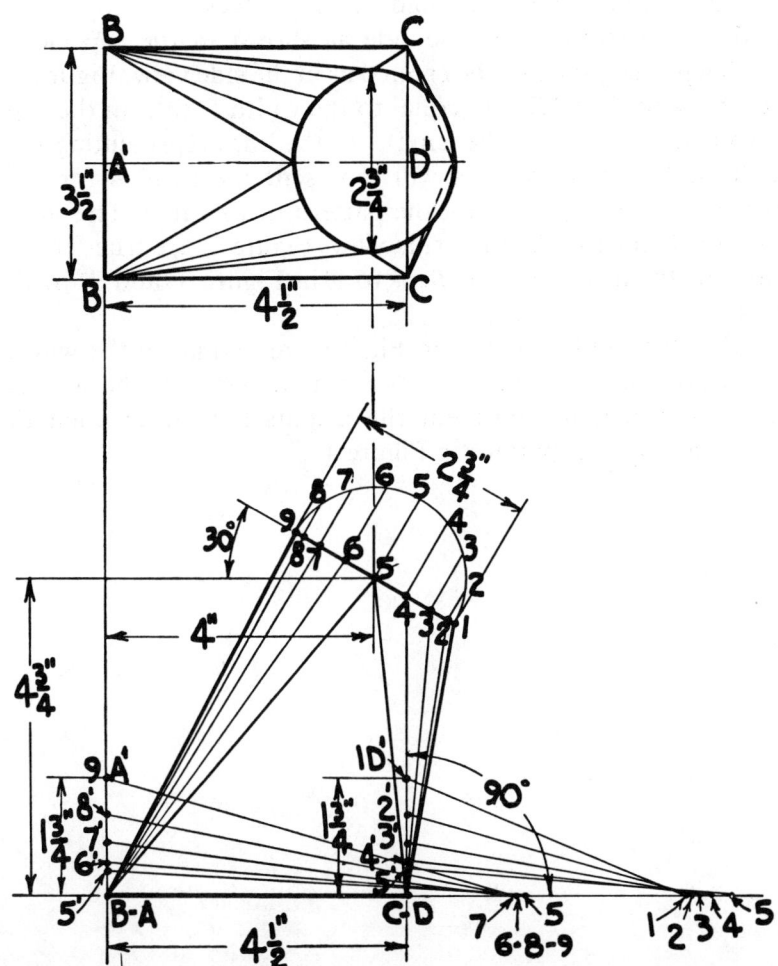

63

PLATE 38 RECTANGULAR-TO-ROUND, STRAIGHT ON ONE SIDE, WITH PITCH AT THE TOP

The procedure for this plate is identical with that for Plate 37, except that, since one side is straight, the patterns are made in two halves.

Because this fitting is straight on one side as shown in the top view, Figure 1, two true-length heights must be erected as in the side view, Figure 2. The distance B to $9C'$ and E to $1D'$ are equal to the width C to B in the top view. To these two heights, transfer the lengths of the lines representing the widths of the half circle. These heights will represent the heights for the true-length lines from points C and D on the pattern in Figure 3. The slant lines from points 9 to B and 1 to E in the side view, Figure 2, are true-length lines for both patterns from points $9'$ to B, $1'$ to E in Figure 3, and $9''$ to B, $1''$ to E in Figure 4.

The distances B to $9A''$ and E to $1F''$ in Figure 2 are equal to the width A to B in the top view, Figure 1. To these two heights transfer the widths of the half circle. These heights represent the heights for the true-length lines from points A and F on the pattern in Figure 4.

PLATE 38

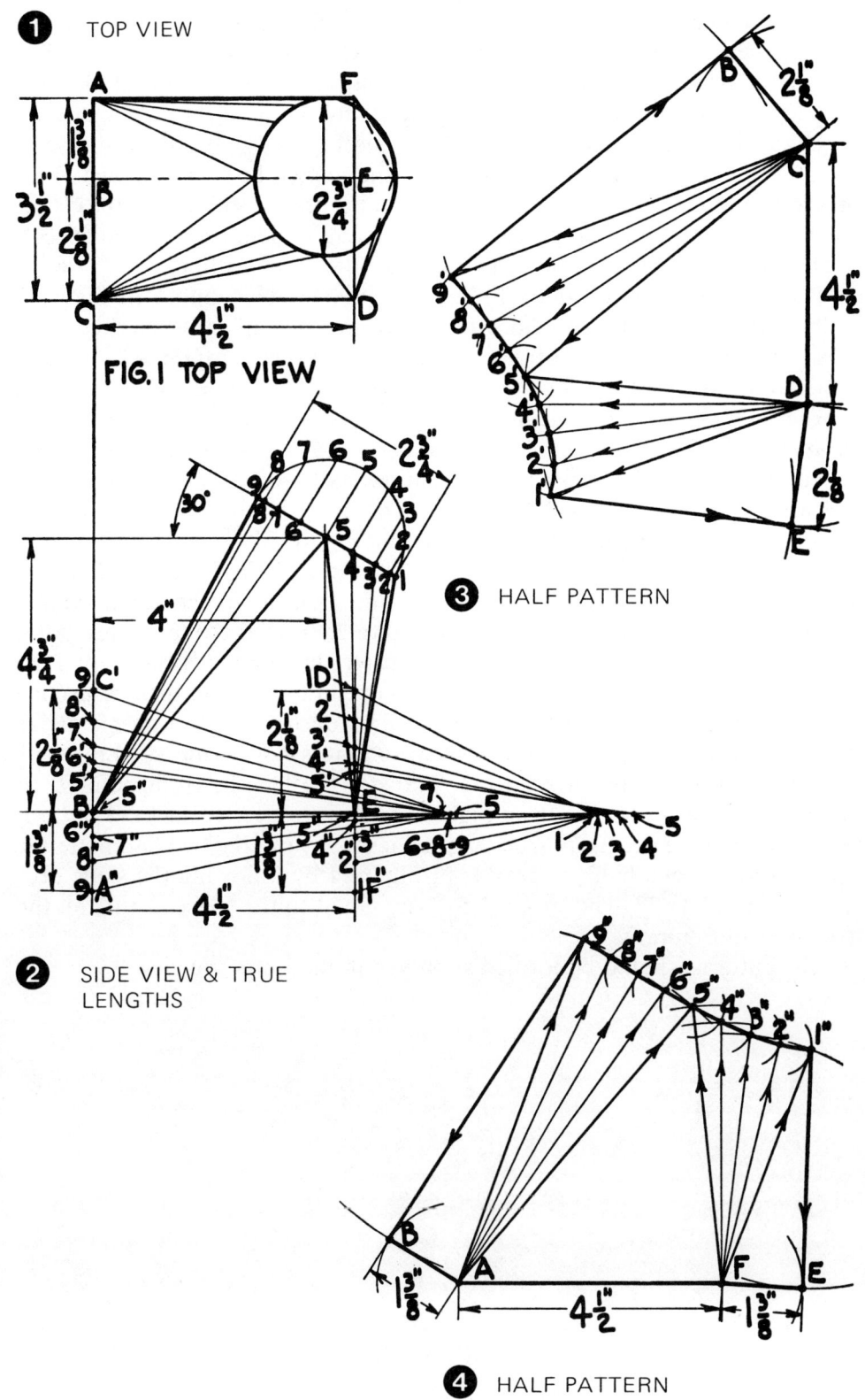

PLATE 39 50-DEG. DOUBLE-OFFSET RECTANGULAR-TO-ROUND

Draw the rectangular base in the top view as represented by points A, $B-C$, and D, then establish points 4 and 10, which represent the double offset and the vertical center line. Also draw the horizontal center line 1–7; then draw the half circle to the given dimension, and divide it into equal spaces.

Draw the base line in the side view equal to the given dimensions, and the 50-deg. angle line as shown. Draw the center line 4–10 down from Figure 1 to intersect the 50-deg. angle line in Figure 2, establishing the center point 4–10. Draw the half circle in Figure 2 and divide it into equal spaces, establishing the various points 1 to 7 on the 50-deg. angle line.

Draw lines up from points 1, 2–12, 3–11, 5–9, 6–8, and 7 in Figure 2 to intersect their respective lines drawn from the half circle in Figure 1, thus obtaining the elliptical freehand curved line through points 1 to 12.

Establish the points in Figure 1 for the kink or bend lines on the patterns, in order to facilitate forming and assembling. As in Plate 18, draw a rectangle parallel to the base lines A, B, C, D to encompass and strike on a tangent the curved lines on the ellipse as shown at points 1, 4, 7, and 10, or the points that are nearest each tangent-line striking point.

To erect the true-length triangles as in Figure 3, draw lines from the points on line 1–7 in Figure 2 to intersect line $F-G$ in Figure 3. Transfer the lengths from point A to points 1, 2, 3, and 4 on the freehand curved line in Figure 1 to their respective lines in Figure 3, from the intersecting point on line $F-G$ to the points as shown. Transfer the lengths from point B to points 4, 5, 6, and 7 on the freehand curved line in Figure 1 to their respective lines in Figure 3, from the intersecting points on line $F-G$ to the points as shown. Transfer the lengths from point C to points 7, 8, 9, and 10 in Figure 1 to their respective lines in Figure 3. Transfer the lengths from point D to points 10, 11, 12, and 1 to E from Figure 1 to their respective lines in Figure 3, thus obtaining the necessary true-length lines to lay out the two half patterns.

Lay out the two half patterns as in Figure 4 by following the same procedure as in Plate 38. The spaces 1 to 7 are equal to the spaces 1 to 7 on the half circle in Figure 2.

The patterns should be formed as shown in the isometric view.

PLATE 39

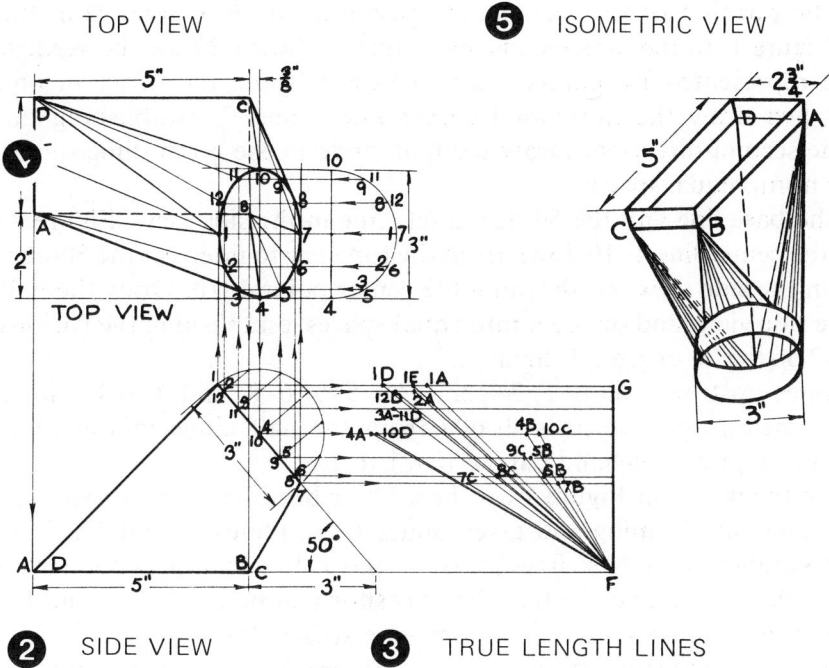

① TOP VIEW
② SIDE VIEW
③ TRUE LENGTH LINES
⑤ ISOMETRIC VIEW

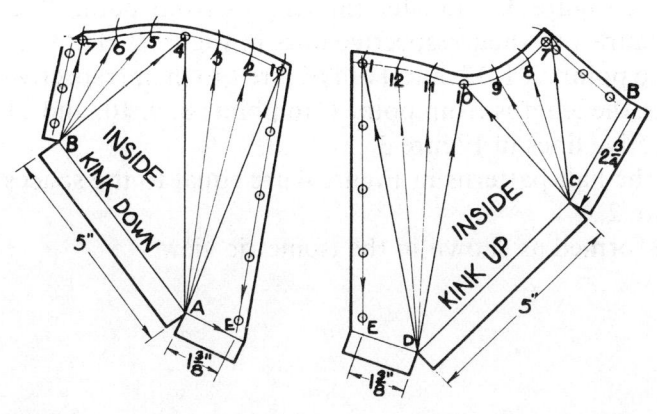

④ TWO HALF PATTERN

67

PLATE 40 50-DEG. TWISTED DOUBLE-OFFSET
RECTANGULAR-TO-ROUND

Draw the position of the rectangular base lines *A*, *B*, *C*, and *D* in the top view, Figure 1, to the dimensions and angle as shown. Draw the vertical center line represented by points 4 and 10 establishing the offset in one direction; then draw the horizontal center line 1 and 7, establishing the offset in the second direction. Draw the half circle to the given dimensions, and divide it into equal spaces.

Draw the base line and the 50-deg. angle line in the side view, Figure 2; then draw the center line 4–10 down from the top view to intersect the 50-deg. angle line in the side view, establishing the center point 4–10. Draw the half circle in the side view, and divide it into equal spaces, establishing the various points 1 to 7 on the 50-deg. angle line.

Draw lines up from points 1, 2–12, 3–11, 5–9, 6–8, and 7 in Figure 2 to intersect their respective lines drawn from the half circle in Figure 1, obtaining the elliptical freehand curved line 1 to 12.

Establish the points in Figure 1 for the kink or bend lines on the patterns in order to facilitate forming and assembling. As in Plates 18 and 39, draw a rectangle parallel to the base lines *A*, *B*, *C*, and *D* to encompass and strike on a tangent the curved lines on the ellipse as shown at points 2, 5, 8, and 11, or the points that are nearest each tangent-line striking point.

To erect the true-length triangles as in Figure 3, draw lines from the points 1 to 7 in Figure 2 to intersect line *F–G* in Figure 3. Transfer the lengths from point *A* to points 2, 3, 4, and 5 on the freehand curved line in Figure 1 to their respective lines in Figure 3. Transfer the lengths from point *B* to points 5, 6, 7, and 8 in Figure 1 to their respective lines in Figure 3. Transfer the lengths from point *D* to points 2, 1, 12, and 11 in Figure 1 to their respective lines in Figure 3. Transfer the lengths from point *C* to points 8, 9, 10, and 11 in Figure 1 to their respective lines in Figure 3.

The spaces 2 to 8 on the half patterns in Figure 4 are equal to the spaces on the half circle in Figure 2.

The patterns must be formed as shown in the isometric view.

PLATE 40

1 TOP VIEW

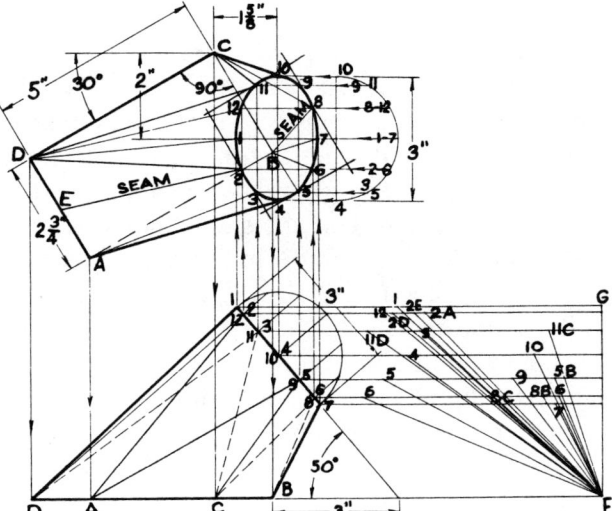

2 SIDE VIEW

3 TRUE LENGTH LINES

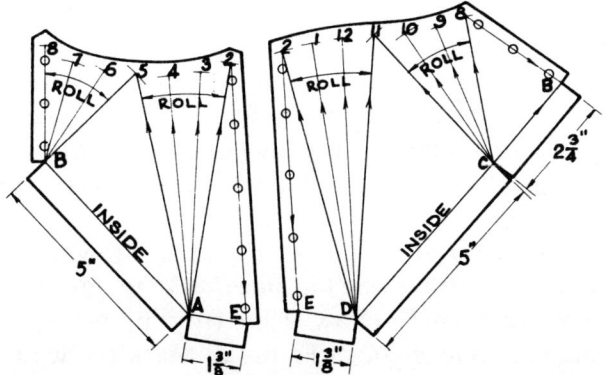

4 TWO HALF PATTERNS

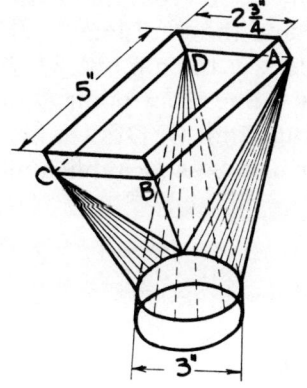

5 ISOMETRIC VIEW

PLATE 41 SQUARE-TO-INTERSECTING HIP ON ROOF

To erect the true-length triangles in Figure 2, transfer the various lengths from the top view of the square-to-round in Figure 1 to their respective base lines in Figure 2.

To lay out the square-to-round pattern as in Figure 4, draw the base line A–C to the dimensions shown. Use the slant true-length line C–F to F in Figure 2 as a radius, and use points A and C as centers to strike arcs to cross each other at point F, Figure 4. Use the slant true-length line C–1 to point L in Figure 2 as a radius, and points C and A in Figure 4 as centers to strike arcs to cross each other at point 1, Figure 4. Use the slant lines C–2 to L and C–3 to L in Figure 2 as radius lengths, and point C in Figure 4 as a center to strike arcs at points 2 and 3. The spaces 1 to 2, 2 to 3, etc., are equal to the spaces 1 to 2 on the circle in Figure 1. Use the slant true-length line G–3 to L in Figure 2 as a radius, and point 3 in Figure 4 as a center to strike an arc at G. Use the slant true-length line C–G to G' in Figure 2 as a radius, and point C in Figure 1 as a center to strike an arc crossing the arc at G. Complete the pattern by transferring the remaining true-length lines from Figure 2 to their respective points in Figure 4.

To obtain the cutout opening in Figure 3, transfer the spaces from the slant hip line N–M in Figure 2 to the center line N–M in Figure 3. Transfer the width A' to A and J' to J from Figure 1 to their respective lines in Figure 3 obtaining the cutting line H, J, A, and F.

To obtain the top view of the square-to-round in Figure 1, draw line D to A to any length desired. Draw line C–D parallel to the hip line M–P to intersect line D–A at point D. Draw line C to A equal to the dimensions shown. The remaining points will be obtained after the side view in Figure 2 has been constructed.

To obtain the side view of the square-to-round in Figure 2, draw a line from point A in Figure 1 to intersect the slant hip line N–M in Figure 2 at point A', establishing the base line A'–K. Draw a line from points B, C, D, and E in the top view to the base line A'–K in Figure 2. Mark the width of the diameter, and draw a slant line from points A' and D to intersect at point O. Draw a slant line from points B and E to point O crossing the hip line at F' and H'. Draw a line from points H', J', and F' to cross line L–K. Point G is established at the crossing point of line J' and the slant line D–O. Draw a line from points F' and H' in Figure 2 to intersect the hip line M–P in Figure 1, establishing points F and H. Also draw a line from point G in Figure 2 to intersect line D–A in Figure 1, establishing point G. Draw a line from point G parallel to D–E to intersect the center line C–J', establishing point J.

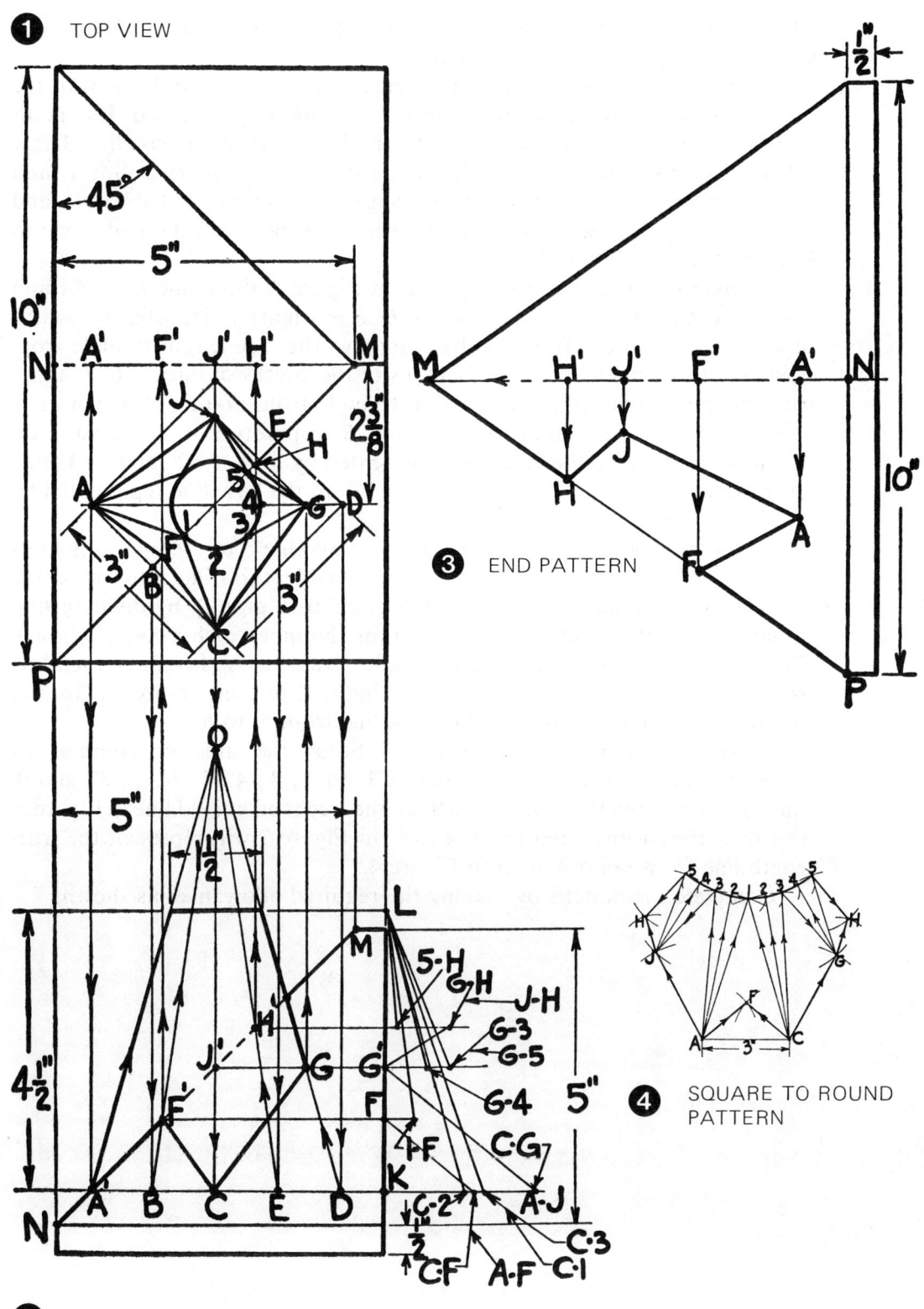

PLATE 42 TWO-PIECE ELBOW BOOT

The method used for this fitting is a combination between the elbow boot and the rectangular-to-round on a pitch.

To construct the side view as in Figure 2, draw line 9 to 9' equal to the dimensions as shown, and draw line 1–1' to any length desired. Use point 9' as a center to draw an arc tangent to line 1–1' toward the base line. Draw a line from point B tangent to the arc just drawn to intersect line 1, and establish point 1'. Then draw a slant from 1' to 9'. Draw the half circle, and divide into equal spaces. Draw a line from each point on the half circle to intersect the slant line 1'–9'.

To erect the true-length triangles as in Figure 2, draw line B to $B'1$ and A to $A'9$ equal to one half of the width $B-B$ in Figure 1. Transfer the widths of lines 5, 4, 3, and 2 from the half circle to the true-length triangle from point $B'1$ to points 5, 4, 3, and 2. Transfer the widths of lines 5, 6, 7, and 8 from the half circle to the true-length triangle from point $A'9$ to points 5, 6, 7, and 8. Transfer the lines from point B to points 5, 4, 3, 2, and 1' on the slant line 1'–9' to the base line represented by 5', 4', 3', 2', and 1'. Transfer the lines from point A to 5, 6, 7, 8, and 9' to the base line represented by 5'', 6', 7', 8', and 9'.

To lay out the pattern as in Figure 3, draw line 5 to 5 equal to the circumference of a $3\frac{1}{2}$-in.-diameter circle. Divide this line into twice the number of equal spaces as on the half circle in Figure 2, by dividing first in halves, quarters, eighths, and sixteenths. Number each point as shown. This will allow the seam to be on the side. Transfer the length of each line from line 1–9 to the slant line 1'–9' in Figure 2 to their respective lines in Figure 3, obtaining the freehand curved line from 5' to 5'.

To lay out the pattern as in Figure 4, follow the same procedure as for the elbow boots, except that the spaces 1' to 2', 3', 4', 5', 6', 7', 8', and 9' must be taken from the spaces 1' to 9' on the freehand curved line in Figure 3. The slant throat line from point A to 9' in Figure 2 will represent the true-length line from point 9' to A in Figure 4.

Complete the pattern by making the required allowances as shown.

PLATE 42

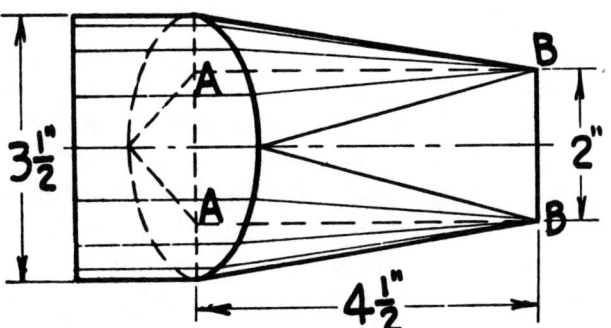

① TOP VIEW

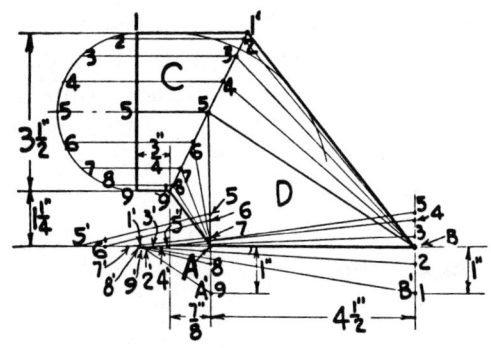

② SIDE VIEW

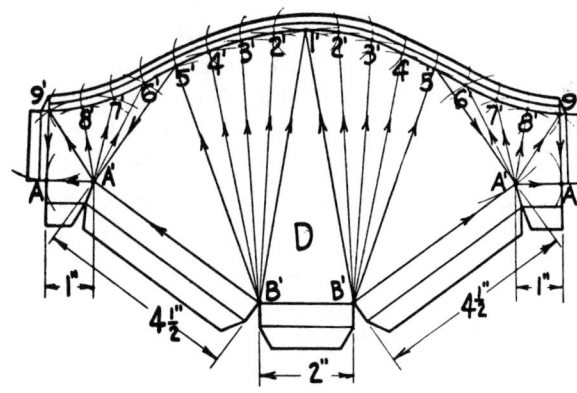

④ FULL PATTERN D

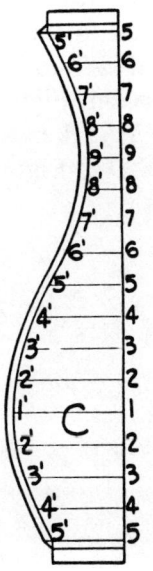

③ FULL PATTERN C

PLATE 43 THREE-PIECE RECTANGULAR-TO-ROUND ELBOW

To obtain the side view as in Figure 2, draw lines 1 to 1' and 9 to 9' to any length desired; also draw lines *A* to *A'* and *B* to *B'* to any length desired. Use the radius point *R* as a center to draw the quarter circle through the center point 5" to the base line. Since this is a three-piece elbow, we treat this the same as a 90-deg. round elbow. The number of pieces times two minus two equals the number of spaces that the center line will be divided. (Thus, $3 \times 2 = 6$, $6 - 2 = 4$.) Therefore, divide the center line into four equal spaces. Then draw a line from point *R* through the first division point crossing lines *A* and *B* to establish points *A'*–*D* and *B'*–*C*, Figure 2. Draw another line from point *R* through the third division point crossing lines 1 and 9 to establish points 1' and 9'. This allows the mid segments to be twice the width at the center than either end segment.

To erect the true-length triangles as in Figure 2, draw a line squaring up from the slant line at *B'*–*C* to 1' and *A'*–*D* to 9'; these are equal to the distance *B* to *C* in Figure 1. Transfer the lines from point *B'*–*C*, Figure 2, to points 1, 2, 3, 4, and 5 on the slant line 1'–9' to the true-length triangle represented by 1', 2', 3', 4', and 5'. Transfer the lines from *A'*–*D* to 5, 6, 7, 8, and 9' to the true-length triangle represented by 5", 6', 7', 8', and 9'.

To lay out the pattern as in Figure 3, draw line 1 to 9 equal to one half of the circumference of a 4-in.-diameter circle. Divide into 8 equal spaces, by dividing in halves, quarters, and eighths. Transfer the length of each line 1 to 9 on section *M* in Figure 2 to their respective lines in Figure 3.

To lay out the pattern as in Figure 4, draw line *C* to *D* equal to the spaces *C* to *D* in Figure 1. The heights are equal to the heights *A* to *A'*–*D* and *B* to *B'*–*C* in Figure 2.

To lay out the pattern as in Figure 5, follow the same procedure as used for Plates 35 and 40, except that the distance *A'* to *B'* is equal to the slant line *A'* to *B'* in Figure 4. The spaces 5' to 4', 3', 2', and 1', also 5' to 6', 7', 8', and 9' are taken from spaces 1' to 9' on the freehand curve in Figure 3. The distances 1' to *C* and 9' to *D* on the pattern in Figure 5 are taken from the slant heel line 1' to *B'*–*C*, and the slant throat line 9' to *A'*–*D* in Figure 2.

The second half pattern is a duplicate of the first.

PLATE 43

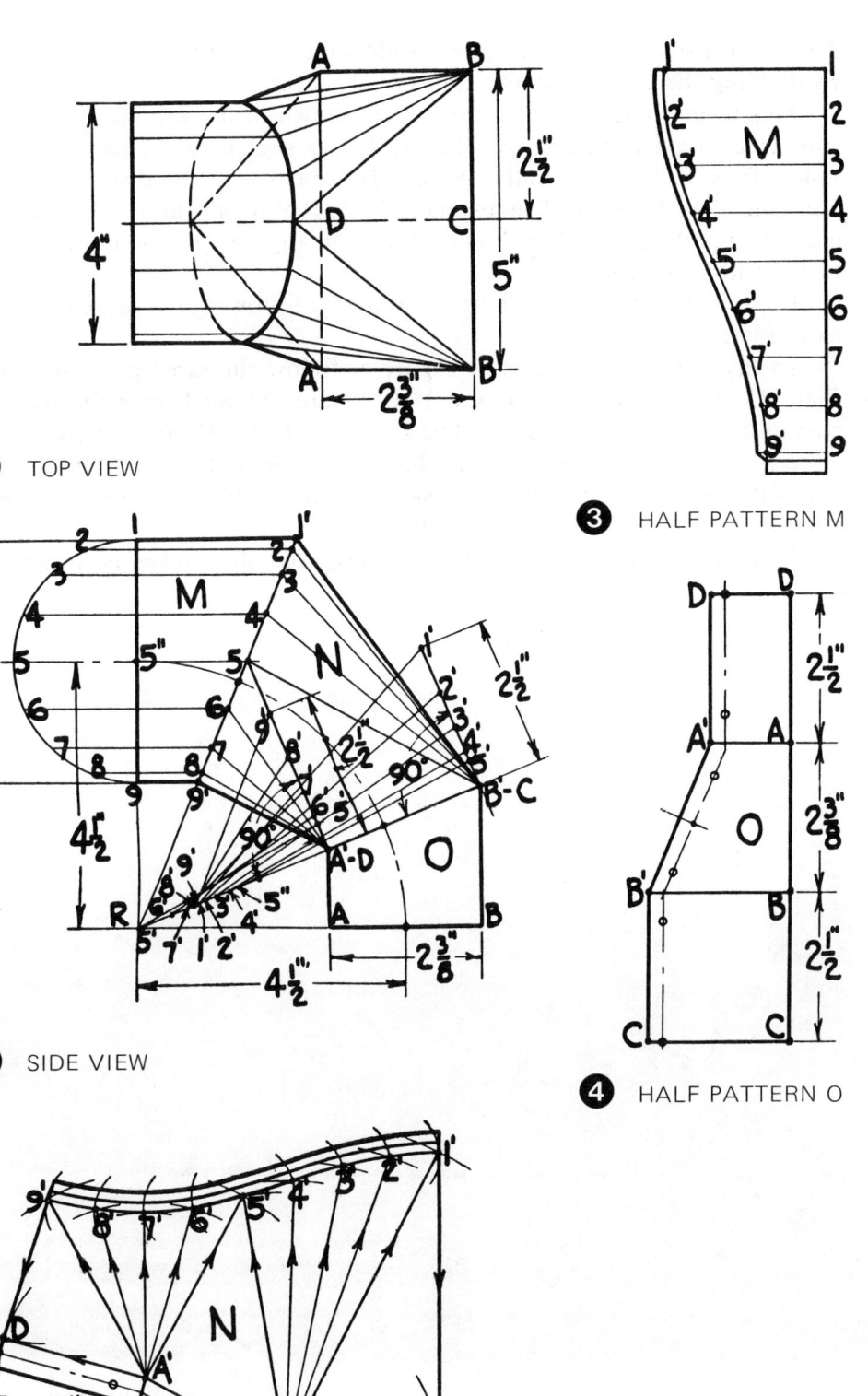

① TOP VIEW

② SIDE VIEW

③ HALF PATTERN M

④ HALF PATTERN O

⑤ HALF PATTERN N

PLATE 44 THREE-PIECE RECTANGULAR-TO-ROUND ELBOW, ONE-SIDE STRAIGHT

To draw the side view as in Figure 2, follow the same procedure as in Plate 43 by dividing the center line into four equal spaces.

This fitting is straight on one side as shown in the top view; therefore, four true-length triangles must be erected as shown in Figure 2. The distances from $E'5'$ to $F1''$ and $E'5'$ to $1'D$ are equal to the distances F to E and E to D in Figure 1. The distances from $B'5'$ to $A9''$ and $B'5'$ to $9'C$ are equal to the distances A to B and B to C in Figure 1, completing the four true-length triangles.

Lay out the patterns in Figures 3 and 4 by following the same procedure as in Plate 43.

To lay out the pattern as in Figure 5, follow the same procedure as in Plate 43. The spaces $1''$ to $9'$ and $1''$ to $9''$ are obtained from the freehand curved line $5''$ to $5''$ in Figure 3. The spaces D' to C', C' to B', F' to A', and A' to B' are obtained from the slant lines D', C', B' and F', A', B' in Figure 4. The slant throat line 9 to B' in the side view in Figure 2 represents the true-length line $9'$ to B' and $9''$ to B' in Figure 5.

Complete the pattern by making the required allowances as shown.

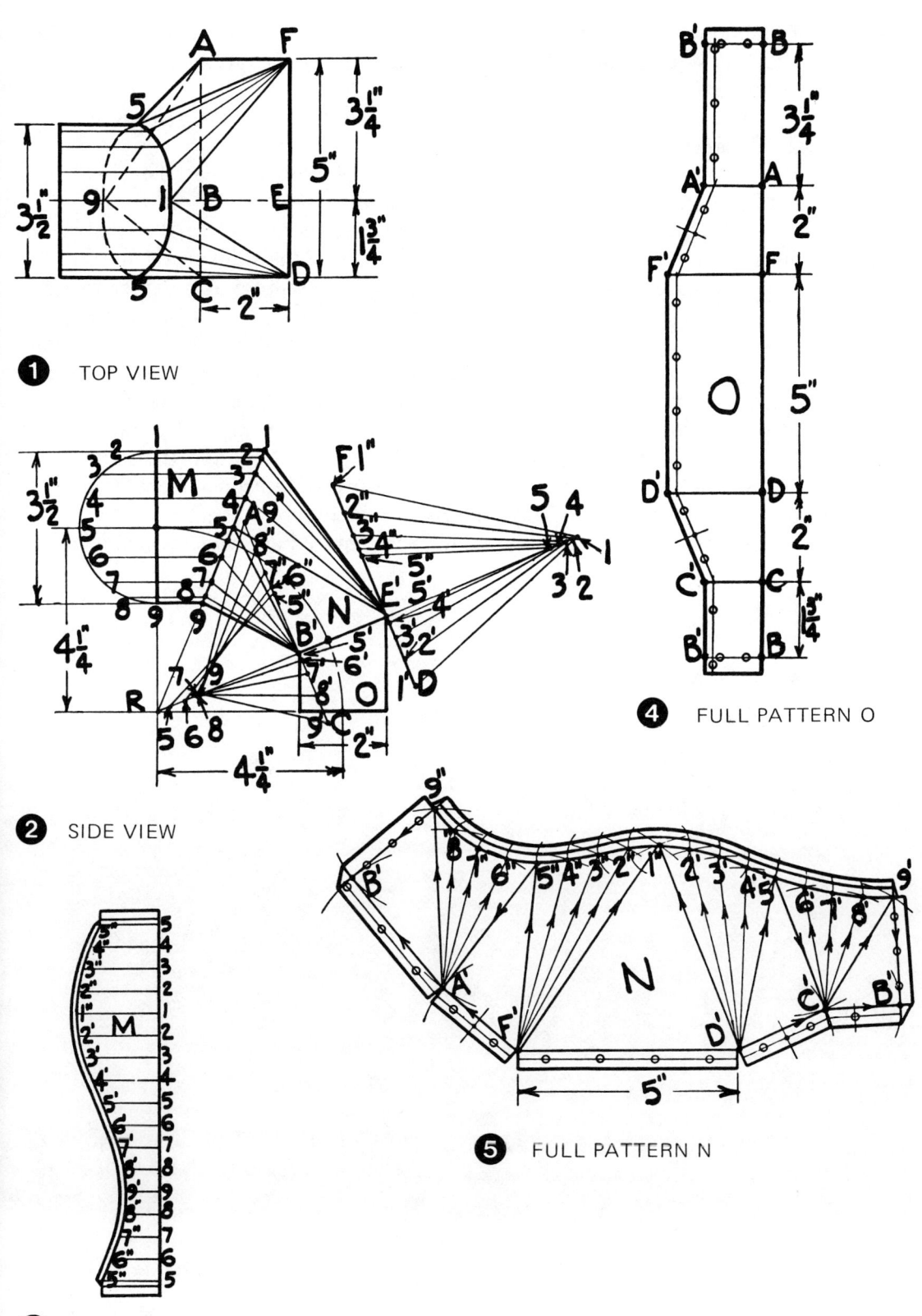

PLATE 44

PLATE 45   THREE-PIECE OBLONG-TO-RECTANGULAR ELBOW

To obtain the side view as in Figure 2, draw the center line and divide it into four equal spaces. Then follow the same procedure as in Plate 43, the rectangular-to-round elbow.

Erect the two true-length triangles B'–C to $1'$ and A'–D to $8'$ equal to the distance B to C in Figure 1.

Lay out the patterns as in Figures 3 and 4. Only half patterns are shown, but full patterns may be laid out if desired.

To lay out the pattern as in Figure 5, draw line A' to B' equal to the distance A'–B' on the pattern in Figure 4. Use the slant true-length line $4'$ to 4 in Figure 2 as a radius, and point B' in Figure 5 as a center to strike an arc at $4'$. Use the slant true-length line from point $5'$–$4'$ to the point marked A' to 4 in Figure 2 as a radius, and point A' in Figure 5 as a center to strike an arc crossing the arc at point $4'$. The remaining pattern may be completed by following the same procedure as in Plate 43. The spaces $4'$ to $5'$, $6'$, $7'$, and $8'$, also $4'$ to $3'$, $2'$, and $1'$ are taken from the freehand curved line $1'$ to $8'$ in Figure 3. The slant heel line 1 to B'–C and the slant throat line 8 to A'–D in Figure 2 represent the true-length lines from point $1'$ to C and from point $8'$ to D on the pattern in Figure 5.

Complete the pattern by making the required allowances as shown.

PLATE 45

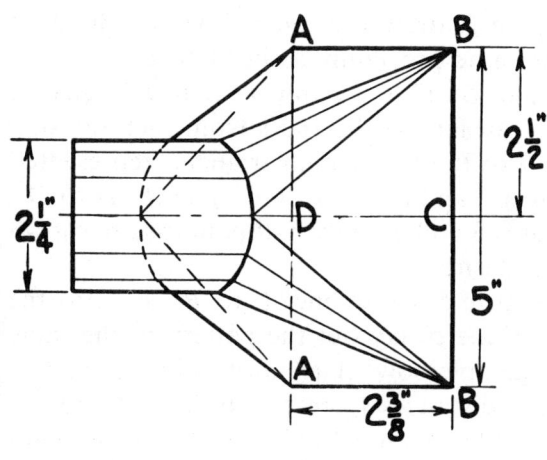

① TOP VIEW

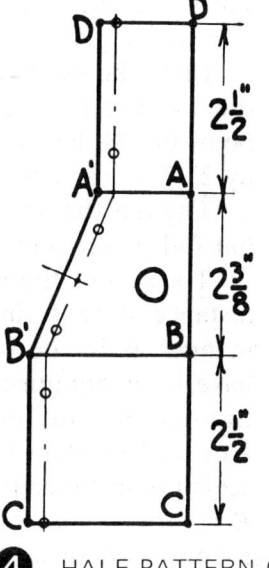

④ HALF PATTERN O

② SIDE VIEW

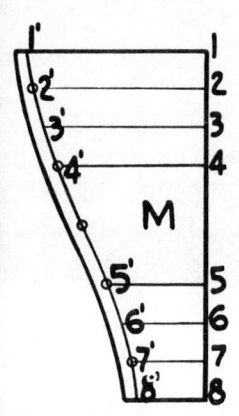

③ HALF PATTERN M

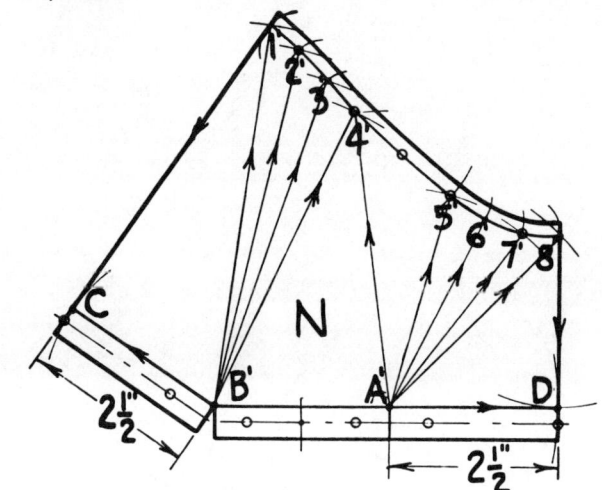

⑤ HALF PATTERN N

PLATE 46 THREE-PIECE OBLONG-TO-RECTANGULAR ELBOW

To obtain the side view as in Figure 2, draw the center line and divide it into four equal spaces. Follow the same procedure as in Plate 43.

Erect the two true-length triangles $B'-C$ to C'' and $A'-D$ to D'' equal to the distance B to C in Figure 1. Transfer the distances from each division point on the half circle to line 1–9 to the true-length triangles represented by 2', 3', 4', and 5' on line $B'-C$ to C'' and 8', 7', 6', 5' on line $A'-D$ to D''.

Lay out the patterns as in Figures 3 and 4. Only half patterns are shown but full patterns may be laid out if desired.

To lay out the pattern as in Figure 5, draw line A' to B' equal to the distance A' to B' in Figure 4. Continue to lay out the pattern in the same manner as in the round-to-rectangular elbow plate until points 2' and 8' have been obtained. To obtain the distances from points 2' to C' to D', transfer the slant heel line $B'-C$ to 1–2 in Figure 2 to line 1–9–R from point 1 to C'. The slant length from point 2 on the half circle to point C' represents the distance from point 2' to C' in Figure 5. The slant heel line from $B'-C$ to 1'–2 in Figure 2 represents the true-lengths line from C' to 1' in Figure 5. Transfer the slant curve line from $A'-D$ to 8–9 in Figure 2 to line 1–9–R from point 9 to point D'. The slant length from point 8 on the half circle to point D' represents the distance from point 8' to D' in Figure 5. The slant throat line $A'-D$ to 8–9 in Figure 2 represents the true-length line D' to 9' in Figure 5.

Complete the pattern by making the required allowances as shown.

PLATE 46

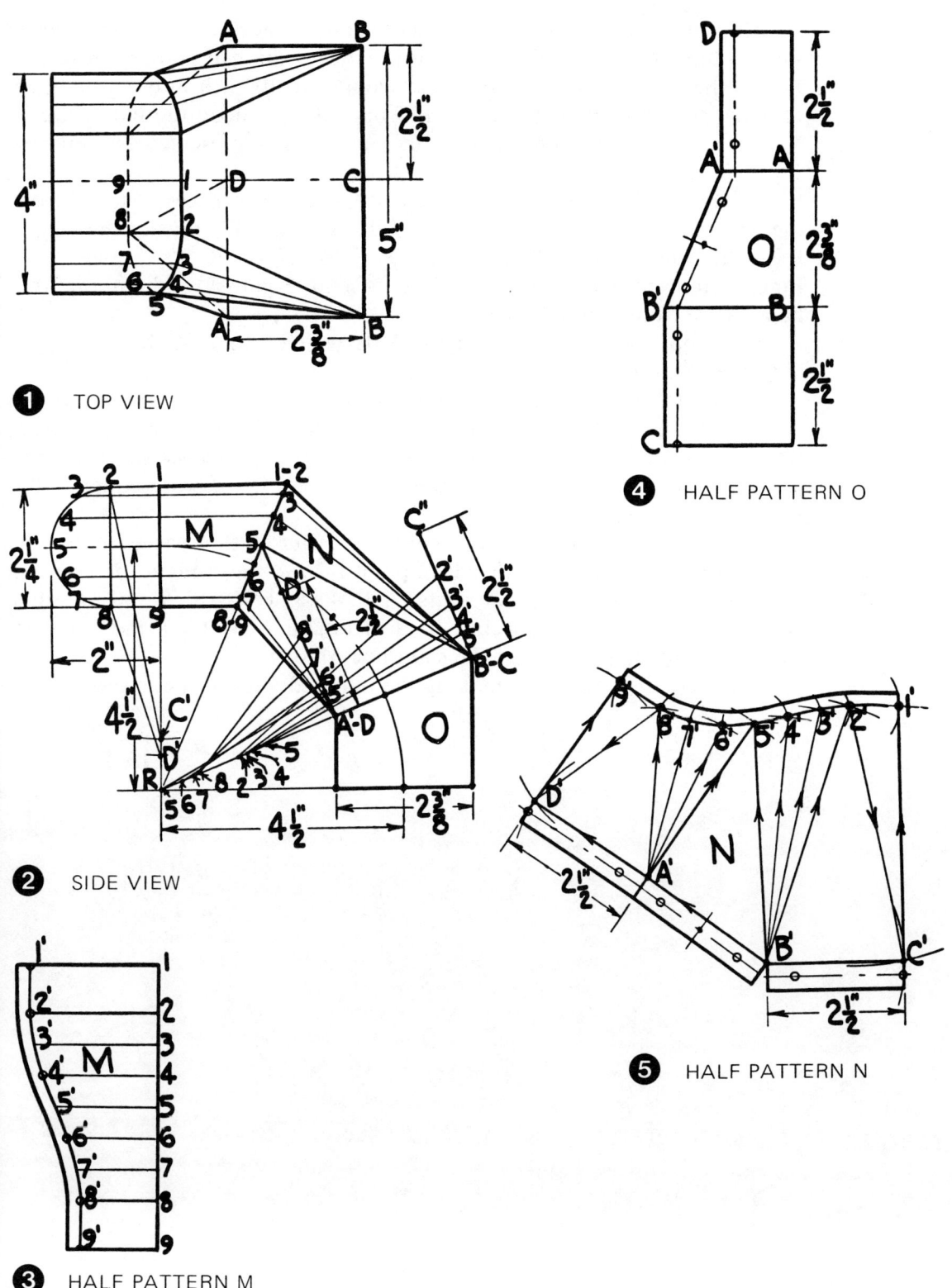

① TOP VIEW

② SIDE VIEW

③ HALF PATTERN M

④ HALF PATTERN O

⑤ HALF PATTERN N

PLATE 47   RECTANGULAR-TO-ROUND WITH CURVED BACK

In this problem only half of the top view and the side view must be drawn. Draw a line from point $A'-B'$ tangent to circle striking at point $6'$, Figure 1. Divide the distance from $1'$ to $6'$ and from $6'$ to 10 each into any number of equal spaces desired. Draw a line from each point on the half circle in Figure 1 to intersect the heel curve in Figure 2 as represented by numbers $1''$ to $10''$.

To lay out the pattern as in Figure 4, transfer the spaces $1''$ to B'' from the heel curve in Figure 2 to line $1''$ to B'' in Figure 4. Transfer the widths from the center line $1'-10$ to points $2'$, $3'$, $4'$, $5'$, $6'$, $7'$, $8'$, $9'$, and $10'$ in Figure 1 to each side of the center line $1''-B''$ in Figure 4 to obtain the freehand curve from point $1''$ to B.

To lay out the pattern as in Figure 5, draw line $6'$ to $6'$ equal to the space $6'$ to $6'$ in Figure 1.

To lay out the pattern as in Figure 6, erect a true-length triangle as AB to C in Figure 1. Lay out the pattern as in Figure 6 by transferring the true-length lines from the triangle in Figure 1, and proceed in the same manner as for a square-to-round. The spaces $6'$ to $6'$ in Figure 6 are equal to the spaces $6'$ to $6'$ in Figure 1.

To lay out the pattern, as in Figure 7, transfer all the spaces from point $1'$ to $A'-B'$ in Figure 1 to each side of point $1'$ to $A'-B'$ in Figure 7. Transfer the heights from line $1''-C'$ to points $2''$, $3''$, $4''$, $5''$, $6''$, $7''$, $8''$, $9''$, $10''$, B'', and A in Figure 2 to their respective lines in Figure 7 to obtain the freehand curve from point $1'$ to B and the slant line $6'$ to A.

Complete the pattern by making the required allowances as shown.

PLATE 47

① TOP VIEW

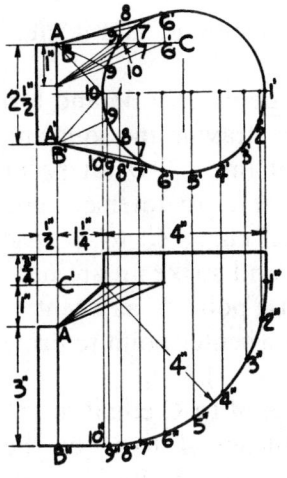

② SIDE VIEW

③ END VIEW

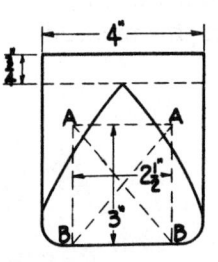

④ BACK PATTERN

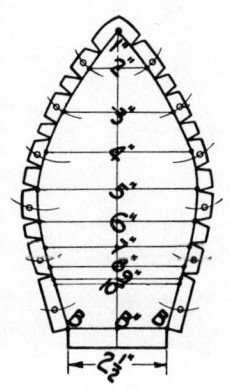

⑤ FRONT COLLAR PATTERN

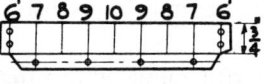

⑥ FRONT TOP PATTERN

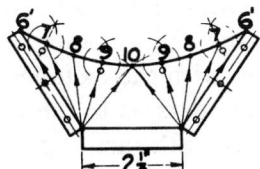

⑦ SIDE PATTERN

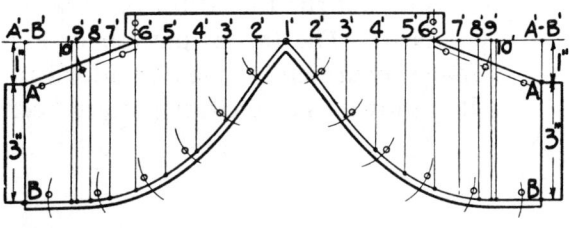

83

PLATE 48 RECTANGULAR-TO-RECTANGULAR TRANSITION ELBOW

This type of fitting generally is used on residential heating and air conditioning.

The top and front views show the position of the openings, and it must be assumed that the heel is curved.

To lay out the patterns as in Figure 3, draw the side pattern M and top pattern N as shown. To lay out the slant side pattern, draw a line squaring up from point B on pattern M to intersect the slant line $C'-D'$ at point E on pattern O. Draw a line squaring up from point E to intersect line $B'-A'$. To obtain the radius point F to draw the heel curve C' to B', use point C' as a center with the distance C' to E radius, and strike an arc near point F. Keep the dividers set to the same radius, and use point B' as a center, to strike an arc crossing the arc at point E, obtaining the center point to draw the heel curve C' to B'.

To lay out the heel pattern as in Figure 4, draw line A to C equal to the distance A to C on the heel line in Figure 3. The slant line A' to C' on the heel pattern, Figure 4, automatically becomes the length of the heel line A' to C' on pattern O in Figure 3.

Lay out the throat for pattern N as in Figure 5.

NOTE: The throat pattern in Figure 5 does not necessarily have to be made separately; it may be added to the throat on the pattern in Figure 3. It is shown as a separate pattern because some shops prefer this method.

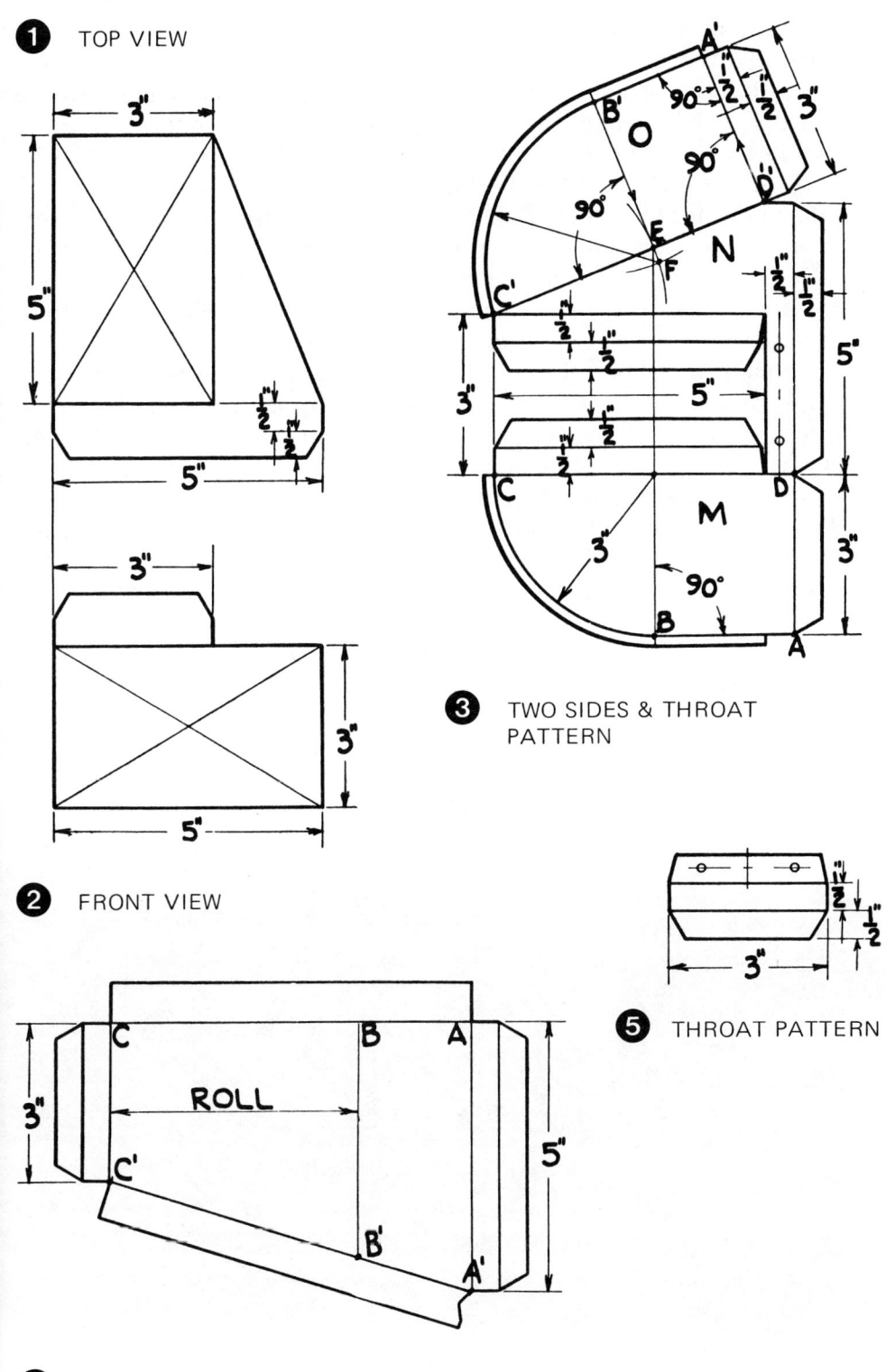

PLATE 48

① TOP VIEW

② FRONT VIEW

③ TWO SIDES & THROAT PATTERN

④ HEEL PATTERN

⑤ THROAT PATTERN

PLATE 49 RECTANGULAR-TO-RECTANGULAR TRANSITION EQUAL TAPER

The top and front views show the position of the openings, and it must be assumed that the heel is curved.

To lay out the top and the two slant cheek patterns as in Figure 3, follow the same procedure as in Plate 48. Draw the top pattern N to the dimensions shown. Draw line E to E parallel to line D' to D'. The distance in from point A'–A' is equal to the height of the cheeks (3 in.) and would also represent the radius for the flat cheek pattern if one cheek was flat as in Plate 46. Use point A' as a center and the distance A' to E as a radius to strike an arc at F. Keep the dividers set, and use point B' as a center to strike an arc crossing the arc at F, thus obtaining the radius point to draw the heel curve from A' to B'.

To obtain the length of the center line A to B on the heel pattern in Figure 4, multiply what would be the radius for the flat cheek 3 in. × 1.57 = 4.71 or $4^{11}/_{16}$ in. plus the straight 2 in. The slant line A' to C' will automatically become the length of the heel line on the slant cheek patterns.

Lay out the throat pattern as in Figure 5.

The throat pattern in Figure 5 does not necessarily have to be made separately; it may be added to the throat on the pattern in Figure 3. It is shown as a separate pattern because some shops prefer this method.

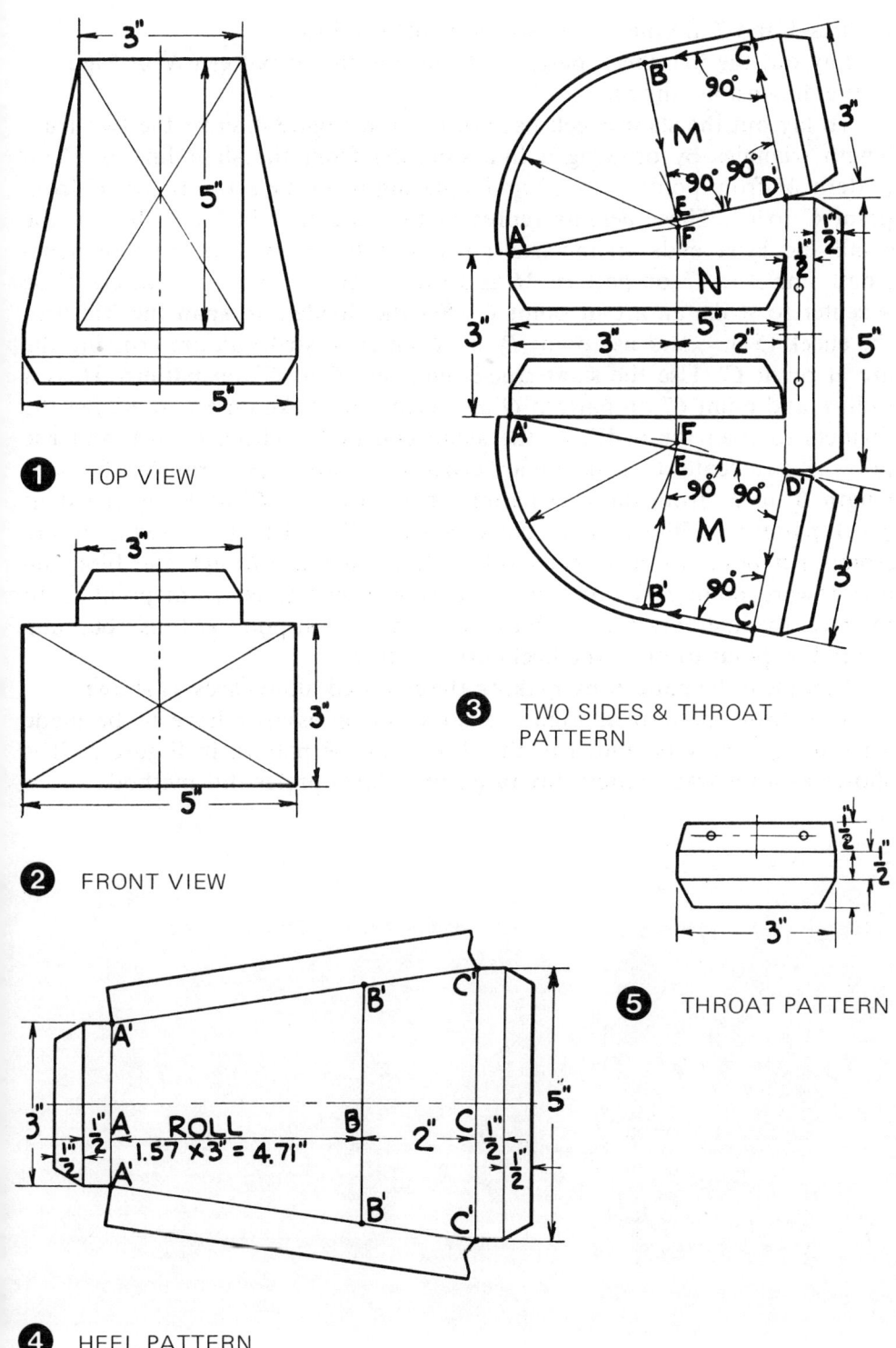

PLATE 49

PLATE 50 TRANSITION ELBOW WITH SLANT THROAT

Figures 1 and 2 are merely illustrations of this fitting.

Lay out the flat cheek pattern M and the throat pattern N in Figure 4 to the dimensions shown.

To lay out the slant cheek pattern O as in Figure 4, erect the two true-length triangles by drawing a line squaring from the slant line $B-C$ on pattern M from point C to C'' and squaring from the slant line $A-C$ from point C to C'. These heights represent the difference in the widths of the small and large ends on the heel pattern in Figure 3. Use the slant true-length line A to C' on pattern M as a radius, and point A' on pattern O as a center to strike an arc at point C'. Set the dividers to span the width of the cheek (5 in.), and use point D' as a center to strike an arc crossing the arc at point C'. Use the slant true-length line B to C'' on pattern M as a radius, and point C' on pattern O as a center to strike on arc at B'. Set the dividers to span the width of the small end of the cheek (3 in.), and use point A' as a center to strike an arc crossing the arc at B'. Transfer the slant length B' to E' from the heel pattern in Figure 3 to B' to E' on the slant cheek pattern O. Bisect the distance between E' and C' by drawing an arc from each point to cross each other. Draw a line through the bisecting arcs toward point F. Draw a line squaring from line $B'-E'$ at point E' to intersect the line drawn through the bisecting arcs at point F, thus obtaining the radius point to draw the heel curve C' to E'.

Complete the pattern by making the required allowances as shown.

The throat pattern in Figure 5 does not necessarily have to be made separately; it may be added to the throat on the pattern in Figure 3. It is shown as a separate pattern because some shops prefer this method.

PLATE 50

① TOP VIEW

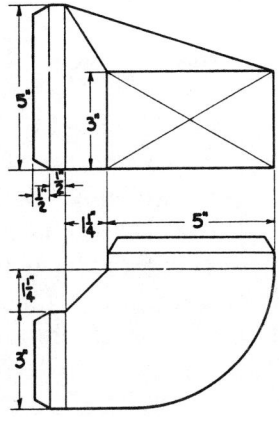

② SIDE VIEW

④ THROAT & TWO SIDE PATTERN

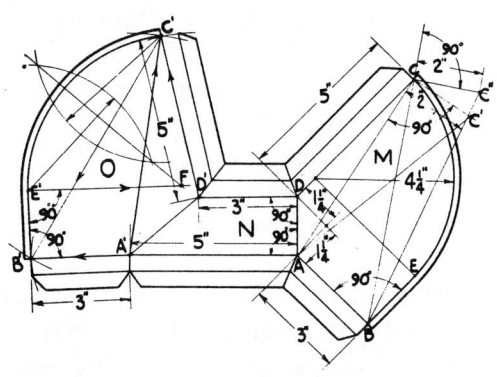

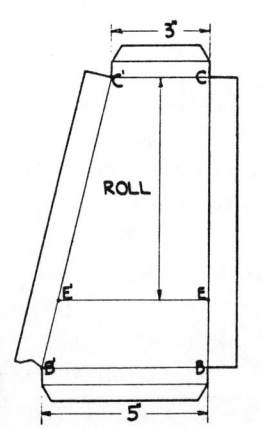

③ HEEL PATTERN

89

PLATE 51 TRANSITION ELBOW WITH SLANT THROAT

This fitting is the same as Plate 50, except that the slant cheek pattern O is laid out with the use of a square to obtain the various points. This method is faster and simpler than that used in Plate 50.

Lay out the flat cheek pattern M and the throat pattern N in Figure 4 to the dimensions shown.

To lay out the slant cheek pattern O as in Figure 4 with the use of a square, set the square so that the point representing $1\frac{1}{4}$ in. on one leg of the square (as presented by D'' to D') rests on point D'; then move the square until the other leg of the square rests on point A'. With the square in its proper position (the $1\frac{1}{4}$-in. mark on the first leg still resting on point D') draw a line from D' to C' to the dimensions given. Set the square so that the $1\frac{1}{4}$-in. point on one leg of the square (as represented by A'' to A') rests on point A'; then move the square until the other leg of the square rests on point C'. With the square in its proper position (the $1\frac{1}{4}$-in. mark on the first leg still resting on point A') draw a line from point A' to B' to the dimensions given. (Note the $1\frac{1}{4}$-in. distances D'' to D' and A'' to A' represent the base and height of the throat.) Draw a line from point B' toward F to any length desired. Set one leg of the square on line $B'-F$, and move the square until the other leg of the square rests on point C'; then draw a line from C' to intersect line $B'-F$ establishing point F.

Lay out the heel pattern and transfer the slant distance $B'-E'$ from the heel pattern in Figure 3 to line $B'-E'$ on the slant cheek pattern O.

Set the dividers to span the distance E' to F, Figure 4; then point E' as a center to draw an arc toward point F'. Use point C' as a center to draw an arc crossing the arc at point F', establishing the center point to draw the heel curve C' to E'.

Complete the pattern by making the required allowances as shown.

PLATE 51

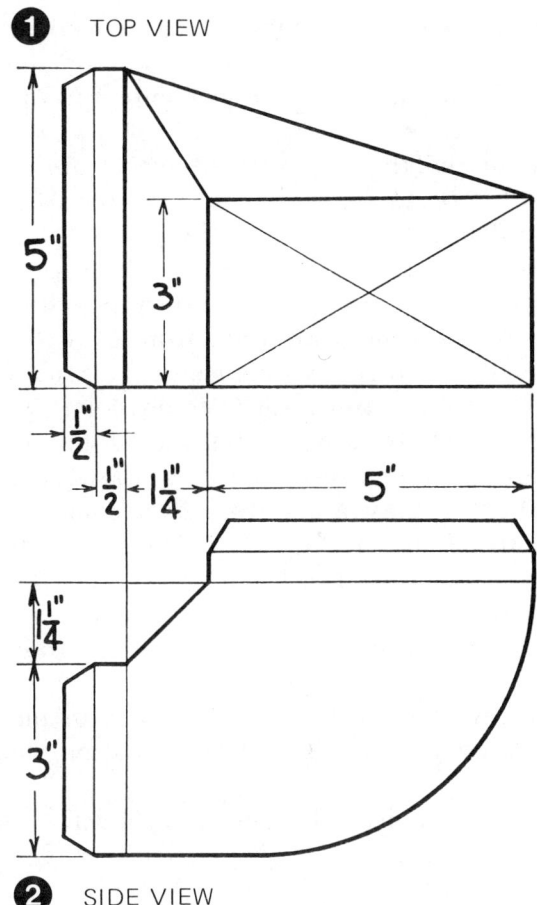

① TOP VIEW

② SIDE VIEW

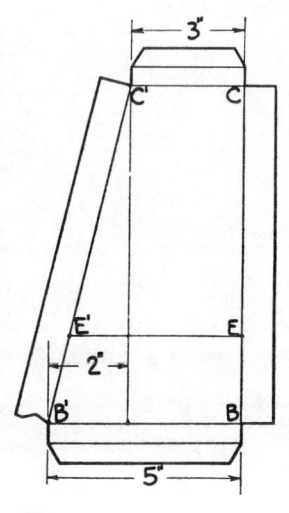

③ HEEL PATTERN

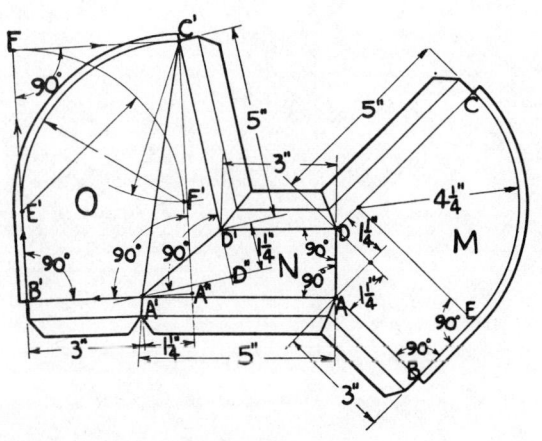

④ THROAT & TWO CHEEK PATTERNS

PLATE 52 TRANSITION ELBOW WITH SLANT THROAT

The method for laying out this slant cheek pattern is identical to the method used in Plate 51.

Lay out the flat cheek pattern and the throat pattern in Figure 4 to the dimensions shown.

Since the dimensions for the base and the height of the throat are of different lengths, these lengths must be considered when laying out the slant cheek pattern O with a square as in Plate 51.

To lay out the slant cheek pattern O, set the square so that the $1\frac{1}{4}$-in. point (E'' to E') on one leg of the square rests on point E'; then move the square until the other leg rests on point A', and draw a line from E' to D'. Set the square so that the $\frac{3}{4}$-in. point (A'' to A') on one leg of the square rests on point A', move the square until the other leg rests on point D', and draw a line from A' to B'. Draw a line from point B' toward F to any length desired. Set one leg of the square on line $B'-F$ and move the square until the other leg rests on point D'; then draw a line from D' to intersect line $B'-F$, establishing point F. Transfer the distance $B'-C$ from the heel pattern in Figure 3 to line $B'-C'$ on the slant cheek pattern O. Use the distance C' to F as a radius; then use points C' and D' as centers to draw arcs to cross each other at point F', establishing the center point to draw the heel curve D' to C'.

NOTE: The distance E'' to E' represents the length of the base of the throat, and the distance A'' to A' represents the height of the throat on the flat cheek pattern M.

Complete the pattern by making the required allowances as shown.

PLATE 52

1 TOP VIEW

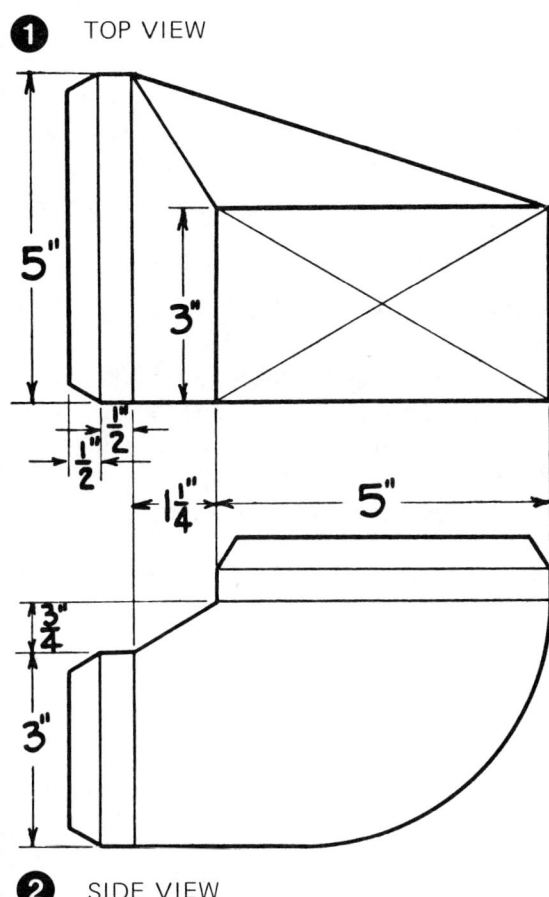

2 SIDE VIEW

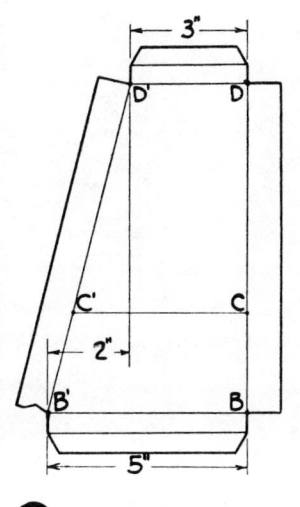

3 HEEL PATTERN

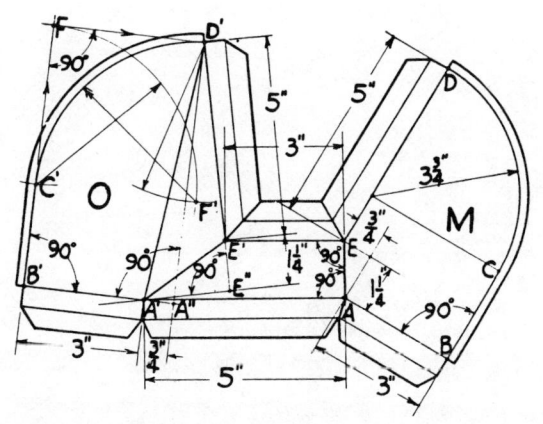

4 THROAT & TWO CHEEK PATTERNS

PLATE 53 DOUBLE-FLARE TRANSITION ELBOW

Lay out the throat pattern in Figure 4 by drawing the center line A to E equal to the length of the slant throat line A to E in the side view in Figure 2. The widths are drawn to the dimensions as shown.

To lay out the two slant cheek patterns as in Figure 4, follow the same procedure as in Plate 51. Set a square so that the $1\frac{1}{4}$-in. point on one leg rests on point E' and the other leg rests on point A', and draw a line from E' to D'. Set the square so that the $\frac{3}{4}$-in. point on one leg rests on point A' and the other leg rests on point D', and draw a line from A' to B'. Set one leg of the square on line $B'-F$ and the other leg resting on point D', and draw a line from D' to F.

Transfer the slant length B' to C' from the heel pattern in Figure 3 to line $B'-C'$ on the slant cheek pattern in Figure 4. Use the distance C' to F as a radius, and points C' and D' as centers to draw arcs to cross each other at point F'. Use F' as a center to draw the heel curve D' to C'.

Complete the patterns by making the required allowances as shown.

PLATE 53

1 TOP VIEW

2 SIDE VIEW

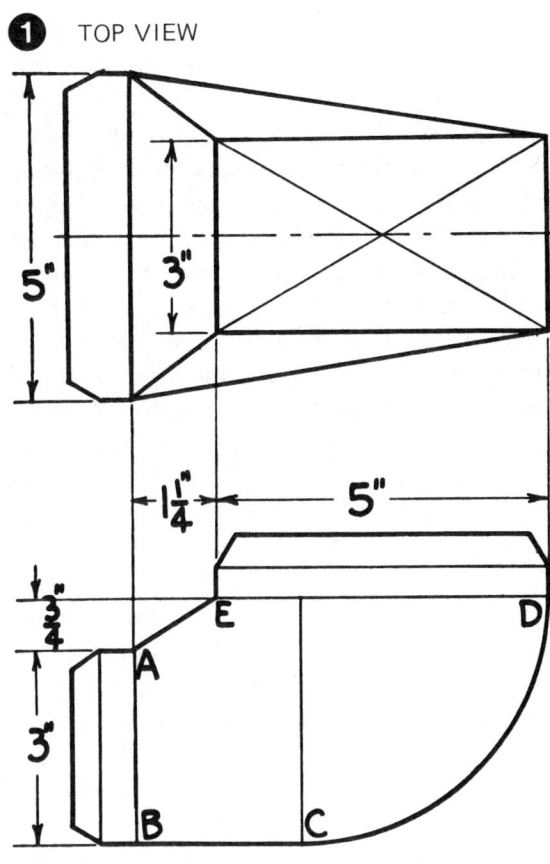

3 HEEL PATTERN

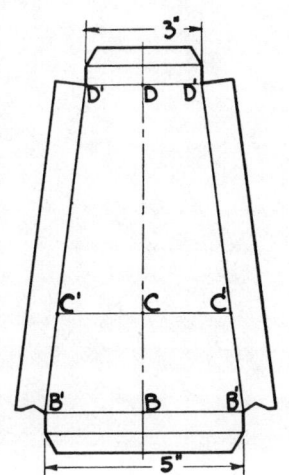

4 THROAT & TWO CHEEK PATTERNS

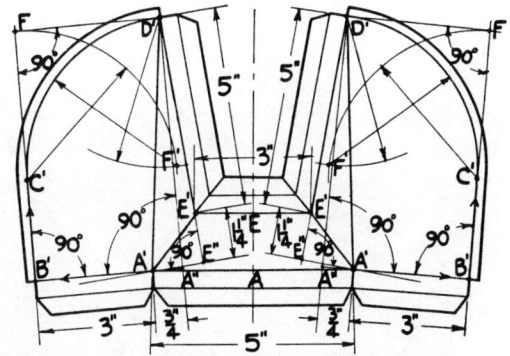

PLATE 54   IRREGULAR-SHAPED ELBOW

The top view shows the difference in the width of the heel and the throat.

Lay out the flat cheek pattern as in Figure 2, and divide the heel and throat curve into four equal spaces.

Erect a true-length triangle as in Figure 2. The distance A to A' represents the difference in the width of the heel and throat patterns in Figures 3 and 4.

To lay out the slant cheek pattern as in Figure 5, draw line A' to B equal to the slant true-length line from point A' to $B-D'$ in Figure 2. The distance A' to C' in Figure 5 is equal to the slant true-length line A' to C' in Figure 2. The distance C' to D' in Figure 5 is the same as the distance A' to B in Figure 5. The distances B to C' and A' to D' in Figure 5 are the same as the distances B to C and A to D in Figure 2. These spaces will not increase because both sides on the heel and the throat patterns in Figures 3 and 4 remain the same in length. The distance D' to 1 is the same length as A' to C' in Figure 5. The distance 1 to 2 is the same as C' to D'. The distances 2 to 3 and 4 to B are the same length as A' to C'. The distances 3 to 4 and B to A' are the same length as A' to B in Figure 5.

To obtain the radius point R to draw the heel and throat curves in Figure 5, bisect the distance between points 2 and A' by drawing an arc from point A' and another from point 2 to cross each other. Draw a line through each of the intersecting points of the arcs to meet at point R, thus obtaining the radius point to draw the heel curve A' to A' and the throat curve B to B.

Complete the pattern by making the required allowances as shown.

PLATE 54

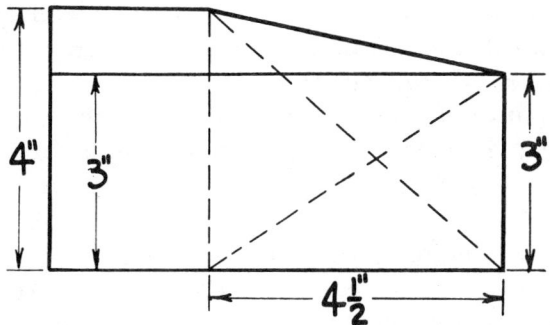

1 TOP VIEW

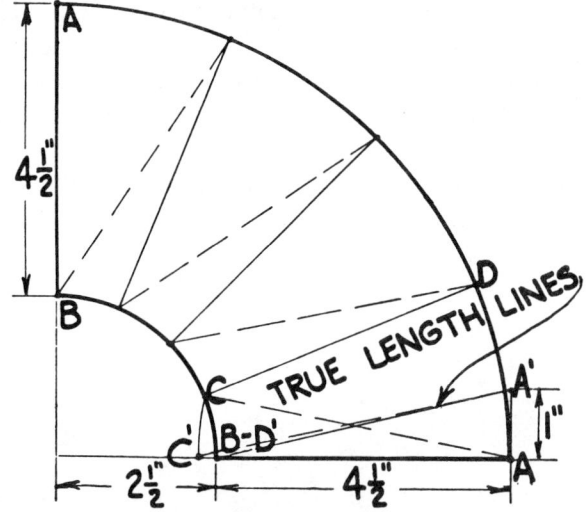

2 FLAT CHEEK PATTERN

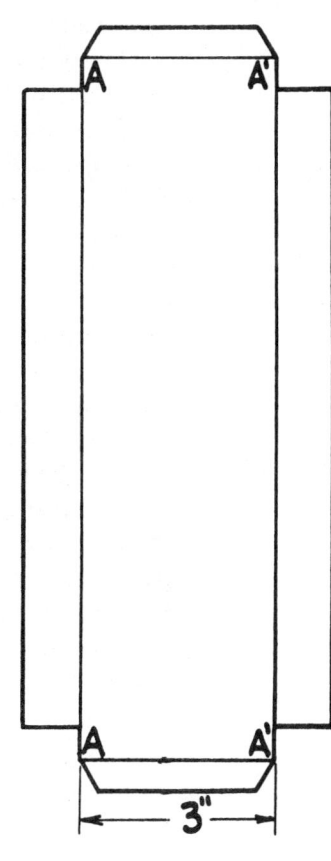

3 HEEL PATTERN

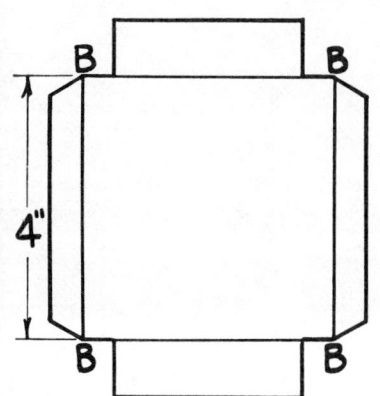

4 THROAT PATTERN

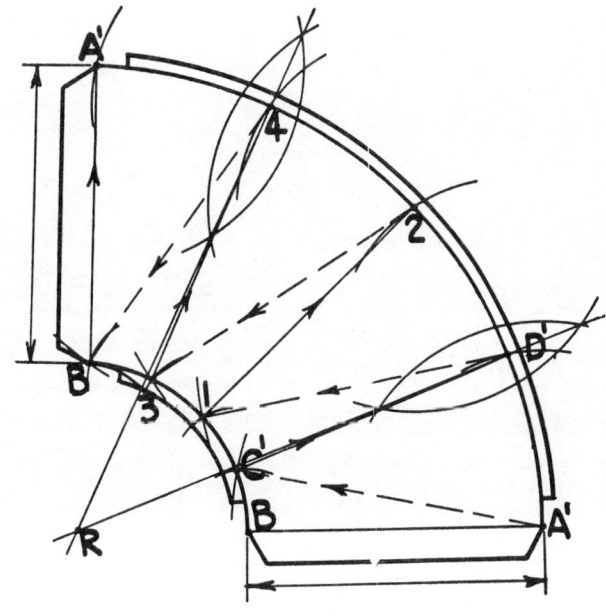

5 SLANT CHEEK PATTERN

PLATE 55 TRANSITION ELBOW IRREGULAR-TO-RECTANGULAR

The front view in Figure 2 shows the irregular shape of the front opening.

Draw the side view as in Figure 1, and divide the heel curves from 2 to 10, and 1 to 9 into the same number of equal spaces.

Erect a true-length triangle as in Figure 4. The height A to B is obtained from the width A to B in Figure 2. Transfer the various slant lengths from the side view in Figure 1 to the base line of the true-length triangle in Figure 4.

To lay out the heel pattern as in Figure 3, draw lines A to B and 1 to 2 to the dimensions shown. The distance 2 to 3 is equal to the slant true-length line from point B to 3 in Figure 4. The distance 1 to 3 is equal to the distance 1 to 3 on the small heel curve in Figure 1. The distance 3 to 4 is equal to the slant true-length line from point B to 4 in Figure 4. The distance 2 to 4 is equal to the distance 2 to 4 on the large heel curve in Figure 1. Continue to lay out the remaining pattern by transferring the true lengths from Figure 4 to their respective places in Figure 3. The spaces 4 to 12 are obtained from the large heel curve 4 to 12 in Figure 1, and the spaces 3 to 11 are obtained from the heel line on the small cheek in Figure 1. Complete the pattern by making the required allowances as shown.

Lay out the throat pattern in Figure 5, and the large and small cheek patterns in Figures 6 and 7 with the required allowances as shown.

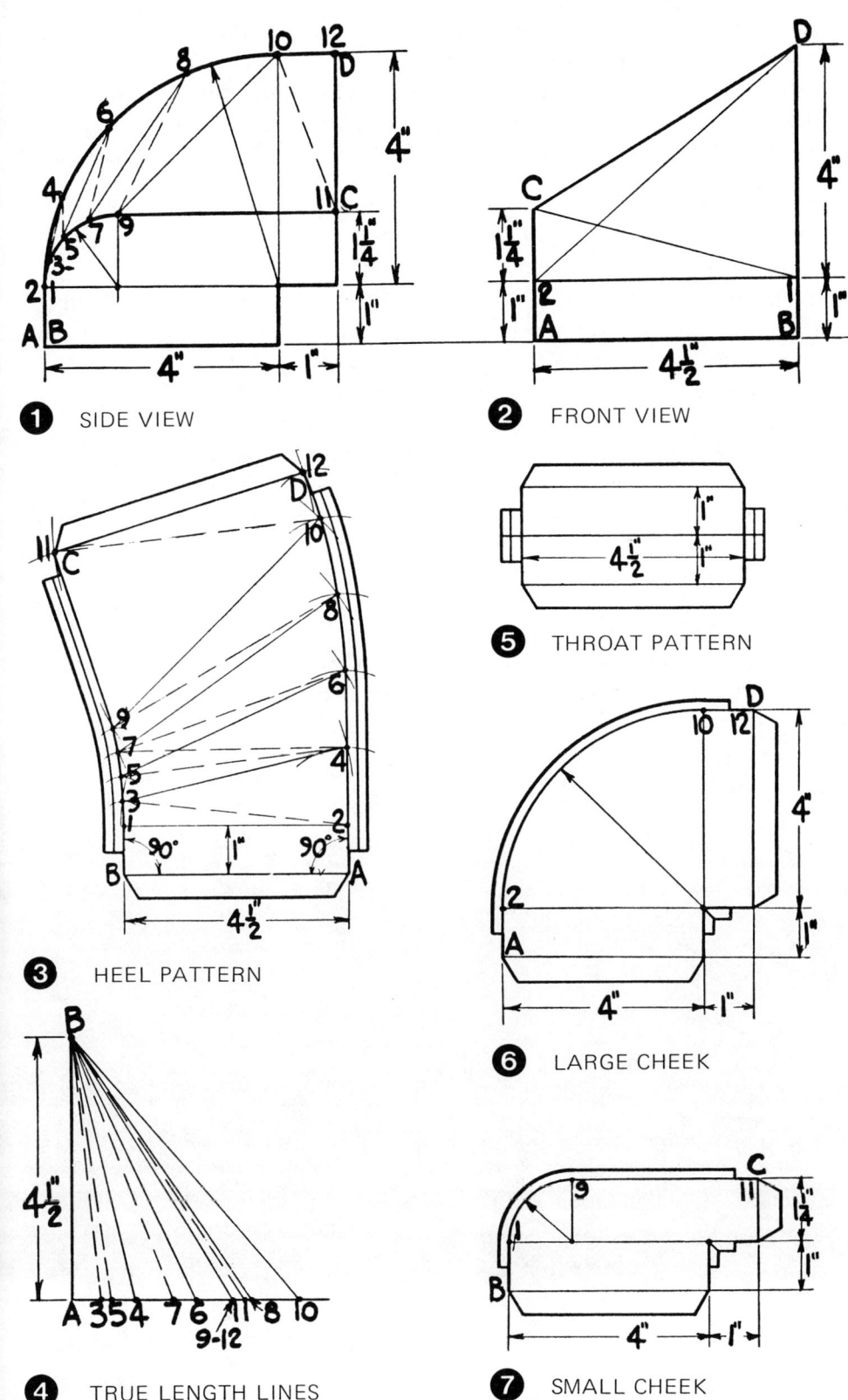

PLATE 56 TRANSITION ELBOW WITH CURVED HEEL AND THROAT PATTERNS

Draw the top view as in Figure 1, and divide the throat curve and the heel into as many equal spaces as desired.

Draw the front view as in Figure 2, and draw lines from the points on the heel and throat curves in Figure 1 to intersect the base line in Figure 2.

Lay out the heel pattern as in Figure 3, by drawing line 1 to 13 equal to the heel length in Figure 1. Transfer the heights from the base line 1–13 to the slant heel line 1'–13' in Figure 2 to their respective lines in Figure 3, obtaining the freehand curved line 1' to 13'.

Lay out the throat pattern as in Figure 4, by drawing line 2 to 10 equal to the length of the throat curve in Figure 1. Transfer the heights from the base line 2–10 to the slant heel line 2'–10' in Figure 2 to their respective lines in Figure 4, obtaining the freehand curved line 2' to 10'.

Draw lines from the points on the slant heel line 1'–13' and from the slant throat line 2'–10' in Figure 2 to intersect line $A-B$, obtaining the various heights for the true-length triangles. Transfer the slant lengths from Figure 1 to their respective lines intersecting line $A-B$ in Figure 2.

Lay out the cheek pattern as in Figure 5, by using the slant true-length lines at the true-length triangle $A-B$ in Figure 2. The spaces 2' to 10' are equal to the spaces 2' to 10' on the freehand curved line on the throat pattern in Figure 4. The spaces 1' to 13' are equal to the spaces 1' to 13' on the freehand curved line on the heel pattern in Figure 3.

The patterns in Figures 3, 4, and 5 are formed to the required shape, then the edges are turned by the use of a hand turning machine.

PLATE 56

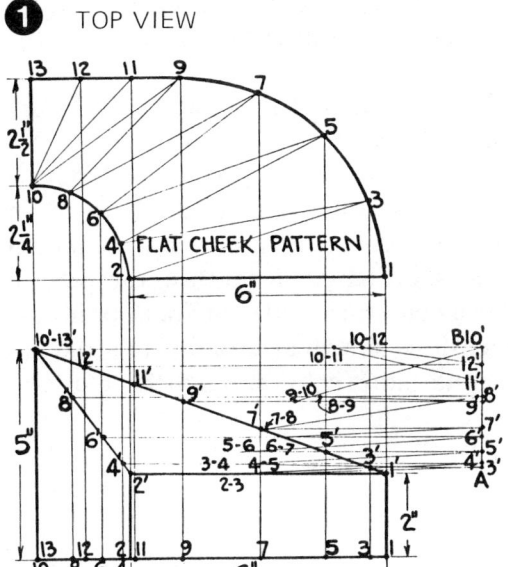

1 TOP VIEW

2 FRONT VIEW

5 DEVELOPED SLANT CHEEK PATTERN

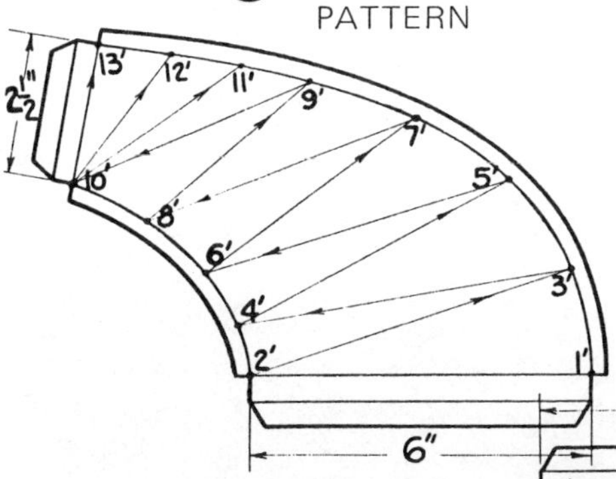

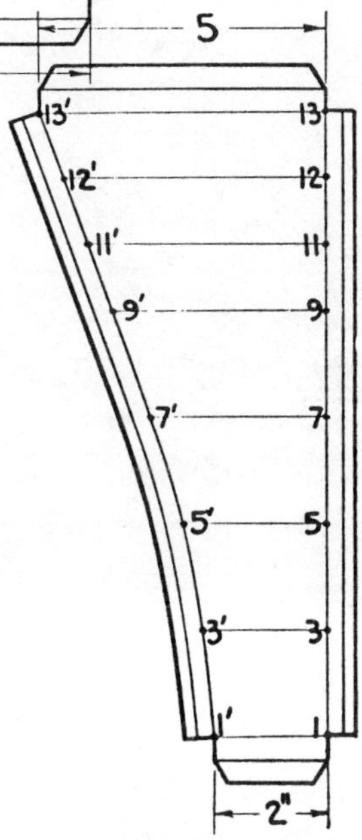

3 HEEL PATTERN

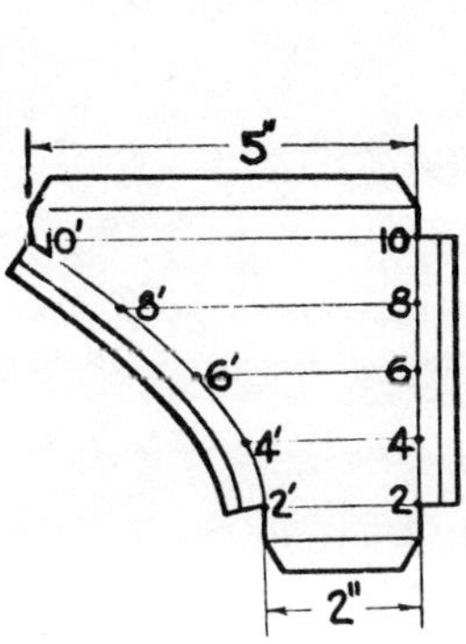

4 THROAT PATTERN

101

PLATE 57   RECTANGULAR-TO-IRREGULAR SHAPED TRANSITION

Draw the top view as in Figure 1, and draw line B' to B'' in Figure 2. Transfer the distance B to C from the top view to the base line B' to C' in the front view Figure 2.

Lay out the pattern in Figure 3. The distances D to C' and B to A are equal to the dimensions in the top view, Figure 1.

Use the distance A to B in Figure 1 as a radius, and point A in Figure 3 as a center, to strike an arc at point B''. Use the slant length C' to B'' in Figure 2 as a radius, and use point C' in Figure 3 as a center to strike an arc to cross the arc at point B''. Use the slant length B to D on the pattern in Figure 3 as a radius, and point B'' as a center, to strike an arc at point D'. Use point C' as a center with the dimensions shown as a radius to strike an arc crossing the arc at D'. Complete the pattern by making the required allowances as shown.

An isometric view is shown in Figure 4.

PLATE 57

① TOP VIEW

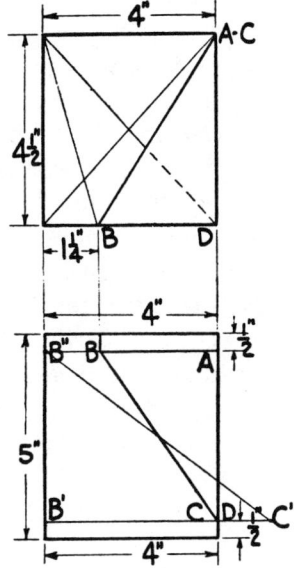

② FRONT VIEW & TRUE LENGTH LINE

④ ISOMETRIC VIEW

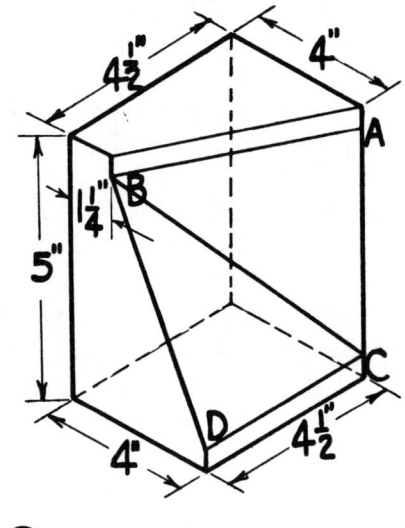

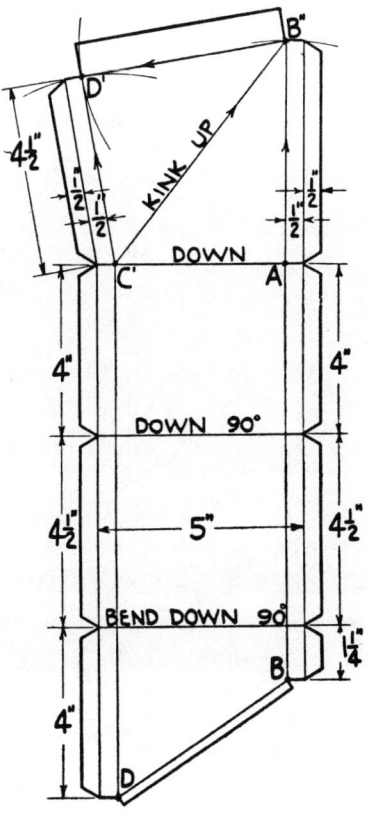

③ FULL PATTERN

103

PLATE 58 RECTANGULAR-TO-TRIANGULAR TRANSITION

Draw the top view as in Figure 1, and the front view as in Figure 2. Transfer the lengths A to C, B to C, B to E, and A to D from the top view to the base line in Figure 2, erecting the true-length triangle as shown.

Lay out the front pattern in Figure 3 to the dimensions shown.

To lay out the pattern as in Figure 5, draw line A to B equal to the dimensions shown. Use the slant line A–C in Figure 2 as a radius, and point A, Figure 5, as a center to strike an arc near point C. Use the slant line B–C in Figure 2 as a radius, and point B, Figure 5, as a center to strike an arc crossing the arc at point C. Use the distance C to E in Figure 1 as a radius, and point C in Figure 5 as a center to strike an arc at point E. Use the slant line B–E in Figure 2 as a radius, and point B in Figure 5 as a center to strike an arc crossing the arc at point E. Use the distance C to D in Figure 1 as a radius, and point C in Figure 5 as a center to strike an arc at point D. Use the slant line A–D in Figure 2 as a radius, and point A in Figure 5 as a center to strike an arc crossing the arc at point D. Use the dimensions shown as a radius, and points A and B as centers to strike an arc at points G and F. Transfer the slant lengths E–F and D–G from Figure 3 to E to F and D to G in Figure 5. Complete the pattern by making the required allowances as shown.

An isometric view is shown in Figure 4.

PLATE 58

① TOP VIEW

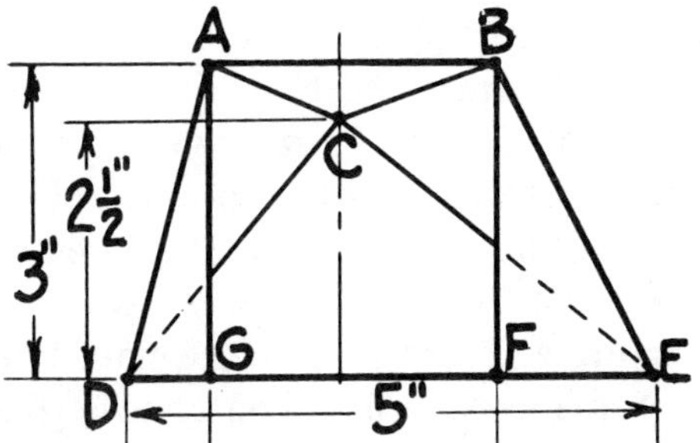

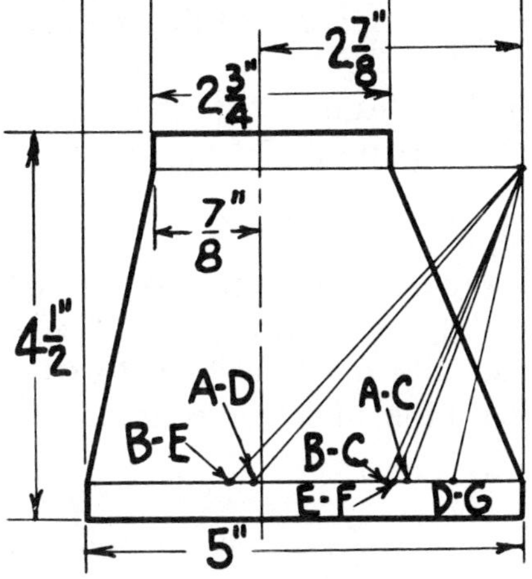

② FRONT VIEW & TRUE LENGTHS

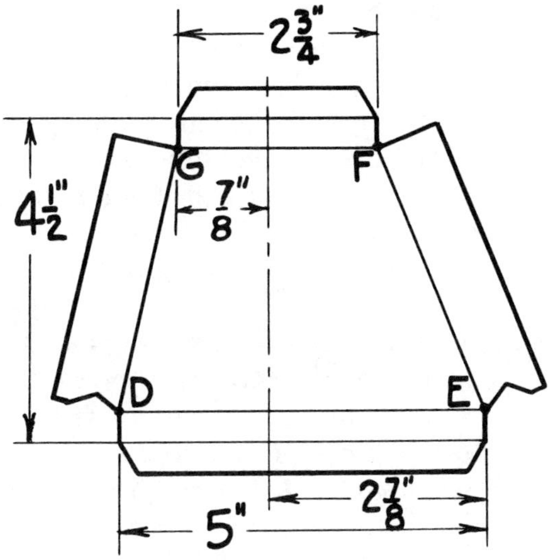

③ FRONT PATTERN

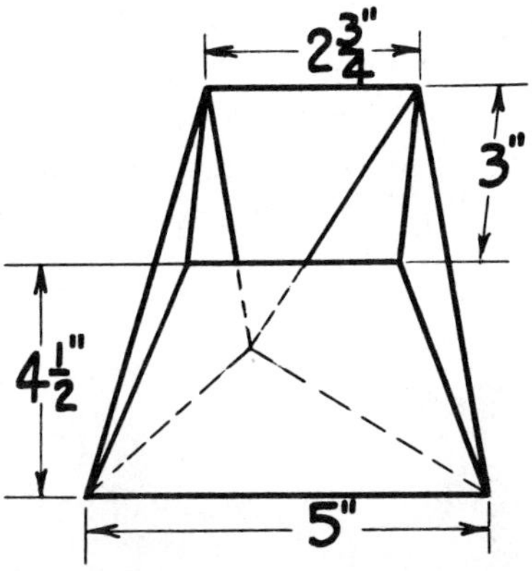

④ ISOMETRIC VIEW

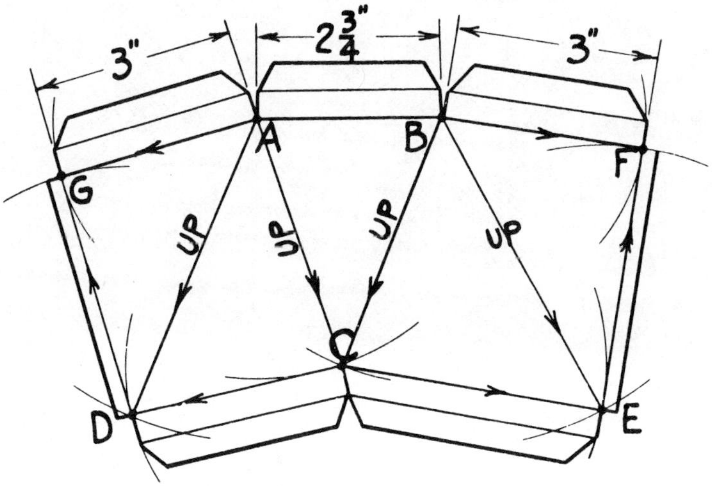

⑤ BACK & TWO SIDE PATTERNS

105

PLATE 59 RECTANGULAR-TO-IRREGULAR-SHAPED TRANSITION

Draw the top view as in Figure 1. Transfer the lengths B to D, A to C, and B to C from Figure 1 to the base line in Figure 2, and the distances E to F and H to G from Figure 1 to the top line in Figure 2.

To lay out the side pattern as in Figure 3, draw the base line A to B equal to the dimensions shown. Use the slant line B–C in Figure 2 as a radius, and point B in Figure 3 as a center to strike an arc at point C. Use the slant line A–C in Figure 2 as a radius and point A in Figure 3 as a center to strike an arc crossing the arc at point C. Use the dimensions shown as a radius, and point C as a center to strike an arc at point D. Use the slant line B–D in Figure 2 as a radius, and point B in Figure 3 as a center to strike an arc crossing the arc at point D.

Lay out the end pattern as in Figure 4. The distance H to G is equal to the slant line H–G in Figure 2.

Lay out the end pattern as in Figure 5. The distance E to F is equal to the slant line E–F in Figure 2. Complete the patterns by making the required allowances as shown.

An isometric view is shown in Figure 6.

NOTE: Two patterns are required as in Figure 3, one left and one right. All markings should be on the inside of the patterns for neatness. In Figure 3 the markings are on the outside to coincide with the bend line B to C on the isometric view, illustrating the kink toward the outside in order to fit the irregular shape of the fitting.

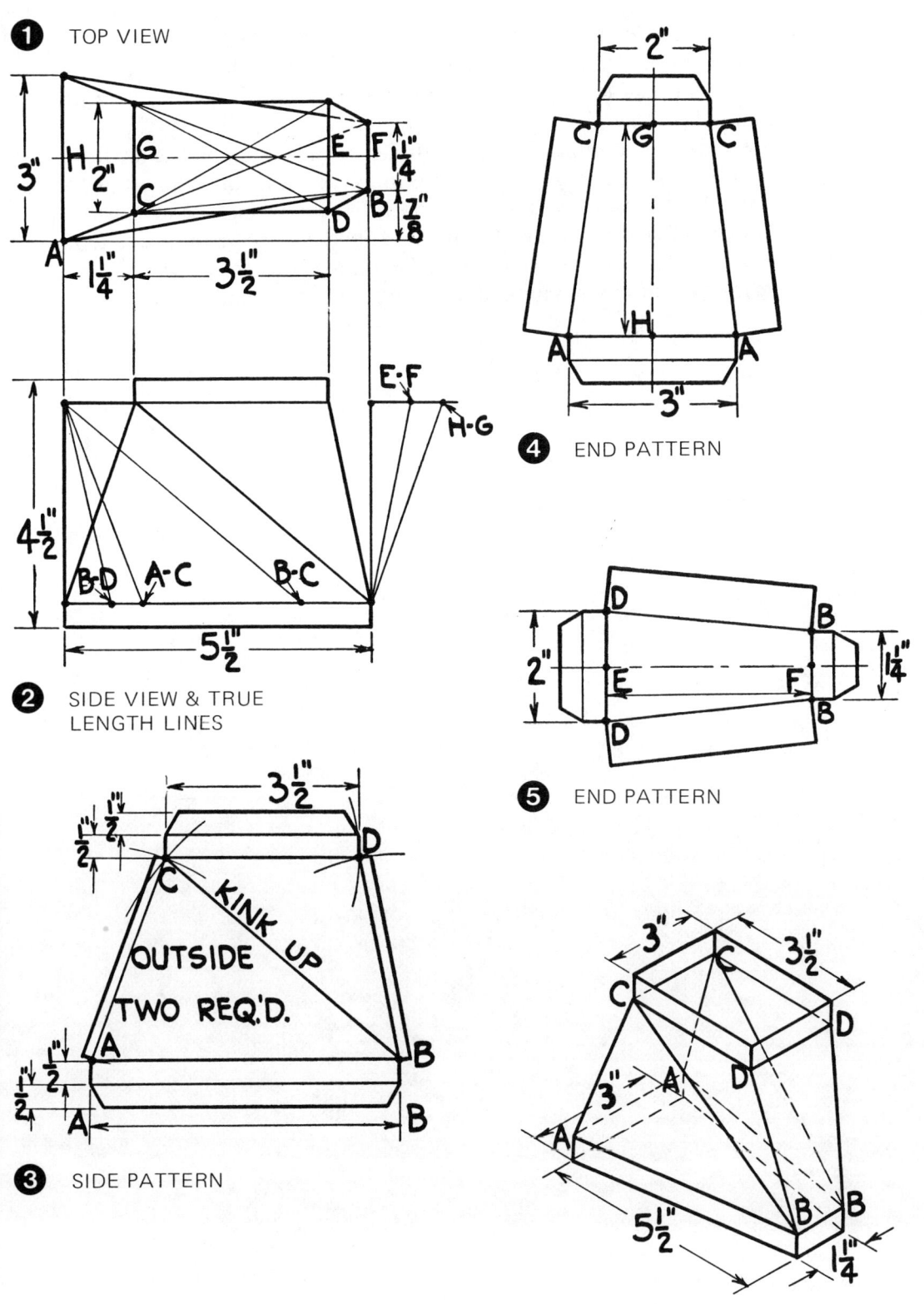

PLATE 59

PLATE 60 TRANSFORMER IN ONE PIECE

When this type of transformer is used for residential heating and air conditioning, the fitting is generally small, and short in length; therefore, the four patterns may be laid out in one continuous piece of metal, and fit on a standard-size sheet.

Since the top and the side are both flat or on a straight line as illustrated in Figures 1 and 2, the two patterns may be laid out in one continuous straight line. The slant side and the bottom pattern may be attached to the top and the flat side pattern by drawing 90-deg. right-angle lines from the slant side of their respective mate to the dimensions as shown.

Allowances for seaming and cleat edges may be made as desired.

PLATE 60

1 TOP VIEW

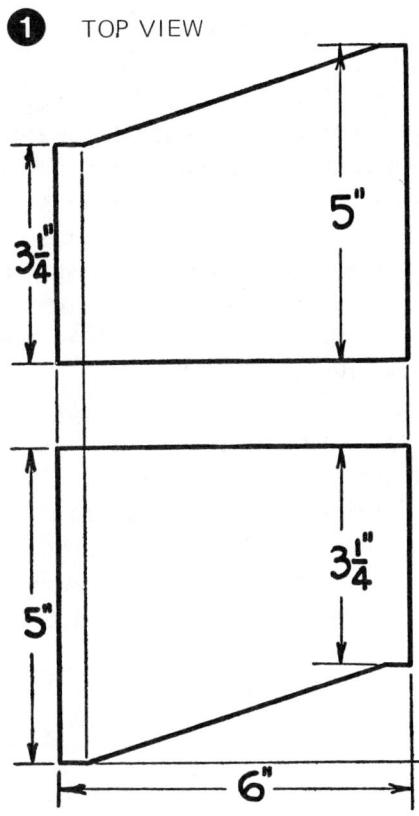

2 SIDE VIEW

4 ISOMETRIC VIEW

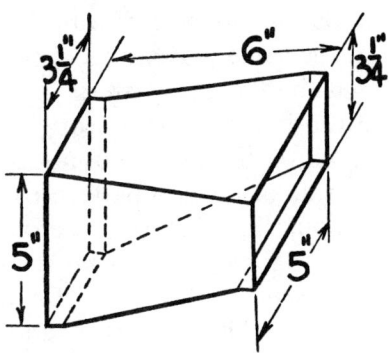

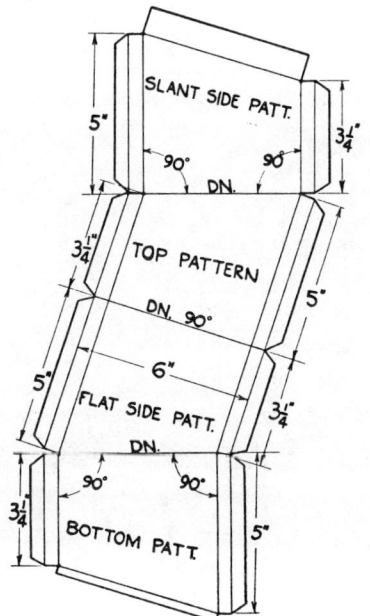

3 FULL PATTERN

PLATE 61 EQUAL-TAPERING TRANSFORMER IN ONE PIECE

The patterns for this type of transformer may be laid out in one continuous pattern in the same manner as in the previous plate.

Lay out the flat top pattern in Figure 3 equal to the given dimensions. Since the top is flat or on a straight line as illustrated in Figure 2, the two slant side patterns may be obtained by drawing 90-deg. right-angle lines from each side of the top pattern to the dimensions, as shown.

To lay out the bottom pattern attached to the slant side of the side pattern, the triangle $A-B-C$ must first be established. NOTE: The distance A to B representing the radius for drawing the arc at point B is equal to the distance A to B in Figure 1, and represents the amount of offset or the taper for the top and bottom pattern as illustrated in Figure 1.

After the arc is drawn by using point A as a center, draw a straight line from point C tangent to the arc at point B. Then draw a 90-deg. right-angle line from point A to B by setting one leg of the square on line $C-B$, and move the square along this line until the other leg of the square rests on point A. Next draw a straight line from point A to B thereby establishing point B. Continue this line to point A' equal to the dimensions as shown. Again set the square on line $C-B$ and draw a 90-deg. line from point C to C' equal to the dimensions as shown.

Allow for seaming and for cleat edges as may be desired.

PLATE 61

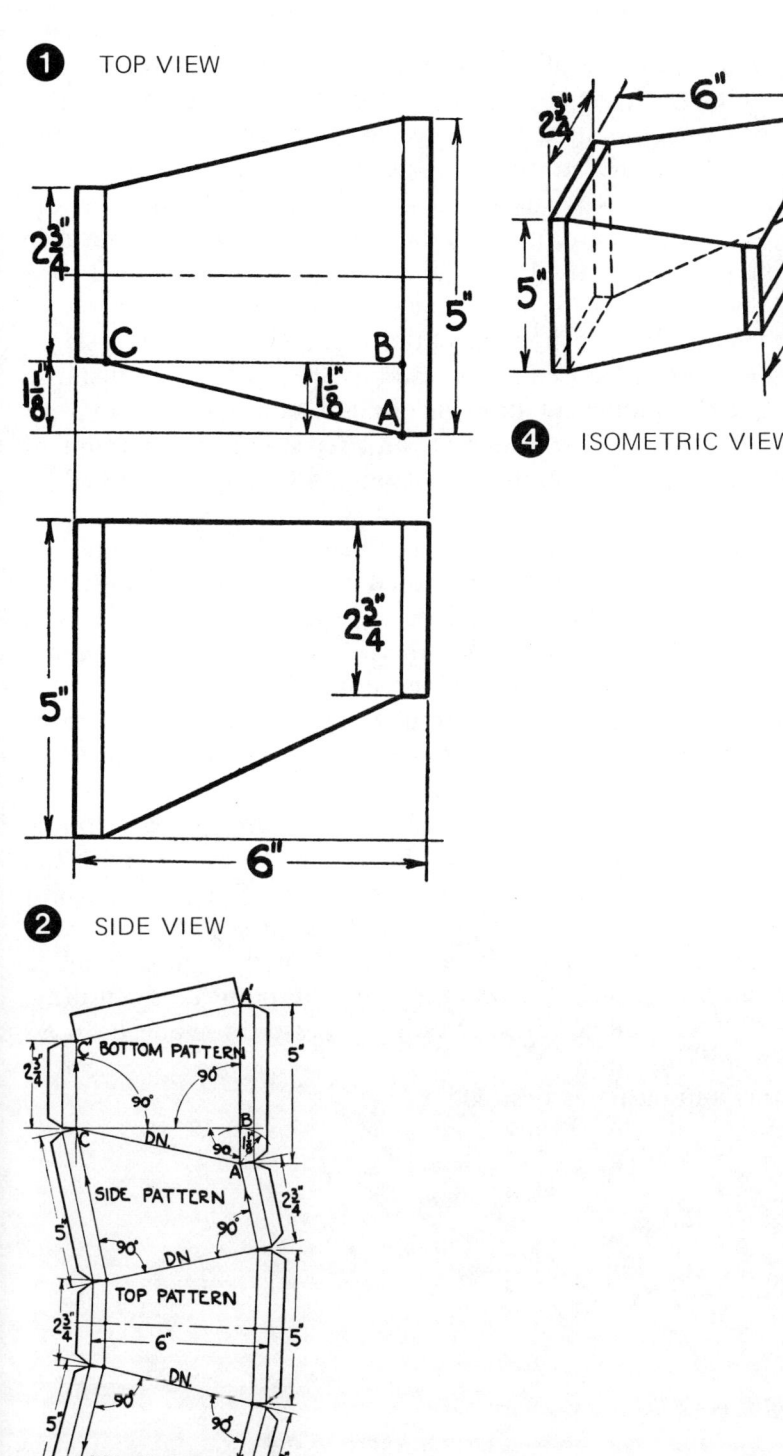

1. TOP VIEW
2. SIDE VIEW
3. FULL PATTERN
4. ISOMETRIC VIEW

PLATE 62   DOUBLE-OFFSET TRANSFORMER IN ONE PIECE

To lay out the patterns in a continuous piece of metal for a transformer that has the two sides, with the top and the bottom offsetting or tapering, follow the same procedure as used in the previous plate for laying out the bottom pattern that is attached to the side pattern.

Select any one of the four patterns and lay it out to its proper length and size. In this plate, side pattern No. 1 has been selected. Lay out side pattern No. 1. The distance A to B is equal to the slant length line A to B in Figure 1.

To lay out the top and the bottom patterns attached to side pattern No. 1, the triangles C–E–F must first be established. NOTE: The distance E to F representing the radius for drawing the arc at point F is equal to the distance E to F or A to C in Figure 1, and represents the amount of offset or the taper for the top and the bottom pattern as illustrated in Figure 1.

After the arc is drawn by using point E as a center, draw a straight line from point C tangent to the arc at point F, then draw a 90-deg. right-angle line from point E to F by setting one leg of the square on line C–F, and move the square along this line until the other leg rests on point E, draw a straight line from point E to F, establishing point F. Then continue this line to point E on the top and bottom patterns to the dimensions as shown. Draw a 90-deg. line from point C to C' on the top pattern, and C to G on the bottom pattern to the given dimensions.

To lay out the side pattern No. 2 attached to the bottom pattern, follow the same procedure as used for the top and bottom patterns. Establish the triangle E–G–H by using the distance G to H in Figure 2 as a radius and point G as a center to draw the arc at point H. Draw a straight line from point E' tangent to the arc at point H. Set one leg of the square on line H–E', and move the square until the other leg rests on point G. Then draw a line from point G to H, and continue to point C''; also draw line E' to E'' equal to the dimensions as shown.

Allow for seams and cleats as desired.

PLATE 62

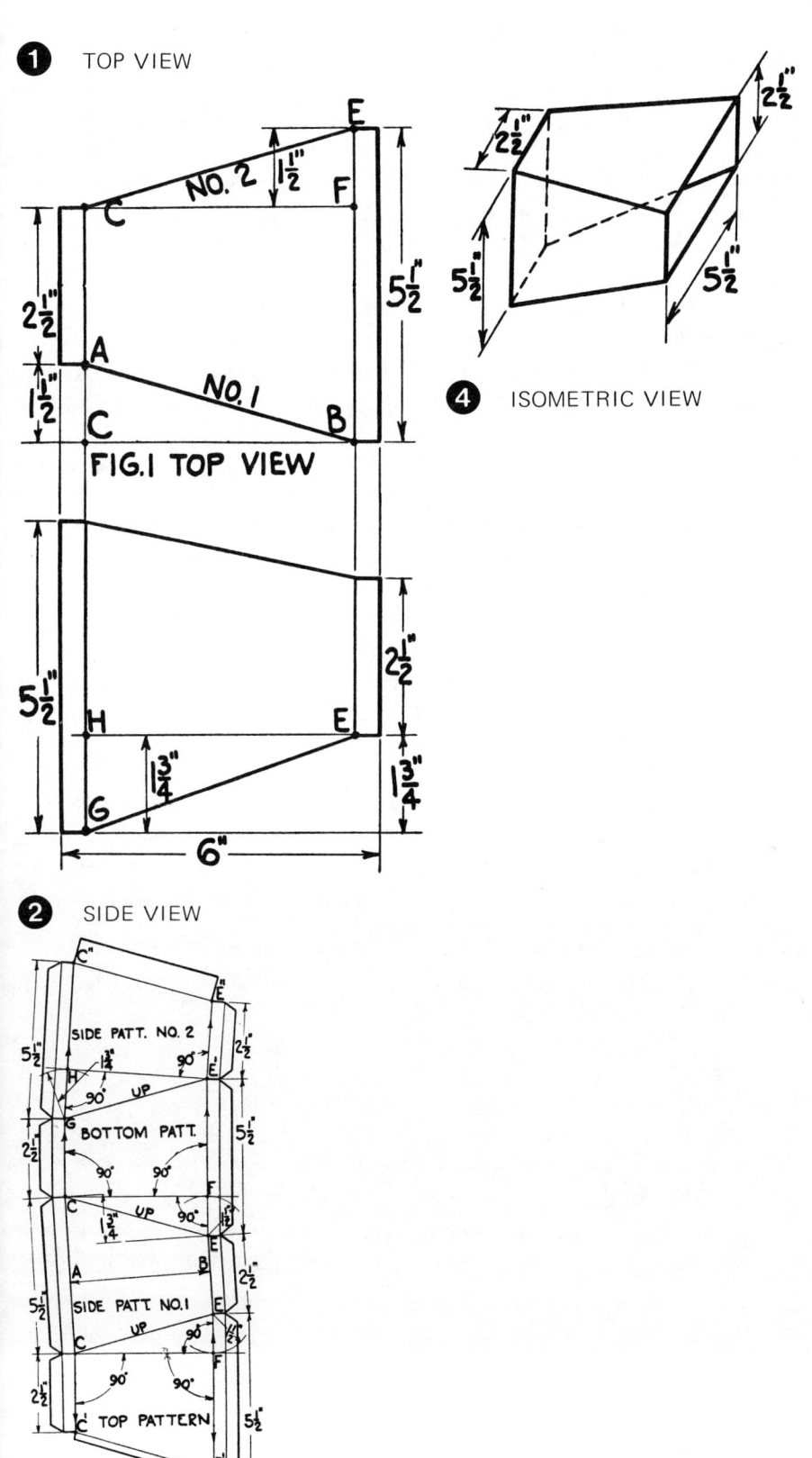

① TOP VIEW

FIG.1 TOP VIEW

② SIDE VIEW

③ FULL PATTERN

④ ISOMETRIC VIEW

PLATE 63 DOUBLE-OFFSET TRANSFORMER IN ONE PIECE

To lay out the patterns for this type of transformer in one continuous pattern, the procedure is the same as that in the previous plate except that the sides on each pattern have different offsets or tapers, as illustrated in Figures 1 and 2.

Select any one of the four patterns and lay it out to its proper length and dimensions. In this plate the top pattern has been selected.

Lay out the top pattern. The line A to B is equal to the slant length line A to B in Figure 2.

The remaining patterns may be laid out by following the same procedure as used in Plate 62.

To lay out the two side patterns Nos. 1 and 2, use the height C to B in the side view, Figure 2, as a radius, and each point C on the top pattern as a center to strike an arc at B. Draw a straight line from point D tangent to the arc B. Set one leg of the square to rest on line $B-D$, move the square until the other leg rests on point C, draw a line from point C to B, and continue to point G equal to the dimensions shown. Also draw a line squaring from line $B-D$ to point E equal to the dimensions shown.

To lay out the bottom pattern, use the height E to F in the top view, Figure 1, as a radius, and point E in Figure 3 as a center to strike an arc at F. Draw a straight line from point G tangent to the arc F. Set one leg of of the square to rest on line $G-F$, and move the square until the other leg rests on point E. Draw a line from point E to F, and continue to point E'. Draw a line squaring from line $F-G$ to point G'.

Allow for seams and cleat the edges as may be desired.

An isometric view is shown in Figure 4.

NOTE: In actual shop practice it is not necessary to draw the top or the side view to obtain the true lengths for any pattern. All true lengths may be obtained by drawing only a triangle such as $A-B-C$ in Figure 2, or by measuring across from the two legs of a square.

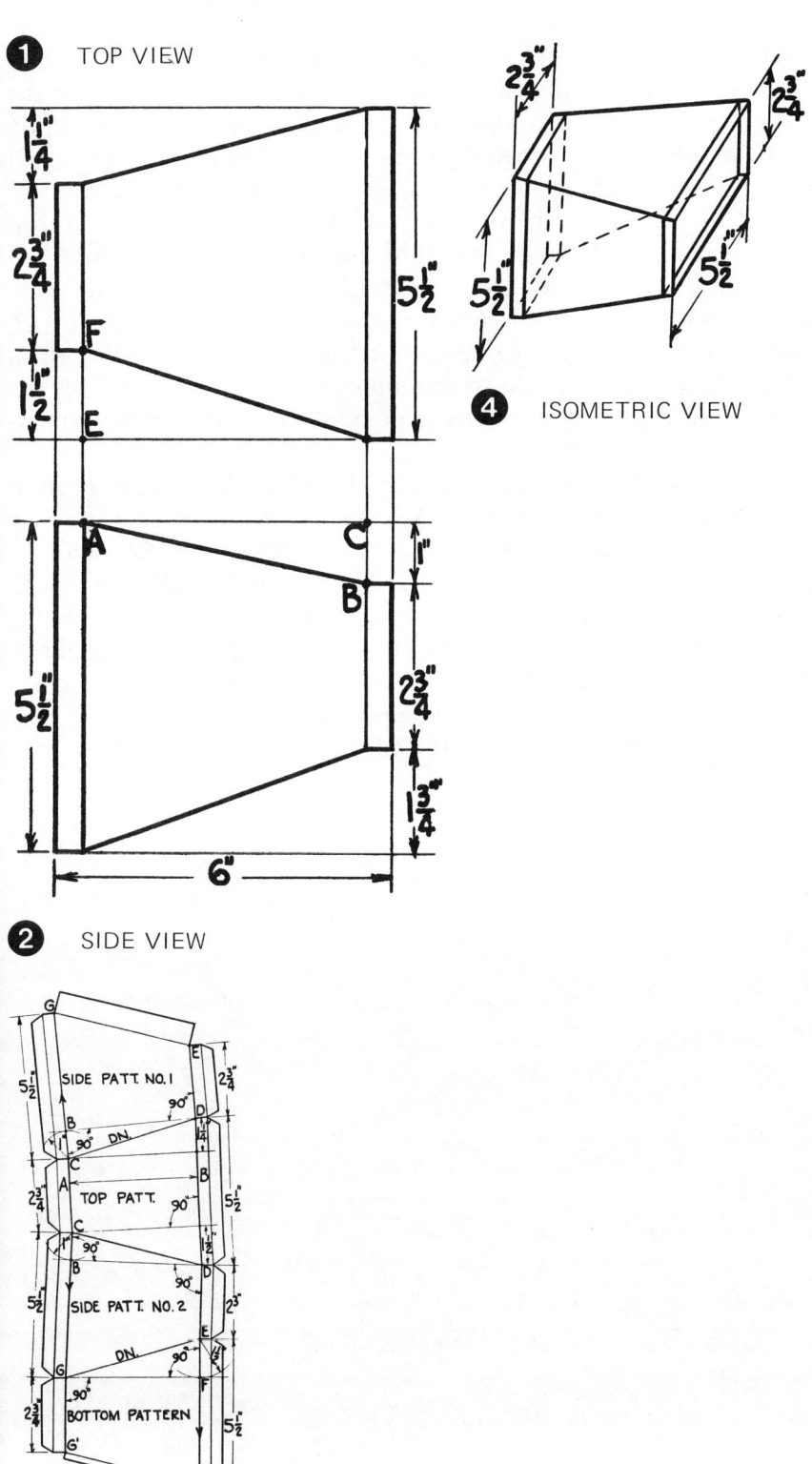

PLATE 63

PLATE 64   DOUBLE-OFFSET TRANSFORMER IN ONE PIECE

The procedure for this plate is the same as that in Plate 62.

Lay out any one of the patterns to their proper length. In this case the top pattern is laid out first. The distance A' to C', Figure 3, representing the length of the top pattern is equal to the slant line A to C in Figure 2.

To lay out the two side patterns attached to the top pattern, transfer the angle A, B, and C from Figure 2 to each side of point C on the top pattern. Use the distance C to B in Figure 2 as a radius, and point C in Figure 3 as a center to strike an arc at point B. Draw a straight line from point A tangent to the arc at point B. Set one leg of the square on line $A-B$, move the square until the other leg rests on point C, and draw a line from C to B, continuing to line D equal to the dimensions shown. Again set the square on line $A-B$, and draw a line from point A to F equal to the dimensions shown.

To lay out the bottom pattern, transfer the angle D, E, and F from Figure 1 to one side of the side pattern No. 1. Use the distance E to F in Figure 1 as a radius, and point F in Figure 3 as a center to strike an arc at point E. Draw a straight line from point D tangent to the arc at point E. Set one leg of the square on line $D-E$, move the square until the other leg rests on point F, and draw a line from point F to E, continuing to point H equal to the dimensions shown. Again set the square on line $D-E$, and draw a line from point D to G equal to the dimensions shown.

Allow for seaming and for cleating the edges as may be desired.

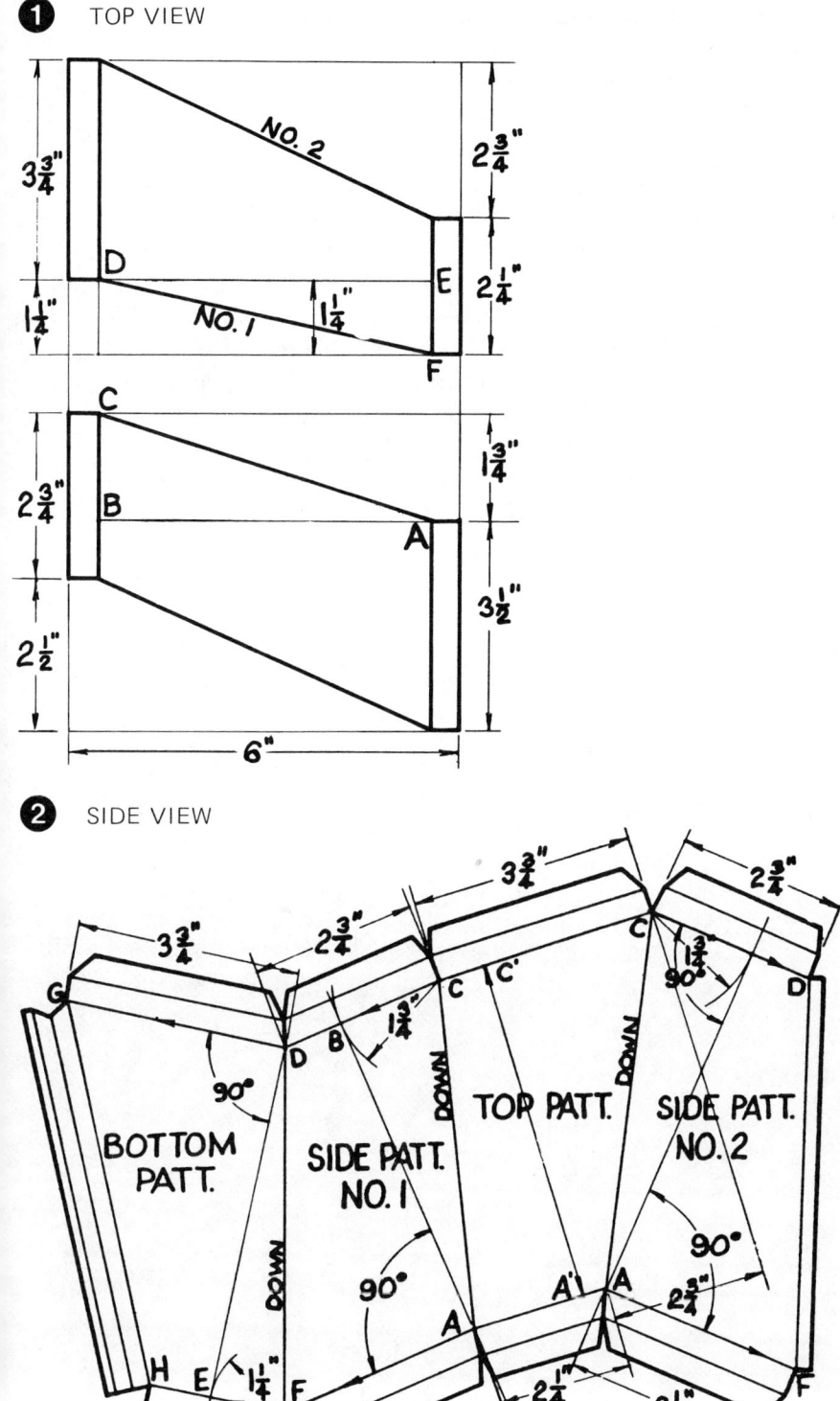

PLATE 64

① TOP VIEW

② SIDE VIEW

③ FULL PATTERN

117

PLATE 65 TWISTED TRANSFORMER

To lay out the patterns for this type of twisted fitting, it is not necessary to draw any of the views in Figures 1, 2, or 3.

Lay out the back and side patterns as in Figure 5 equal to the dimensions shown.

To lay out the side, center, and front patterns as in Figure 4, draw line K to J equal to the slant line K to J on the back pattern in Figure 5.

To lay out the center pattern in Figure 4, use the distance E to F (or the difference in the length and width) in the top view in Figure 1 as a radius, and points H and F in Figure 4 as centers to strike arcs at points G and E. Use the distance B to A in Figure 5 as a radius, and point F in Figure 4 as a center to strike an arc to cross arc G. Use the distance J to K in Figure 5 as a radius, and use point G as a center to strike an arc to cross arc E. Draw lines squaring from line E–G to points C and D equal to the dimensions as shown.

Allow for seams and cleat the edges as may be desired.

An isometric view is shown in Figure 6.

PLATE 65

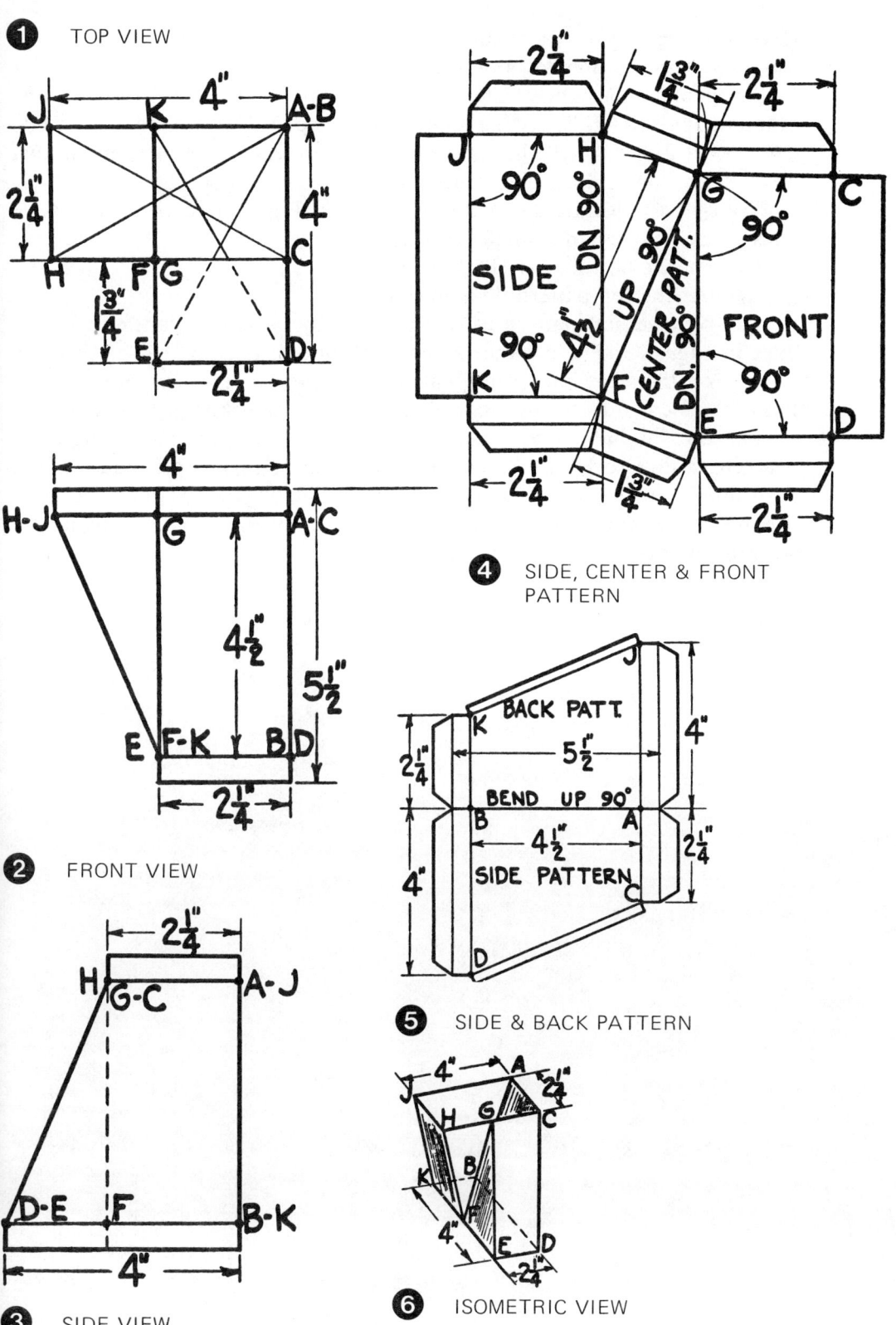

PLATE 66 TWISTED RECTANGULAR OFFSET

Draw the top view as in Figure 1. To draw the side view as in Figure 2, use point 5 as a center to draw the arc at the heel. Draw a line from point 2–8 tangent to the arc to intersect the line drawn up from the base line obtaining point 1–7. Bisect the angle of this line at the top, and draw a line from point 2–8 through the bisecting arcs to intersect the heel line drawn down from the top line, obtaining point 4–6.

Transfer the lengths from the top view in Figure 1 to their respective lines at the true-length triangles in Figure 3.

Lay out the bottom pattern *A* as in Figure 4.

Lay out the top pattern *C* as in Figure 5.

Lay out the half pattern as in Figure 6, by using the slant true-length lines in Figure 3. The distance 1' to 3', Figure 6, is equal to the distance 1' to 3' on the slant line on the pattern in Figure 4. The distance 2' to 4', Figure 6, is equal to the distance 2' to 4' on the slant line in Figure 5.

Lay out the half pattern as in Figure 7, by using the remaining slant true-length lines in Figure 3. The distance 5' to 7', Figure 7, is equal to the slant line 5' to 7' on the pattern in Figure 4. The distance 6' to 8', Figure 7, is equal to the slant line 6' to 8' on the pattern in Figure 5, completing the patterns as shown.

PLATE 66

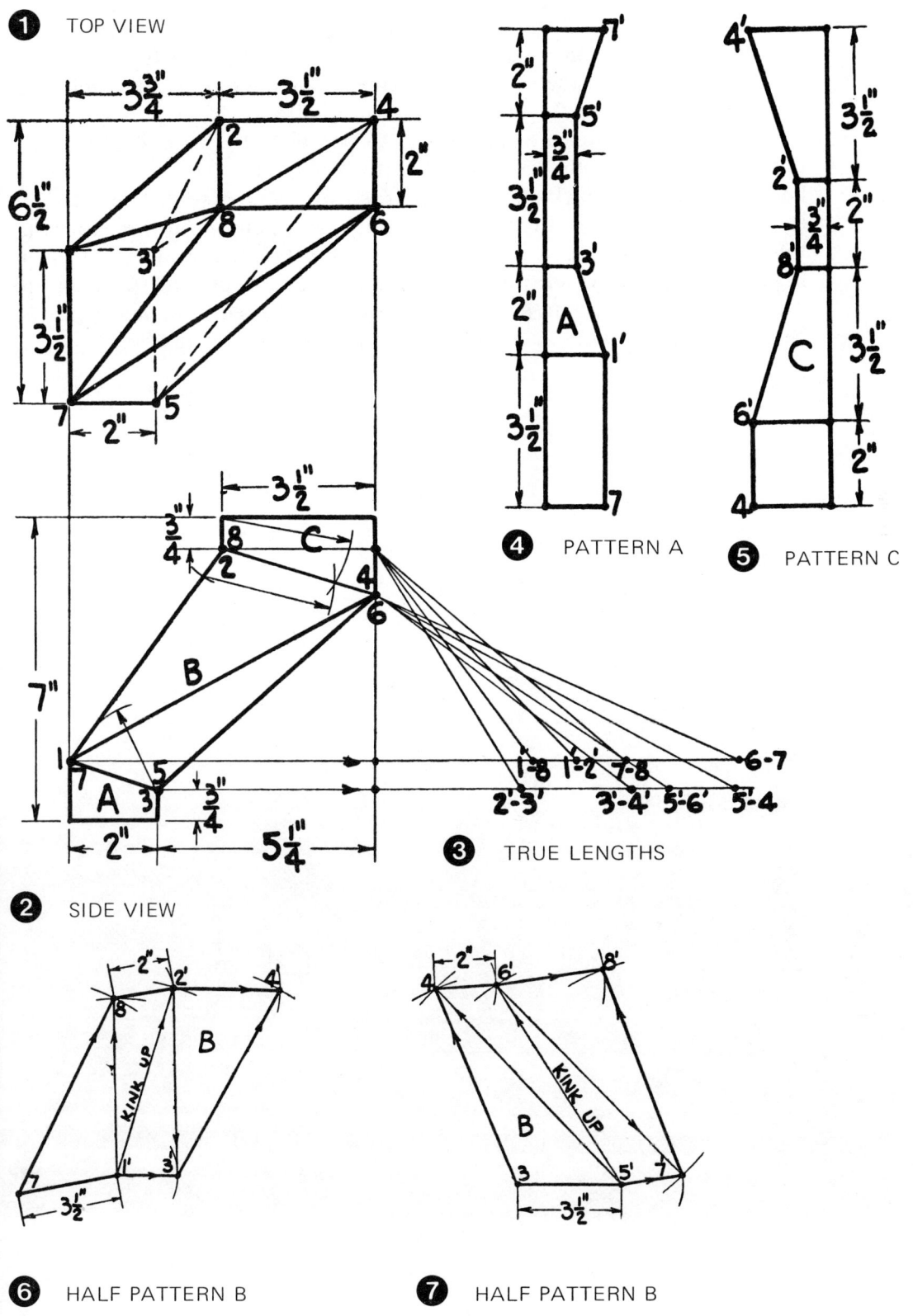

PLATE 67 TRANSFORMER QUARTER ROUND-TO-RECTANGULAR

Draw the top view and the side view as in Figures 1 and 2.

Transfer the distance 4 to *B*, and *B* to points 6, 7, and 8 from Figure 1 to the base line *B* on the triangle in Figure 3, thus obtaining the slant true-length lines to lay out the throat pattern in Figure 5.

Transfer the distance from point *A* to points 1, 2, 3, and 4 in Figure 1 to the base line *A* on the triangle in Figure 3, obtaining the slant true-length lines to lay out the back and side patterns from point 1 to 4. Use the slant true-length line 4–*B* in Figure 3 as a radius, and point 4 on the pattern in Figure 4 as a center, to strike an arc at point *B*. Use the distance *A* to *B* in Figure 2 as a radius; use point *A* in Figure 4 as a center to strike an arc crossing the arc at point *B*. The height *B* to 5 in Figure 4 is equal to the height *B* to 5 in Figure 2, completing the patterns as shown.

PLATE 67

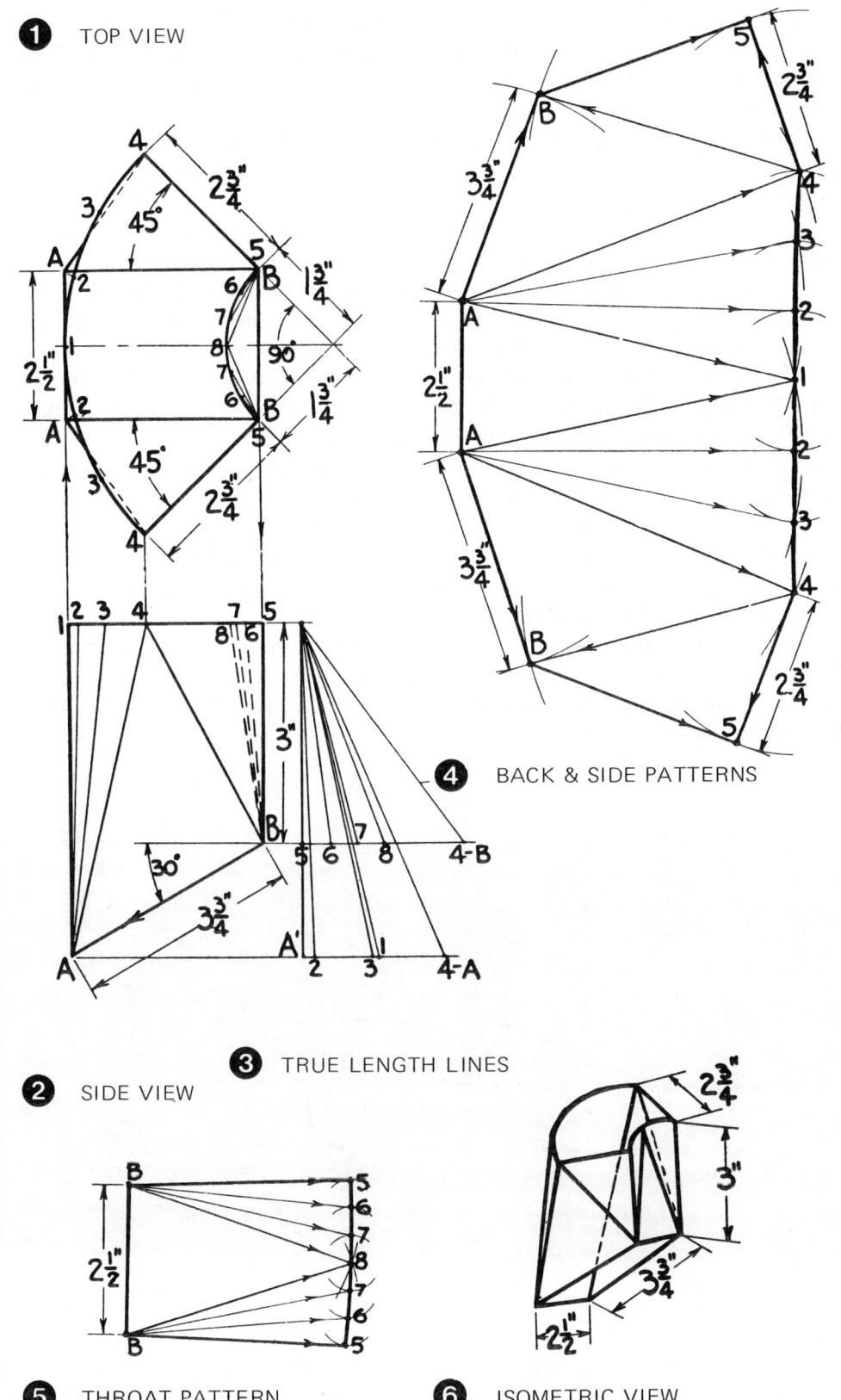

PLATE 68 ROUND TAPER WITH BASE MITERED 30 DEG.

This plate illustrates the conventional method for finding the true-length lines in triangulation for many irregular-shaped round tapers. This is a long method and is used by some sheet metal workers who are not familiar with the short cut method.

NOTE: The plate should be completed so that the student will easily absorb and understand the short cut methods in the next and following plates.

Draw the side view as in Figure 1, then transfer the distance 2 to 3' from Figure 1 to the base line 2 to 3' in Figure 2A, then transfer the height 3' to 3 from Figure 1 to erect the height of triangle 3'–3 in Figure 2.

NOTE: The distance 1 to 2 and 13 to 14 on the half pattern Figure 5 is equal to the distance 1 to 2 and 13 to 14, in the side view Figure 1. These are true-length lines. The distance 2 to 3 is equal to the slant length line 2 to 3 in Figure 2A. Transfer the distance 3' to 4' in Figure 1 to the base line 3'–4' in Figure 2B, then transfer the height 3'–3 from Figure 1 to one end of the base line in Figure 2B, then transfer the height 4'–4 from Figure 1 to the opposite end of the base line in Figure 2B obtaining the slant true-length line 3 to 4 on the half pattern in Figure 5.

Follow this same procedure to form the base lines and the end heights to obtain the remaining slant true-length lines in Figure 2, C, D, E, F, G, H, I, J, and K, thereby obtaining all the true length lines for the half pattern in Figure 5. The spaces on the curve at the top and bottom may be obtained by drawing a straight line equal to one quarter circumference of the small and large diameter in Figure 1 and divide each into equal spaces as illustrated in Figures 3 and 4.

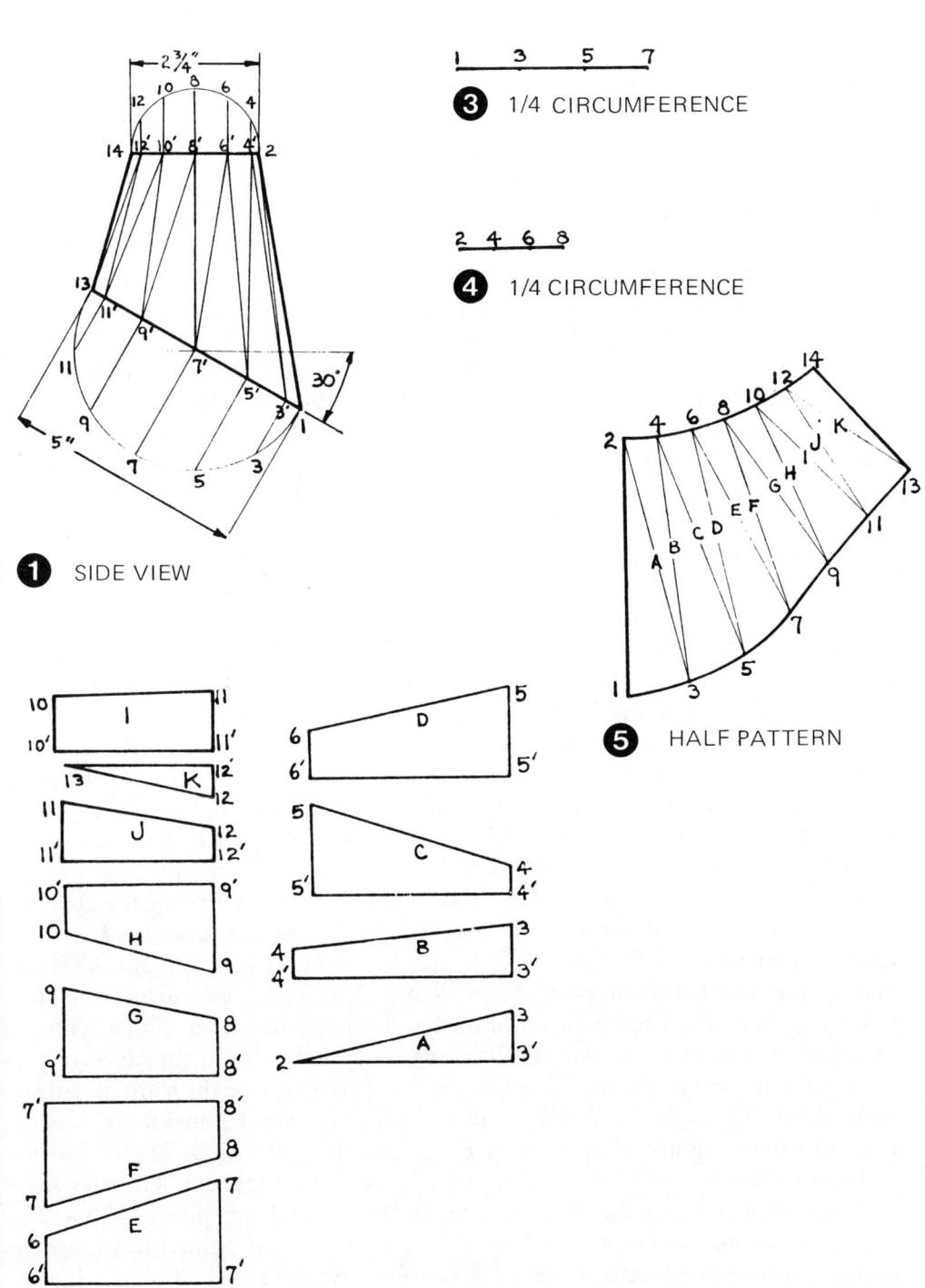

PLATE 69 ROUND TAPER WITH BASE MITERED 30 DEG.

Plate 69 illustrates a short, modern method in triangulation, as used in industry, to lay out a pattern for any irregular-shaped round fitting. Many unnecessary steps are eliminated in comparison to the old-fashioned long method used by many layout men as shown in Figure 2. This long method is obsolete and should not be used because of the waste of time involved in the erection of the many true-length triangles.

Draw the side view as in Figure 1. To use the short, modern method efficiently, transfer all the heights from the small diameter to their respective lines on the large diameter at the base. The distance 1 to 2 is a true-length line and does not have to be increased in length. The next line is 2 to 3′. The length of this line must be increased since point 2 has no height. Then the full height of line 3 on the large circle will represent the height of the true-length triangle as represented by 2-3 to 3′. The next line is 3′ to 4′. The length of this line must be increased according to the difference in the heights of lines 4-4′ on the small diameter and 3-3′ on the large diameter. Therefore, transfer the height of line 4-4′ to line 3-3′ from point 2-3 to point 3-4. The distance from 3-4 to 3′ represents the difference in height of the two lines, thereby obtaining the height of the raise for the true-length line 3′ to 4′. The next line is 4′ to 5′. Transfer the height of line 4-4′ to line 5 from point 5 to point 4-5. The next line is 5′ to 6′. Transfer the height of line 6-6′ to line 5 from point 5 to point 5-6. Continue to transfer the heights of the remaining lines 8-8′, 10-10′, and 12-12′ to their respective lines on the large diameter as shown. Since point 13 has no height, then the height 12-12′ is transferred from point 13 to point 12-13. This full height will represent the height of the true-length triangle. Now that all the heights for the triangles have been established, the pattern may be laid out.

To lay out the pattern as in Figure 3, draw line 1 to 2 equal to line 1 to 2 in Figure 1. Following the procedure in Plate 24, use any one of the spaces 1 to 7 on the ¼ circumference line in Figure 4 as a radius, and point 1 as a center, to strike an arc at point 3. Transfer the distance 2 to 3′ from the side view to the base line from point 3′ to point 2-3. Use the slant distance from point 2-3 to 2-3 in Figure 1 as a radius, and point 2 in Figure 3 as a center, to strike an arc to cross the arc at point 3. Transfer the distance 3′ to 4′ from the side view to the base line from point 3′ to 3-4. Use the slant distance from point 3-4 to 3-4 in Figure 1 as a radius, and point 3 in Figure 3 as a center to strike an arc at point 4. Use any one space 2 to 8 in Figure 5 as a radius to draw an arc to cross the arc at point 4 in Figure 3. Transfer the distance 4′ to 5′ from the side view to the base line from point 5′ to 4-5. Use the slant line 4-5 to 4-5 in Figure 1 as a radius, and point 4 in Figure 3 as a center to strike an arc at point 5. Continue this same procedure to obtain the remaining true-length lines to complete the pattern as in Figure 3. The distance 13 to 14 in Figure 3 is equal to line 13 to 14 in Figure 1.

PLATE 69

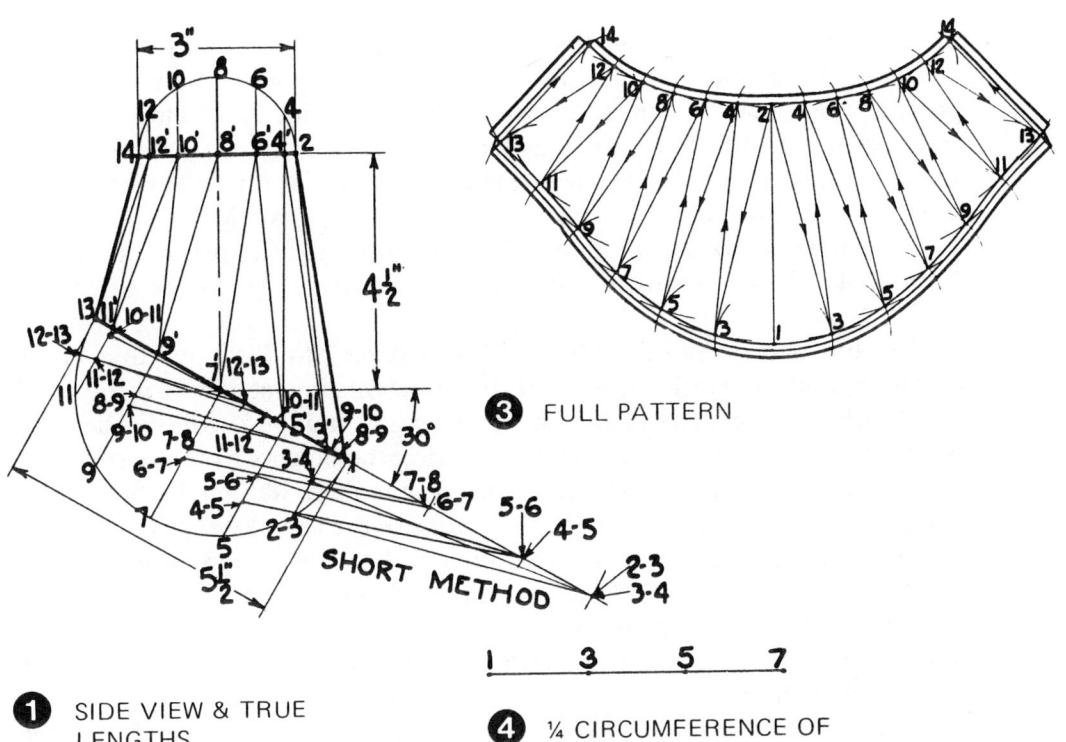

① SIDE VIEW & TRUE LENGTHS

③ FULL PATTERN

④ ¼ CIRCUMFERENCE OF 5 INCH DIAMETER

⑤ ¼ CIRCUMFERENCE OF 3 INCH DIAMETER

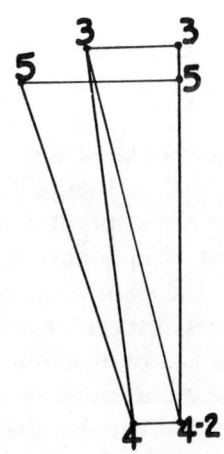

② LONG METHOD FOR FINDING TRUE LENGTHS

PLATE 70 ROUND TAPERING OFFSET

To draw the side view as in Figure 1, use the distance 2 to 14 as a radius, and point 14' as a center to draw an arc toward point 1'. Draw a line from point 1' tangent to the arc to intersect line 2 at point 2'. Draw a slant line from point 2' to 14'. Use 1 to 13 as a radius, and point 1 as a center to draw an arc toward point 14'. Draw a line from point 14' tangent to the arc to intersect line 13 at point 13'. Draw a slant line from 1' to 13'.

Lay out pattern O as in Figure 2. The distance 8 to 8 is equal to the circumference of a 3½-in. diameter, and the heights are equal to the heights on section O in Figure 1.

Lay out pattern M in Figure 3. The distance 7 to 7 is equal to a 2½-in. diameter, and the heights are equal to the heights on section M in Figure 1.

To obtain the heights for the true-length triangles, follow the same procedure as used in Plate 69, by transferring the heights from the small half circle, Figure 1, to their respective lines on the large half circle. Since point 2' has no rise, then the distance 2' to 3' will rise the full height of line 3 on the small half circle. This height 3 is transferred to line 2, represented by the height from point 2 to 2'–3', also to line 4 from point 4 to 3'–4'. Transfer the height of line 5 from the small half circle to the large half circle from point 4 to 4'–5'. Continue transferring the remaining heights from the small half circle to their respective lines on the large half circle as shown.

Transfer the distance 2' to 3' from section N to the base line from point 2 to point 2'–3'. Transfer the distance 3' to 4' to the base line from line 4 to point 3'–4'. Transfer the distance 4' to 5' to the base line from line 4 to point 4'–5'. Continue transferring the remaining lengths of the lines from section N to the base line as shown.

NOTE: The distances 1' to 2' and 13' to 14' in Figure 1 do not have to be increased in length; they are the true lengths of lines 1' to 2' and 13' to 14' on the pattern in Figure 4.

To lay out the pattern in Figure 4, draw line 1' to 2' equal to the length of line 1' to 2' in Figure 1. Use the slant length 2'–3' to 2'–3' in Figure 1 as a radius, and point 2' in Figure 4 as a center to strike an arc at point 3'. Use the distance 1' to 3' in Figure 3 as a radius, and point 1' in Figure 4 as a center to strike an arc crossing the arc at point 3'. Use the slant length 3'–4' to 3'–4' in Figure 1 as a radius, and point 3' in Figure 4 as a center to strike an arc at point 4'. Use the distance 2' to 4' in Figure 2 as a radius, and point 2' in Figure 4 as a center to strike an arc crossing the arc at point 4'. Continue this procedure by transferring the remaining slant true-length lines from Figure 1 to the pattern in Figure 4. The spaces 1' to 13' are equal to their respective spaces on the curved line 1' to 13' in Figure 3. The spaces 2' to 14' are equal to their respective spaces on the curved line 2' to 14' in Figure 2.

Complete the pattern by making the required allowances as shown.

PLATE 70

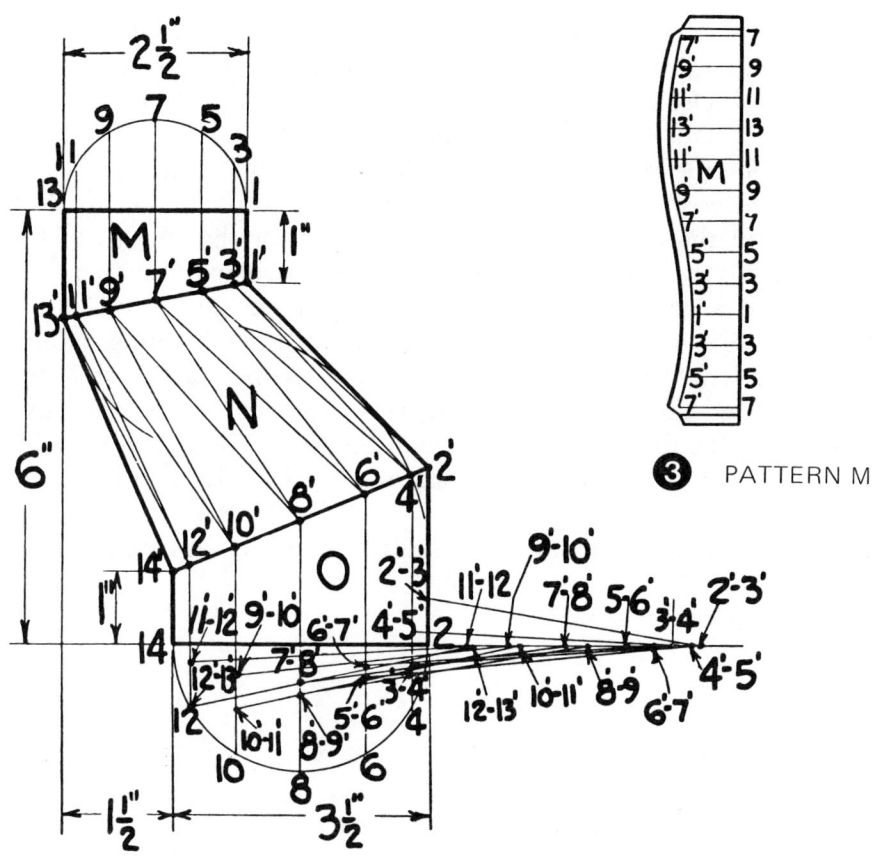

① SIDE VIEW & TRUE LENGTH LINES

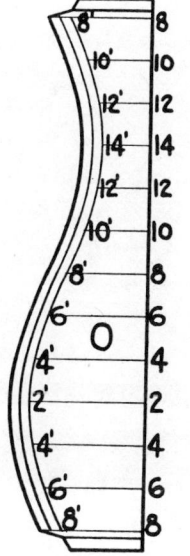

② PATTERN O

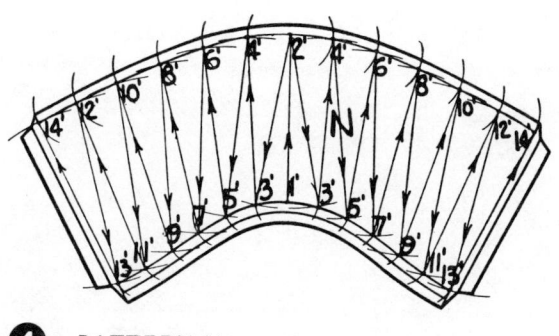

③ PATTERN M

④ PATTERN N

PLATE 71   OBLONG-TO-ROUND OFFSET

Draw the side view in the same manner as in Plate 70, by using point 1' as a center to draw the arc at line 15. Draw a line from point 1' tangent to the arc to point 15. Use point 14' as a center to draw an arc at line 2. Draw a line from point 1' tangent to the arc to point 2'.

To obtain the true-length triangles, follow the same procedure as in Plate 68, by transferring the heights from the oblong at the top to their respective lines on the half circle at the base. Transfer the lengths of the lines in section N to their respective positions on the base line as shown.

To lay out the pattern as in Figure 4, draw the center line 1' to 2' equal to line 1' to 2' in Figure 1; then continue laying out the remaining pattern by using the true-length triangles in Figure 1 in the same manner as in the previous plate. The spaces 2' to 14' in Figure 4 are equal to the respective spaces 2' to 14' on the freehand curved line in Figure 2. The spaces 1' to 15' in Figure 4 are equal to their respective spaces from 1' to 15' on the curved line in Figure 3.

Complete the pattern by making the required allowances as shown.

PLATE 71

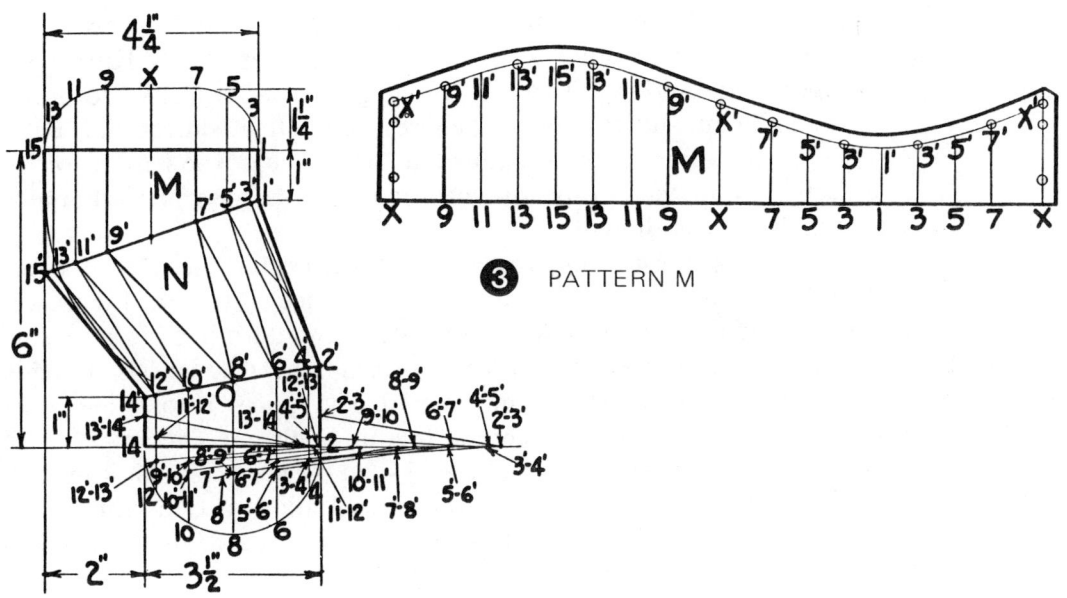

1. SIDE VIEW & TRUE LENGTH LINES
2. PATTERN O
3. PATTERN M
4. PATTERN N
5. ISOMETRIC VIEW

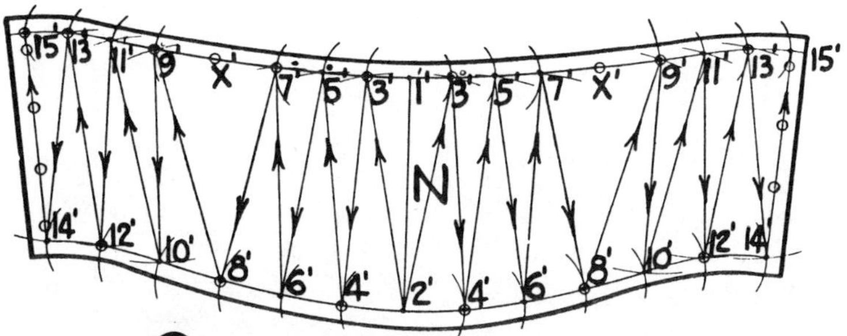

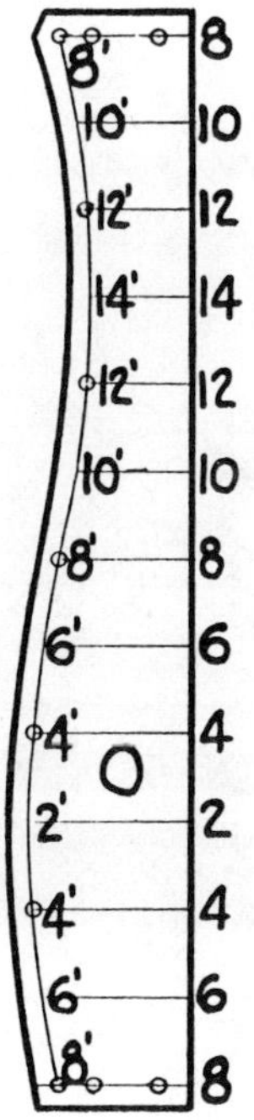

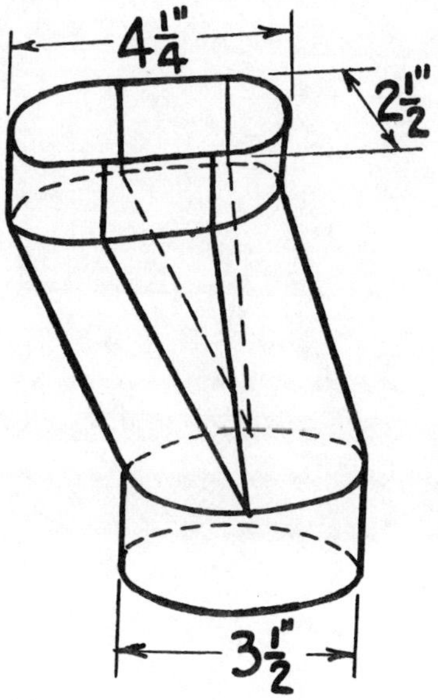

131

PLATE 72 OBLONG-TO-ROUND OFFSET

Draw the side view, Figure 1, in the same manner as in Plate 70. The $2\frac{1}{8}$-in. height from line $A-B$ to point 7 on the half circle represents half of the $4\frac{1}{4}$-in. length of the oblong. Transfer the heights from line $A-B$ to points 3, 5, 7, 9, and 11 on the oblong to their respective lines on the half circle at the base, in the same manner as in Plate 70. Transfer the lengths of the lines in section N to their respective positions on the base line, obtaining the true-length lines as shown.

To lay out the pattern as in Figure 4, draw the center line A' to $2'$ equal to the length of line A' to $2'$ in Figure 1. The distance A' to $1'$ in Figure 4 is a straight line equal to the distance A' to $1'$ on the pattern in Figure 3. Continue laying out the remaining pattern by using the true-length triangles in Figure 1 in the same manner as in Plate 70. The spaces $1'$ to B' are equal to their respective spaces on the curved line in Figure 3. The spaces $2'$ to $14'$ are equal to their respective spaces $2'$ to $14'$ on the curved line in Figure 2.

Complete the pattern by making the required allowances as shown.

NOTE: The line drawn from $13'$ to B' in pattern N is a straight line equal to the distance $13'$ to B' on pattern M in Figure 3.

PLATE 72

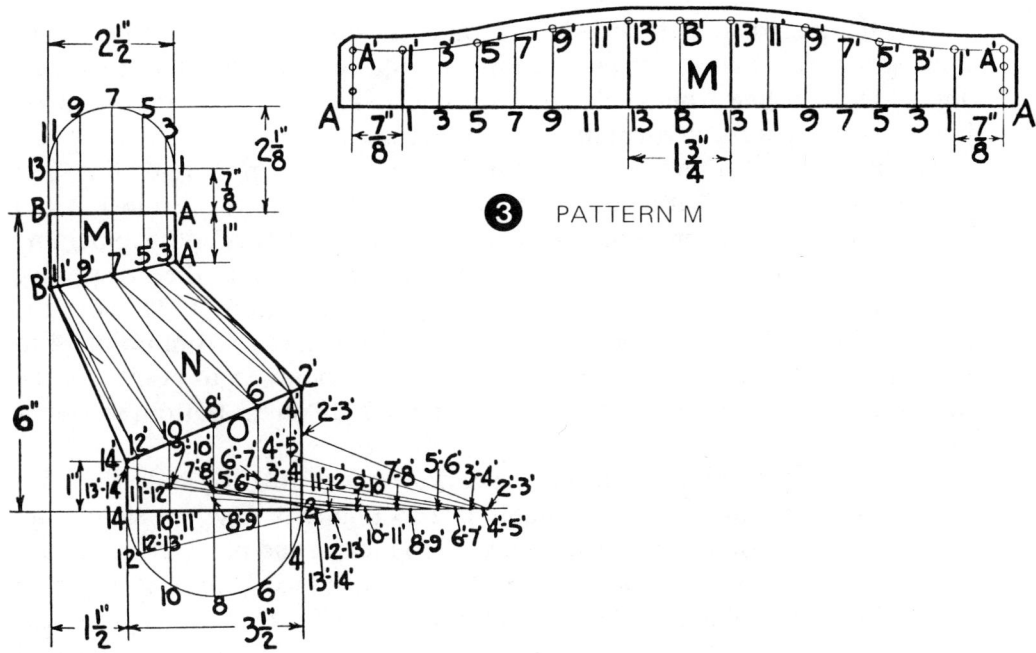

① SIDE VIEW & TRUE LENGTH LINES

③ PATTERN M

④ PATTERN N

⑤ ISOMETRIC VIEW

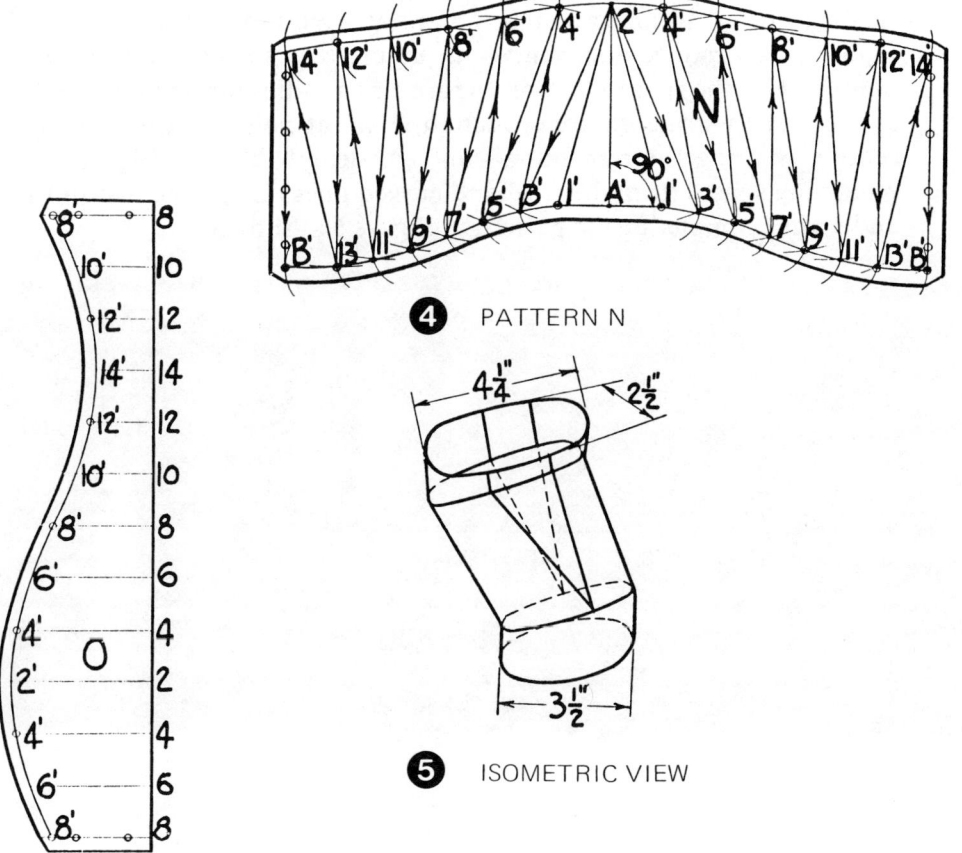

② PATTERN O

133

PLATE 73   TAPERING DOUBLE OFFSET

To obtain the true-length lines for a tapering double-offset fitting of this type, two views must be drawn, the top and side views as shown in Figures 1 and 2. Draw a line from each point on line 1–13 and from each point on line 2–14 to cross the vertical line $A-B$ in Figure 3.

Transfer the distance 1 to 2 from the top view, Figure 1, to line 1 in Figure 3 from point 1 at line $A-B$ to point 1–2. Transfer the distance 2 to 3 from Figure 1 to line 3 in Figure 3 represented by 2–3. Transfer the remaining lengths from point 3 to point 14 in the top view, Figure 1, to their respective lines drawn from line 1–13 on the top section B in Figure 3. Transfer the distance 14 to 15 from the top view to line 14 in Figure 3, from point 14 to point 14–15. Transfer the remaining lengths from point 15 to point 1 in the top view to their respective lines drawn from line 2–14 on the bottom section A in Figure 3.

To lay out the pattern as in Figure 6, draw line 1' to 2' equal to the slant true-length line from point 2 to point 1–2 in Figure 3. Use the slant true-length line from point 2 to 2–3 in Figure 3 as a radius, and point 2' in Figure 6 as a center to draw an arc at point 3'. Use the distance 1' to 3' in Figure 5 as a radius, and point 1' in Figure 6 as a center to draw an arc crossing the arc at point 3'. Use the slant true-length line from point 4 to 3–4 in Figure 3 as a radius, and point 3' in Figure 6 as a center to strike an arc at point 4'. Use the distance 2' to 4' in Figure 4 as a radius, and point 2' in Figure 6 as a center to draw an arc crossing the arc at 4'. Use the distance 4 to 4–5 in Figure 3 as a radius, and point 4' in Figure 6 as a center to strike an arc at point 5'. Use the spaces 3'–5' in Figure 5 as a radius, and point 3' in Figure 6 as a center to strike an arc crossing the arc at point 5'. Continue this procedure by using the remaining slant true-length lines in Figure 3 and the remaining spaces from the freehand curved lines on the patterns in Figures 4 and 5, thus completing the pattern in Figure 6 as shown.

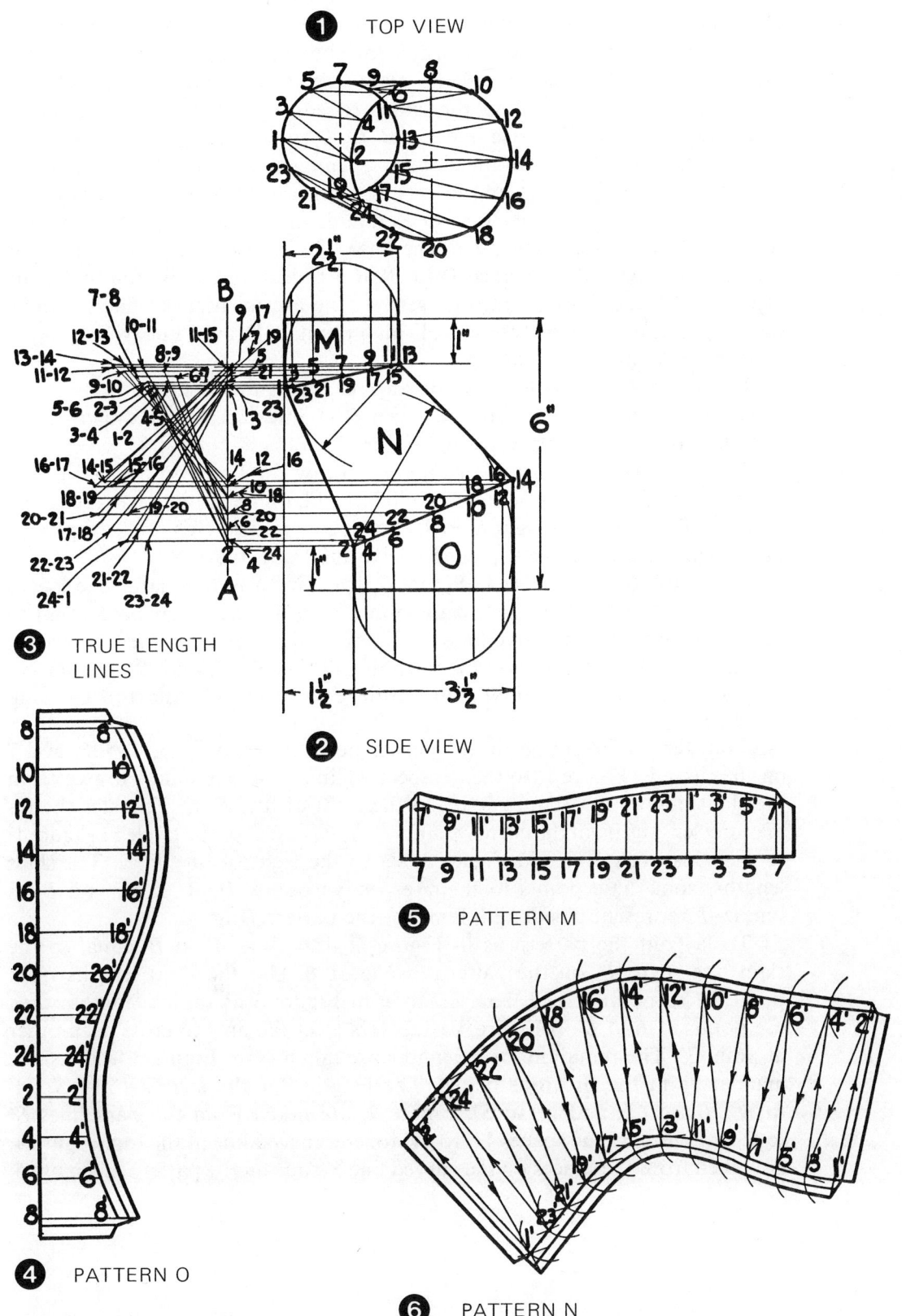

PLATE 74   RECTANGULAR-TO-ROUND DOUBLE OFFSET

Draw the position of the rectangular base line A, B, C, D, and the circular opening in Figure 1 to the dimensions as shown. Now establish the points for the kink or bend lines and the seam line to facilitate forming and assembling the patterns. Follow the same procedure as in Plate 18 by drawing a square parallel to the base lines A, B, C, D to encompass and strike on a tangent the curved lines on the circular opening as shown at points 3, 7, 11, and 15, or the points that are nearest to each tangent-line striking point.

To lay out the front view as in Figure 2, draw lines down from points A, B, C, D, and S in Figure 1 to the base line in Figure 2. Use point D as a center, and with the distance D to B' as a radius, draw an arc toward the top. Draw a line from point 9 tangent to the arc to intersect line B drawn down from the top view, thus establishing point B. Draw a line from point D to B' crossing lines A, C, and S.

Use point 9 as a center to draw an arc. Bisect this arc, and draw a line from the bisecting point through point 9 to intersect line 1 drawn down from the top view, thus completing Figure 2.

To lay out the pattern as in Figure 5, draw the line 5 to 5 equal to the circumference of a 3-in.-diameter circle, and divide it into the required number of spaces. The height of each line is equal to the height of its respective line in Figure 2 from the top edge line to the slant line 1–9.

To lay out the pattern as in Figure 6, draw line S to S equal to the spaces S, B, A, D, C, and S on the rectangle in Figure 1. The heights are equal to the height of their respective lines in Figure 2 from the base line to the slant line D – B'. Draw lines across from points D, A, C, S, and B' to intersect line D–E in Figure 3. Draw a line from each point on line 1–9 crossing line E–D.

Transfer the lengths from the corner point A to points 3, 4, 5, 6, and 7 on the circle in Figure 1, to their respective lines 3, 4, 5, 6, and 7 drawn from the slant line 1–9 in Figure 3, set to the right of line E–D. Transfer the remaining lengths from the corner points B, S, C, and D to the circle in Figure 1, to their respective lines in Figure 3, set to the right of line E–D. The slant lengths from these points to their respective points D, A, C, S, and B on line D–E represent true-length lines for the pattern (Fig. 4).

To lay out the pattern as in Figure 4, draw line A' to B' equal to the slant line A' to B' on the pattern in Figure 6. Use the slant lengths from point $7A$ to point A and from $6B$ to B in Figure 3 as radius lengths; then use points A' and B' in Figure 4 as centers to strike arcs to cross each other at point $7'$. The remaining true lengths are taken from Figure 3 in the same manner as in the previous plates. The lengths for the spaces B' to S', A' to D', D' to C', and C' to S', Figure 4, are taken from the slant lines S' to S' in Figure 6. The spaces 11′ to 11′ for the curved line at the top, Figure 4, are taken from the spaces on the curved line 5′ to 5′ on the pattern in Figure 5.

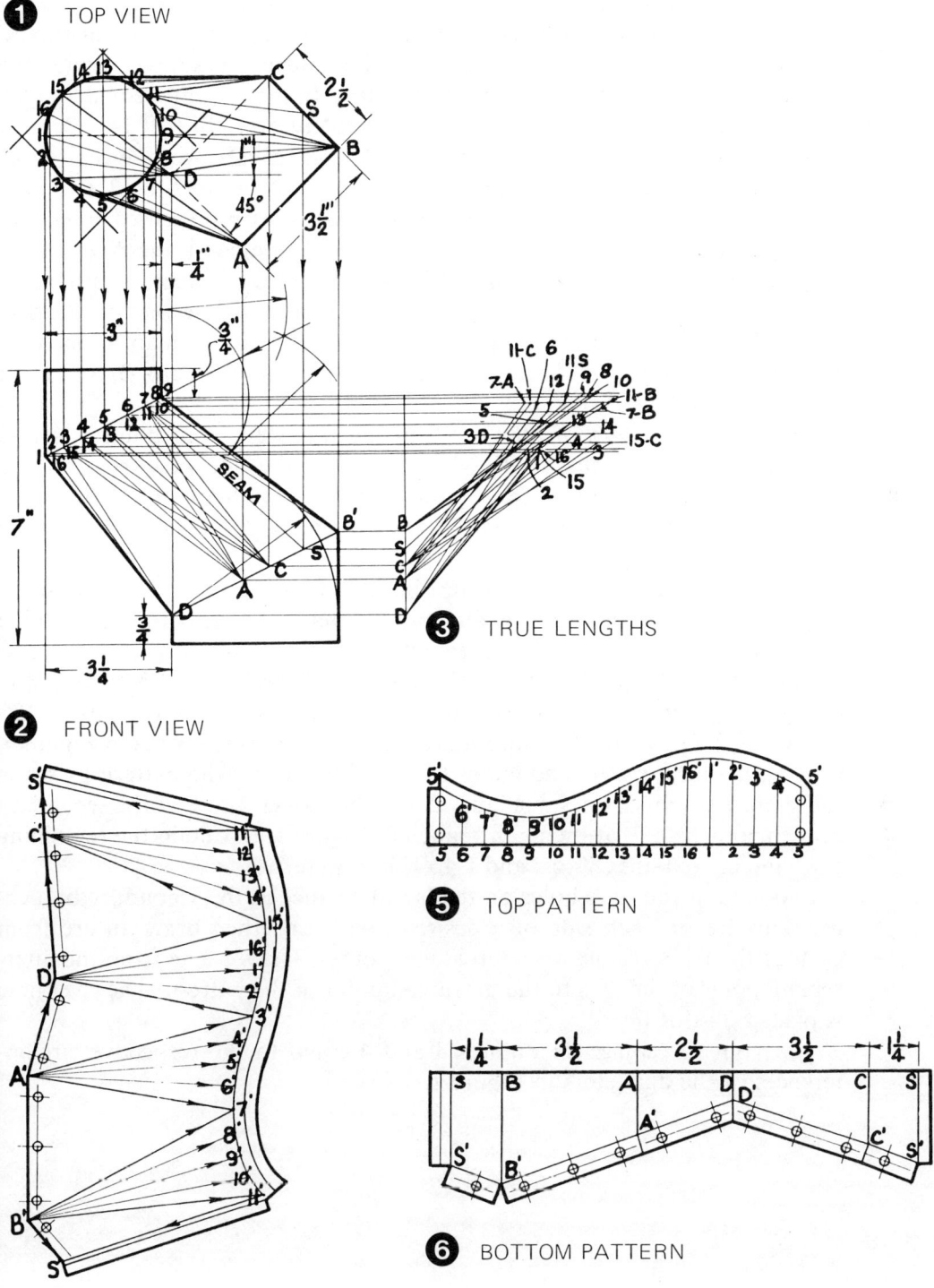

PLATE 74

PLATE 75 BULLHEAD Y BRANCH

This type of Y branch is not recommended for blowpipe systems because of the flat crotch, but it will be practical for discharging air, smoke, or fumes.

Draw the side view as in Figure 1 to the dimensions shown. Divide the half circle 2 to 14′, at one of the secondary branches, into equal spaces, and draw a line from each division point to intersect line 2–14′. Draw a line from points 10′ and 12′ parallel to line 8′–8″ to intersect the center line 7–14″. Transfer the widths of lines 8 to 8′, 10 to 10′, and 12 to 12′ from the half circle to their respective lines on the crotch, as represented by 8″–8″, 10″–10″, and 12″–12″, thus obtaining the freehand curved line 8″ to 14″.

To obtain the various heights for the true-length lines, transfer the heights of lines 4, 6, and 8 from the half circle at the secondary branch, Figure 1, to their respective lines on the half circle at the main branch, as shown from point 3′ to 4–3, and from 5′ to 5–4, and 5′ to 6–5, and from 7′ to 7–6, and 7′ to 7–8″A.

Transfer the length of line 3 to 2 in the side view from point 3 to 3–2 in Figure 2. Transfer the length of line 4 to 3 in the side view from point 3 to 4–3. Transfer line 5 to 4 from point 5 to 5–4, and line 6 to 5 from point 5 to 6–5. Transfer line 7 to 6 and 7 to 8″A from point 7 to point 7 to points 7–6 and 7–8″A in Figure 2.

To lay out the half pattern as in Figure 5, draw line 8′ to 8′ equal to the length of line 8′–8′ in Figure 1. Draw the center line 8″A to 14″ equal to the spaces 8″ to 14″ on the freehand curved line at the crotch in Figure 1. Draw the center line 8″A to 7 equal to the slant true-length line 7–8″A to 7–8″A in Figure 2. Transfer the widths of the lines at the crotch 8″A to 8′, 10″ to 10′, 12″ to 12′, and 14″ to 14′ from Figure 1 to their respective lines in Figure 5, thus obtaining the freehand curve 14′ to 8′. Complete the pattern by using the slant true-length lines in Figure 2 and the spaces 2 to 8 at the secondary branch, and 1′ to 7′ at the main branch in Figure 1, to obtain the remaining freehand curved lines 8′ to 2 and 1′ to 1′ in Figure 5.

To lay out the rivet holes on the lap allowance at the secondary branch, draw an arc on each side of a desired point, and then draw an arc from each of the two arcs drawn to cross each other. Draw a line from the intersecting point of the arcs to the presumed point of the pattern. The rivet hole is placed on this line.

Lay out the collars in Figures 3 and 4 equal to the respective circumferences of the diameters in Figure 1.

PLATE 75

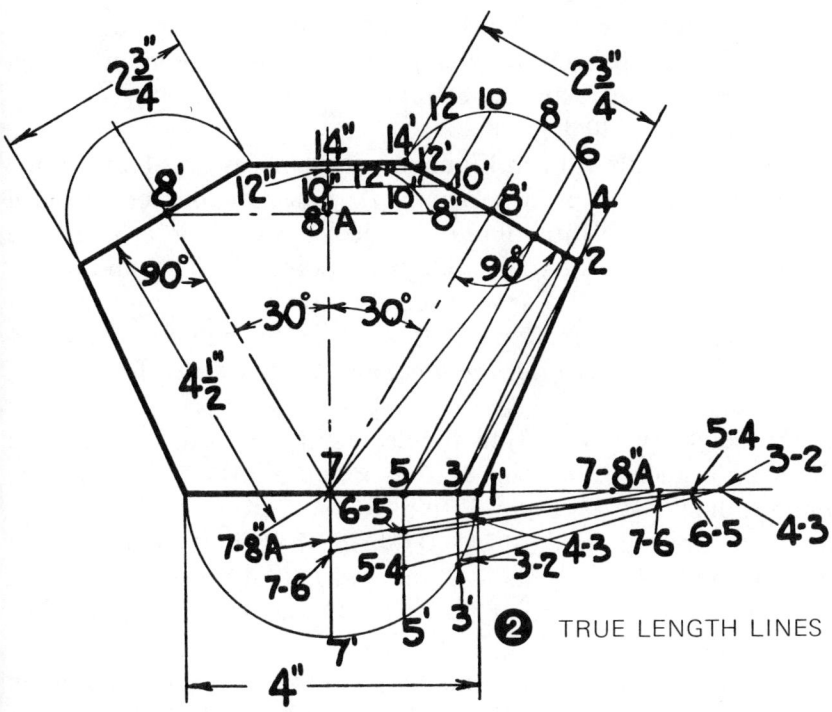

② TRUE LENGTH LINES

① SIDE VIEW

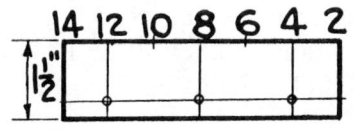

③ HALF PATTERN

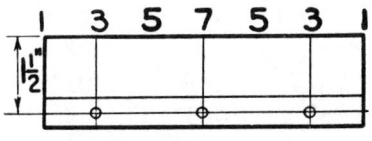

④ HALF PATTERN

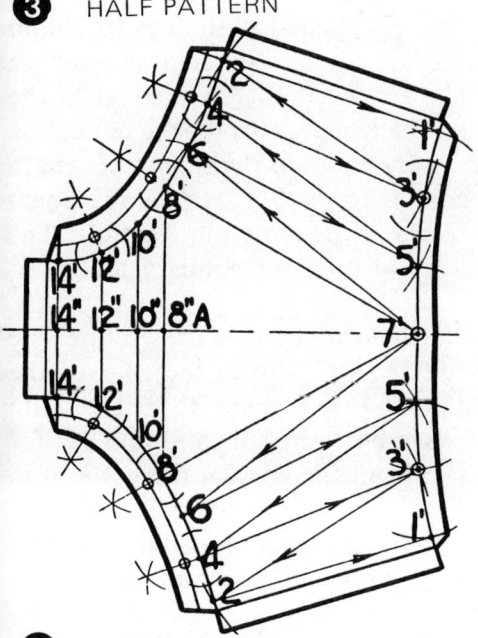

⑤ HALF PATTERN

⑥ ISOMETRIC VIEW

139

PLATE 76   Y BRANCH WITH EQUAL SPREAD

The length of the crotch 7 to 13' shall be at least equal to three quarters of the diameter of the main branch, or greater, if possible. The main branch shall be of a diameter to contain an area equal to the area of the two secondary branches and at an angle of 30 deg.

Only one half of the side view in Figure 1 will be necessary. Draw a line from point 14 at the secondary branch to point 7" at the main branch, crossing the center line at point 13', thus obtaining the length of the crotch. Use point 7 as a center to draw a quarter circle from point 13' to 7", and divide it into equal spaces. Draw a line from each division point to intersect the crotch line 7–13'. Draw a line from points 3 and 5 on the half circle to intersect lines 9 and 11, thus obtaining the freehand crotch curve from point 13' to 7".

Transfer the height of line 4 from the secondary branch, Figure 1, to line 3' on the main branch from point 3' to 3–4, and 5' to 4–5. Transfer the height of line 6 to line 5', from point 5' to 5–6, and 7' to 6–7. Transfer the height of line 8 to 7' from point 7' to 7–8. Transfer the heights of lines 8 and 10 to line 9, from point 9' to points 8–9 and 9–10. Transfer the heights of lines 10 and 12 to line 11, from point 11' to 10–11 and 11–12. Transfer the height of line 12 from 13' to 12–13'.

Transfer the length of lines 2 to 3 and 3 to 4 to the base line of the main branch from point 3 to point 2–3 and 3–4. Transfer the length of lines 4 to 5 and 5 to 6 from point 5 to 4–5 and 5–6. Transfer lines 6 to 7 and 7 to 8 from point 7 to 6–7 and 7–8.

To obtain the true lengths for the crotch, transfer lines 8 to 9 and 9 to 10 from point 9 to points 8–9' and 9'–10. Transfer lines 10 to 11 and 11 to 12 from point 11 to points 10–11' and 11'–12. Transfer 12 to 13' from point 13' to 12–13'.

To lay out the pattern as in Figure 2, draw line 1' to 2 equal to line 1' to 2 in the side view in Figure 1. The remaining true-length lines to obtain the points from 7' to 7' and 8 to 8, Figure 2, are equal to the slant true-length lines at the main branch in Figure 1. The spaces 1' to 7' are equal to the spaces 1' to 7' on the half circle. To obtain the true lengths for the remaining points 7' to 13' and 8 to 14, Figure 2, they are equal to the slant true-length lines at the crotch in Figure 1. The spaces 7' to 9' to 11' to 13' are equal to the spaces 7" to 13' on the freehand curve at the crotch in Figure 1. The spaces 2 to 14 are equal to the spaces 2 to 14 on the secondary branch in Figure 1.

Since both patterns are the same, a lap allowance is made on one side only; the rivet holes on the opposite side are set in from the edge of the pattern. To lay out these rivet holes, use point 7' as a center to strike an arc between points 7' and 9'; also, strike an arc between points 9' and 11' for the next rivet hole. Place the rivet holes in from the edge of the pattern on these arcs on both sides as shown.

At the lower portion of the crotch it is practically impossible to flatten a rivet. This portion, therefore, is notched to allow one branch to be clenched to the other.

NOTE: Both patterns must be formed the same.

PLATE 76

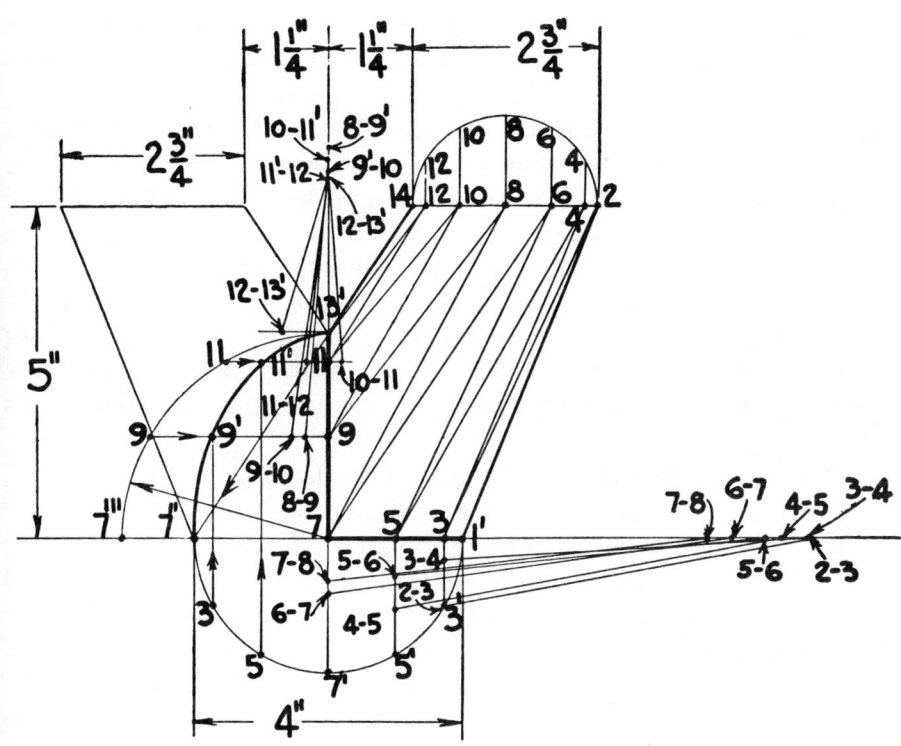

① SIDE VIEW & TRUE LENGTH LINES

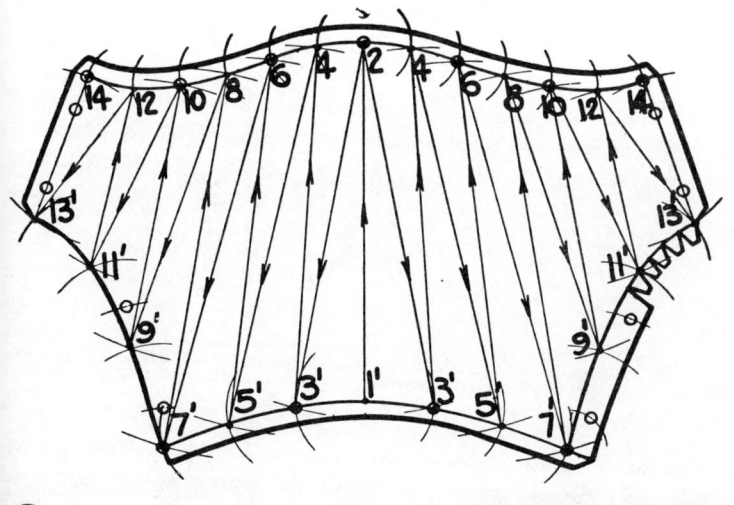

② FULL PATTERN — TWO REQUIRED

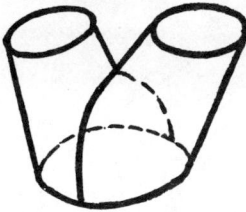

③ ISOMETRIC VIEW

141

PLATE 77 SHORT METHOD: Y BRANCH EQUAL SPREAD

The Y branch in this plate is the same fitting as in Plate 74. This method for laying out the pattern is shorter and more simplified than the method shown in Plate 76.

Draw only a portion of the side view as in Figure 3. Draw a line from point 14 to 7′ crossing the center line at point 13′ obtaining the height of the crotch. Use point 7 as a center to draw the quarter circles 13′ to 7″ and 7′ to 13. Draw a line from points 9″ and 11″ to intersect the center crotch line. Draw a line from points 9 and 11 to intersect the lines drawn from points 9″ and 11″, thus obtaining the freehand curved crotch line 13′ to 7′.

Draw the half bottom view as in Figure 1. Divide both the small half circle 2 to 14, and the large half circle 1 to 13 into equal spaces; draw a line from points 9 and 11 to intersect the center crotch line, thus obtaining points 9′ and 11′.

Erect the true-length triangle by transferring the lengths 1 to 2, 2 to 3, 3 to 4, 4 to 5, 5 to 6, 6 to 7, and 7 to 8 from Figure 1 to the base line in Figure 2. These will represent the slant true-length lines to develop the pattern 8 to 8 and 7′ to 7′ in Figure 4. The remaining true-length lines may be obtained by transferring the lengths 8 to 9′, 9′ to 10, 10 to 11′, 11′ to 12, and 12 to 13′ from the half bottom view in Figure 1 to line C–D in Figure 3, from point C toward point D. The slant lines from points 8–9′ to 9, 9′–10 to 9, 10–11′ to 11, 11′–12 to 11, and 12–13′ to 13′ in Figure 3 represent the slant true-length lines to develop the remaining pattern 8 to 14 and 7′ to 13′ in Figure 4, following the same procedure as in Plate 76.

To lay out the rivet holes on the crotch, use point 7′ as a center to draw the arc between 7′ and 9′, and between 9′ and 11′ as shown.

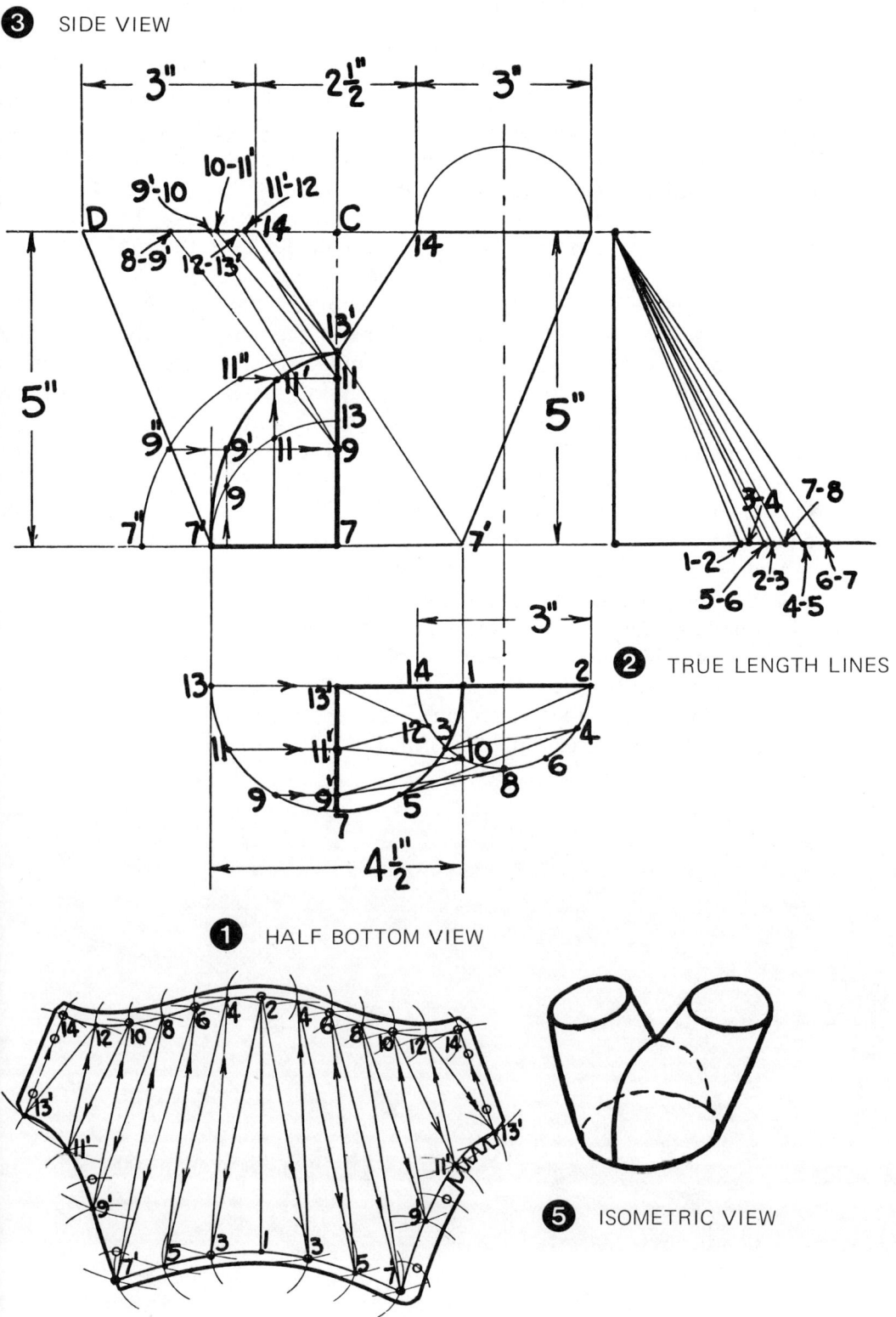

PLATE 77

PLATE 78   Y BRANCH WITH DIFFERENT SPREADS

Each secondary branch has a different spread or angle of direction; therefore, the full side view must be drawn and two different patterns must be laid out.

Draw the side view as in Figure 1. (When two branches are of different angles, draw a line from the secondary branch to the main branch from the one that will allow the longest crotch.) Use point 7 as a center to draw the quarter circle from 13' to 7", and follow the same procedure as in Plate 76, to obtain the freehand crotch curve from 13' to 7'.

Transfer the heights of lines 4, 6, and 8 from any one of the secondary branches to their respective lines on the half circle at the main branch; also lines 8, 10, and 12 to lines 9', 11', and 13' at the crotch.

To obtain the true slant-length lines, transfer the slant lines from the side view of both branches to the base line of the main branch, and to the center crotch line in the same manner as in Plate 76. Note that no slant lines are drawn to the respective heights representing the true slant lines for the crotch. These lines are eliminated to avoid confusion.

To lay out the two branch patterns as in Figures 2 and 3, the slant true-length lines are obtained from Figure 1. The spaces 7' to 7' are equal to the spaces 1 to 7 on the half circle on the main branch. The spaces 2 to 14 are equal to spaces 2 to 14 on the half circle on either one of the two secondary branches. The spaces 7' to 13' on both patterns are equal to the spaces 7' to 13' on the freehand crotch curve in Figure 1.

To lay out the rivet holes on the crotch, use point 7' as a center to draw an arc between 7' and 9' on each side of both patterns with the same radius, and draw an arc between 9' and 11' on each side of both patterns. Notch the lower portion of the crotch, and mark the rivet holes as shown.

NOTE: Both patterns must be formed in the same manner.

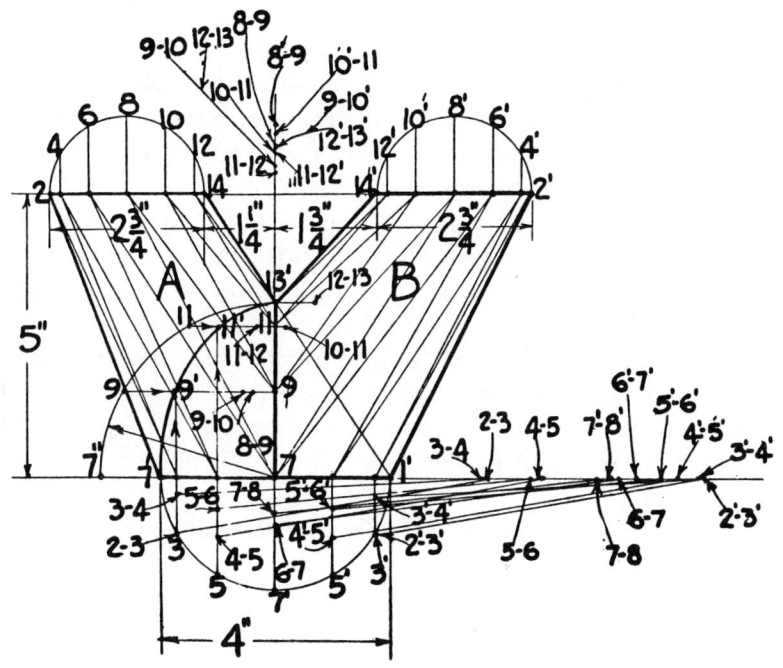

1 SIDE VIEW & TRUE LENGTH LINES

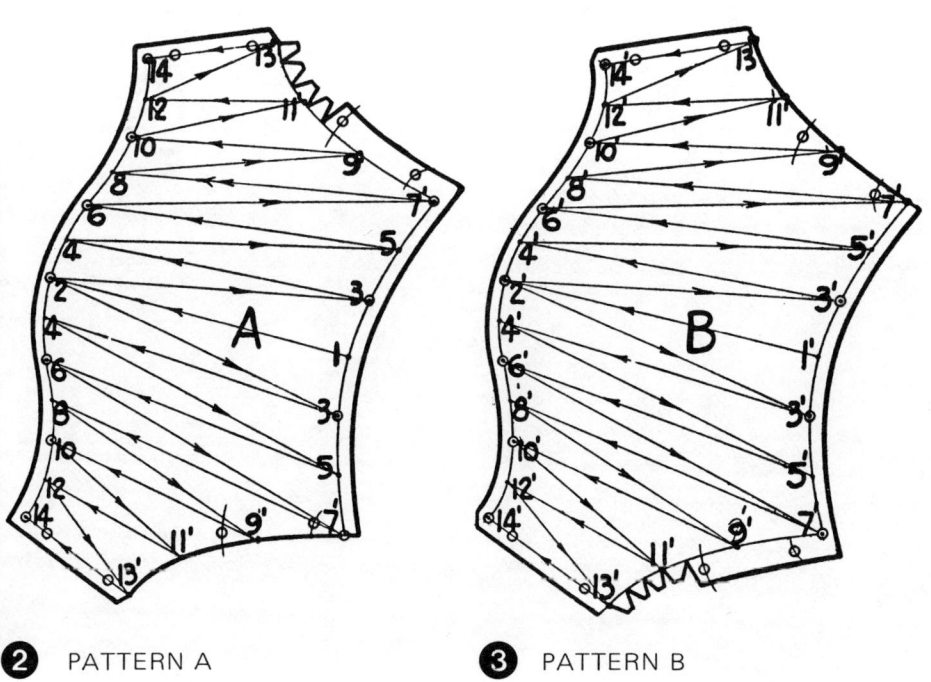

2 PATTERN A

3 PATTERN B

PLATE 79 Y BRANCH OF DIFFERENT DIAMETERS WITH EQUAL SPREAD

Due to the different diameters of the two secondary branches, the full side view must be drawn and two different patterns must be laid out.

Draw the full side view as in Figure 1, and transfer the heights from the half circles on the two secondary branches to their respective lines on the half circle at the main branch and on lines 9′, 11′, and 13′ at the crotch as shown.

Transfer the various length lines from branches A and B to the base line of the main branch and to the center crotch line, in the same manner as in Plate 76. Notice that no lines are drawn to represent the slant true-length lines at the crotch; they have been eliminated to avoid confusion.

To lay out the patterns as in Figures 2 and 3, follow the same procedure as in Plate 76, by transferring each of the slant true-length lines in Figure 1 to their respective patterns in Figures 2 and 3. The spaces 7′ to 13′ at the crotch on both patterns are equal to the spaces 1′ to 13′ on the freehand curved line in Figure 1.

Lay out the rivet holes at the crotch, and complete the patterns as shown.

PLATE 79

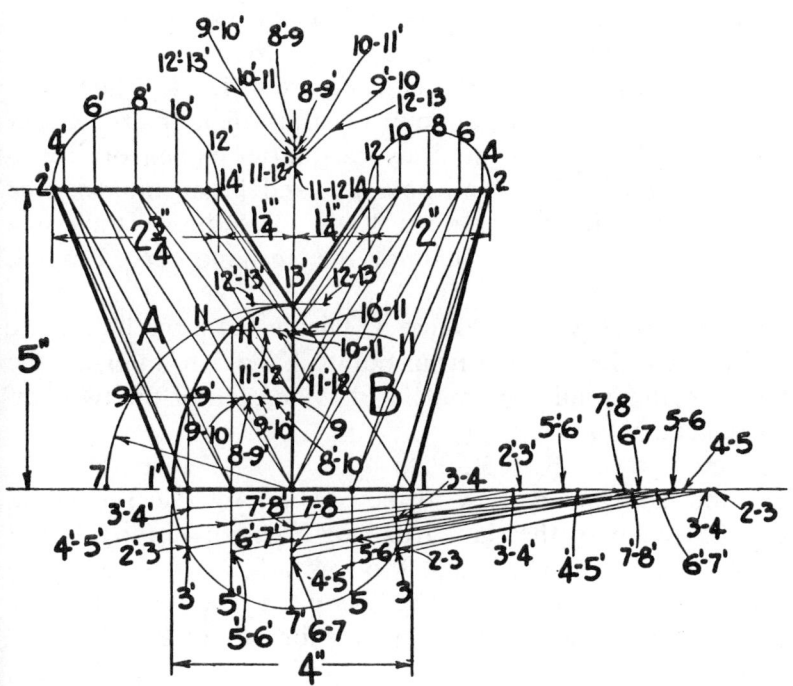

① SIDE VIEW & TRUE LENGTH LINES

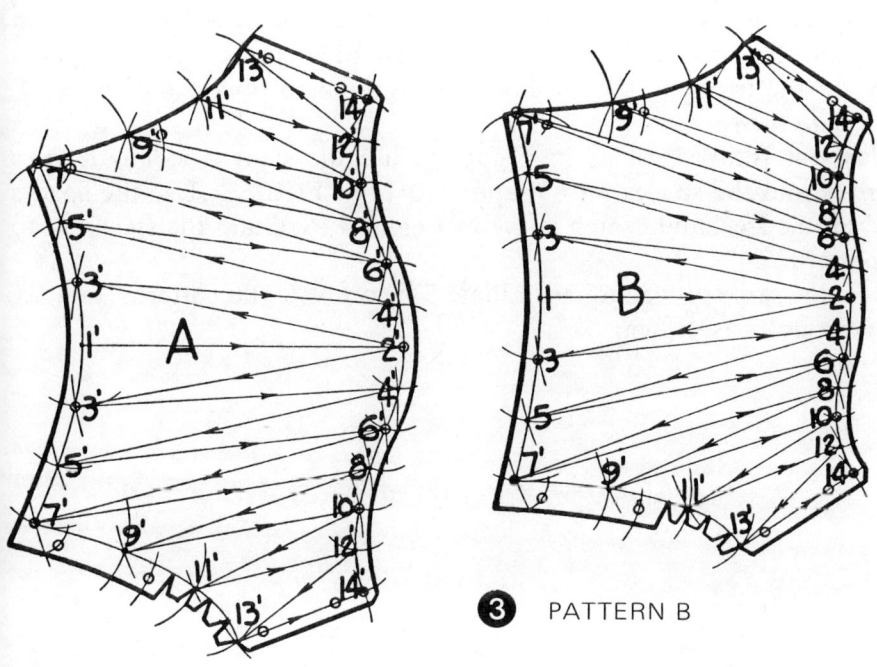

② PATTERN A

③ PATTERN B

PLATE 80   SHORT METHOD: Y BRANCH OF DIFFERENT DIAMETERS WITH EQUAL SPREAD

This Y branch is the same fitting as in Plate 79, but this method for laying out the pattern is shorter and more simplified.

Draw only a portion of the side view as in Figure 4. Draw a line from point 14' to 7', crossing the center line at point 13', thus obtaining the height of the crotch. Use point 7 as a center to draw the quarter circle from 13' to 7", and from 7' to 13. Draw a line from points 9" and 11" to intersect the center crotch line. Draw a line from points 9 and 11 to intersect the line drawn from points 9" and 11", thus obtaining the freehand curved crotch line 13' to 7'.

Draw the half bottom view as in Figure 1. Divide the large half circle 1 to 1' into equal spaces. Draw a line from points 3 and 5 to intersect the center crotch line, obtaining points 9' and 11'. Divide each of the two half circles into equal spaces.

Transfer the lines 1' to 2', 2' to 3', 3' to 4', 4' to 5', 5' to 6', 6' to 7', and 7' to 8' from the half bottom view to the base line of the true-length triangle in Figure 2, obtaining the slant true-length lengths for pattern *A* in Figure 5. Transfer the line 1 to 2, 2 to 3, 3 to 4, 4 to 5, 5 to 6, 6 to 7', and 7' to 8 from the half bottom view to the base line of the true-length triangle in Figure 3, thus obtaining the slant true-length line for pattern *B* in Figure 6. Transfer lines 8' to 9', 9' to 10', 10' to 11', 11' to 12', and 12' to 13' from Figure 1 to line *C–D* in Figure 4, from point *C* toward *D*, obtaining the slant true-length lines for the crotch on pattern *A*. Transfer lines 8 to 9', 9' to 10, 10 to 11', 11' to 12, and 12 to 13' from Figure 1 to line *C–E* in Figure 4, from point *C* toward *E*, obtaining the slant true-length lines for the crotch on pattern *B*.

To lay out the pattern as in Figure 5, use the slant true-length lines in Figure 2, and the spaces 1' to 7', and 2' to 14' in Figure 1. Use the spaces 7' to 13' on the freehand crotch curve in Figure 4 to obtain the spaces 7' to 13' on pattern *A*.

To lay out the pattern as in Figure 6, use the slant true-length lines in Figure 3, and the spaces 1 to 7' and 2 to 14 in Figure 1. Use the spaces 7' to 13' on the freehand crotch curve in Figure 4 to obtain the spaces 7' to 13' on pattern *B*.

Follow the same procedure as in Plate 77; complete the patterns, and lay out the rivet holes as shown.

PLATE 80

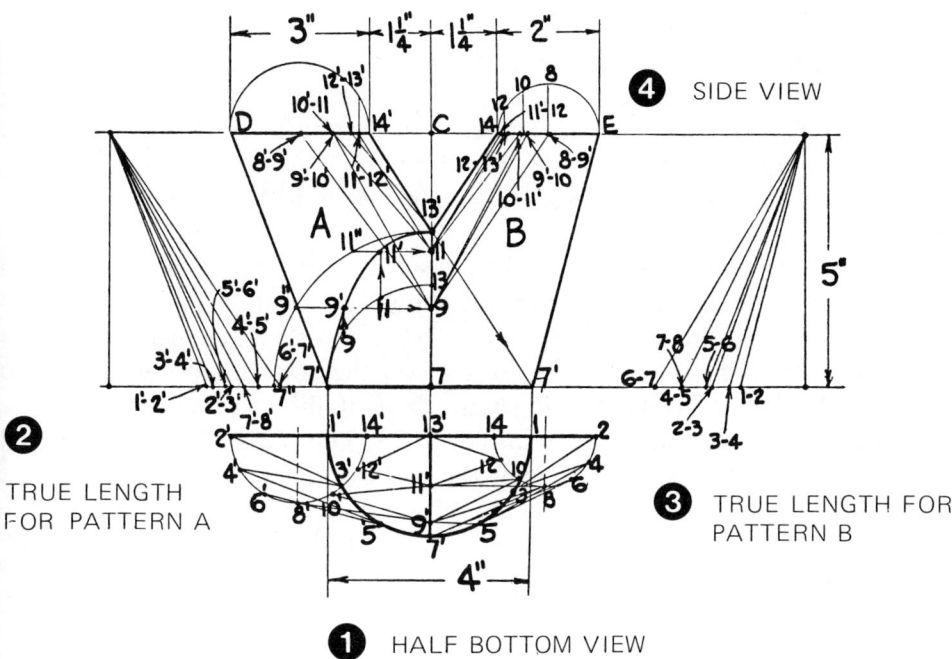

① HALF BOTTOM VIEW
② TRUE LENGTH FOR PATTERN A
③ TRUE LENGTH FOR PATTERN B
④ SIDE VIEW

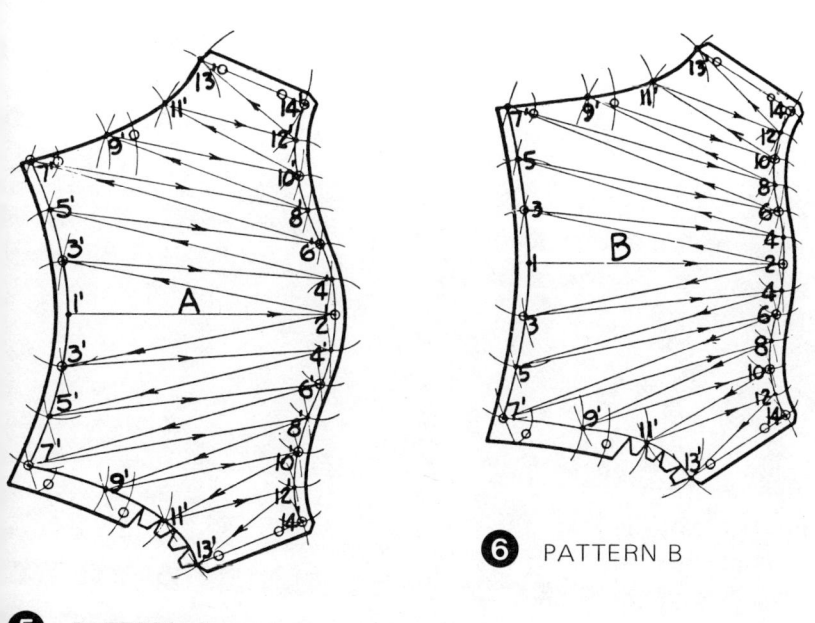

⑤ PATTERN A
⑥ PATTERN B

149

PLATE 81 Y BRANCH WITH ROUND MAIN AND OBLONG SECONDARY BRANCHES

The procedure for this type of oblong-to-round Y branch is the same as for the round-to-round Y branch in Plate 80.

Draw a line from point 14 to 7′ crossing the center line, thus obtaining point 13′. Transfer the height of lines 4, 6, 8, and 8′ from the secondary branch to lines 3, 5, and 7 at the main branch. Transfer the height of lines 8′, 10, and 12 to lines 9′, 11′, and 13′ at the crotch in the same manner as in Plate 76.

Transfer the slant lines from the side view to the base line of the main branch and to the crotch center line as shown.

To lay out the pattern as in Figure 2, use the slant true-length lines at the base line of the main branch to develop the pattern from 7′ to 7′ and 8′ to 8′. Then use the slant true-length lines at the crotch to develop the remaining pattern from 7′ to 13′ and 8′ to 14. The spaces 7′ to 7′ are equal to spaces 1 to 7 on the half circle at the main branch, Figure 1; the spaces 14 to 14 are equal to the spaces 2 to 14 at the secondary branch, and the spaces 7′ to 13′ are equal to the spaces 7′ to 13′ at the freehand curved crotch line.

Lay out the rivet holes at the crotch, and complete the pattern in the same manner as in Plate 76.

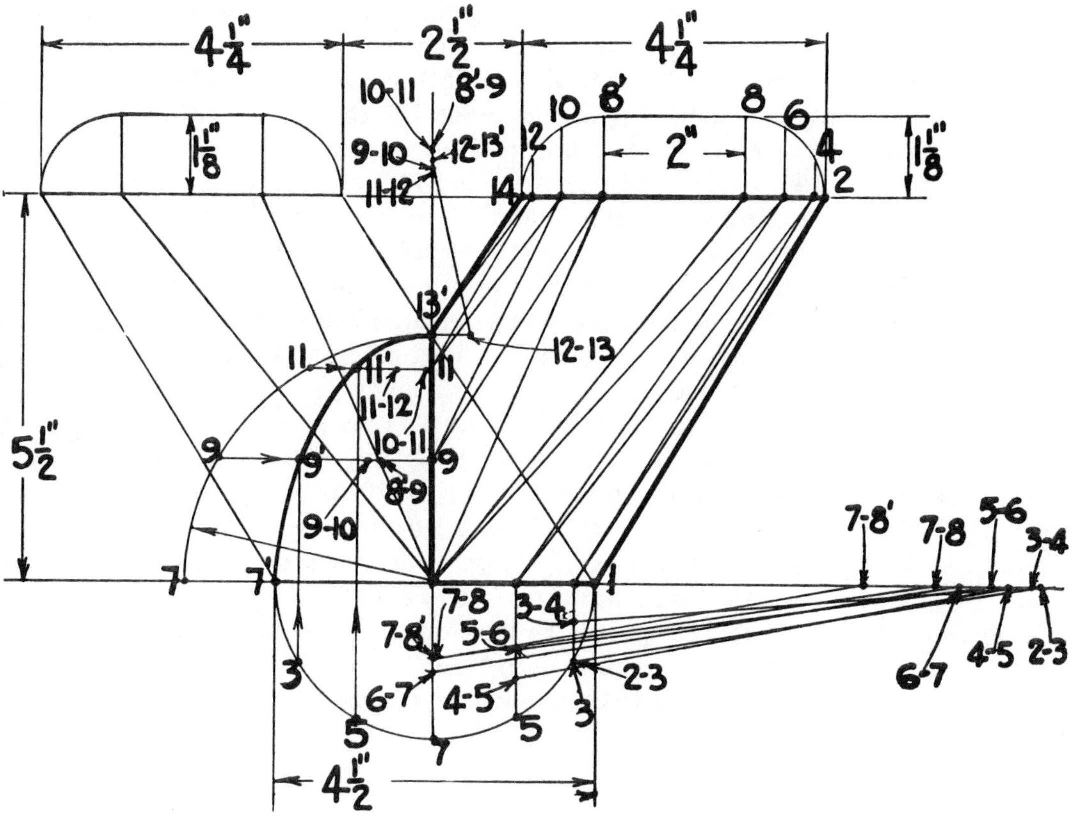

PLATE 81

① SIDE VIEW & TRUE LENGTH LINES

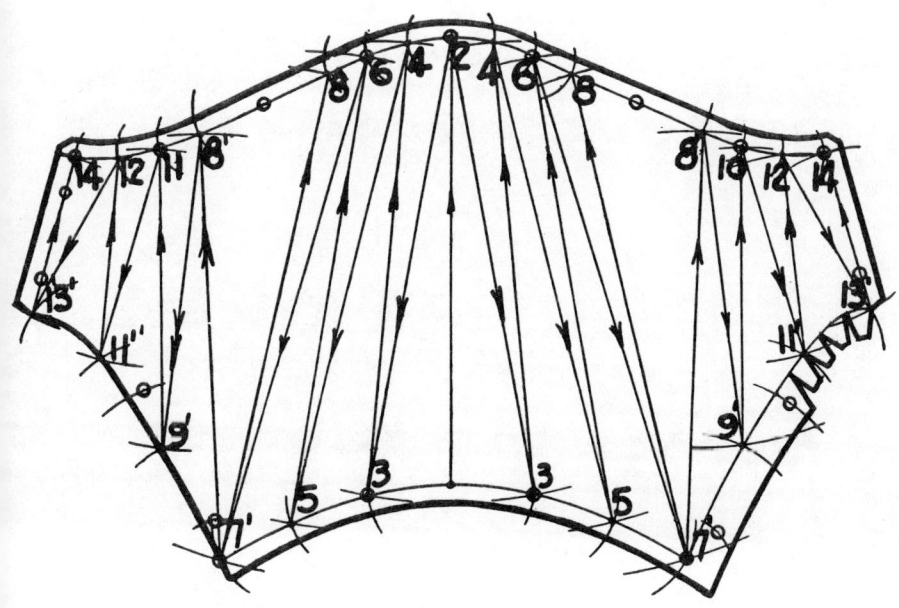

② FULL PATTERN — TWO REQUIRED

PLATE 82   SHORT METHOD: ROUND-TO-OBLONG Y BRANCH

This Y branch is the same fitting as in Plate 81, but this method for laying out the pattern is shorter and more simplified.

Draw a portion of the side view as in Figure 3. Draw a line from point 14 to 7' crossing the center line at point 13', thus obtaining the height of the crotch. Use point 7 as a center to draw the quarter circles 13' to 7", and 7' to 13. Draw a line from points 9" and 11" to intersect the center crotch line. Draw a line from points 9 and 11 to intersect the lines drawn from points 9" and 11", thus obtaining the freehand curved crotch line 13' to 7'.

Draw the half bottom view as in Figure 1. Divide the two quarter circles on the oblong into equal spaces, and the large half circle 1 to 13 into equal spaces. Draw a line from points 9 and 11 to intersect the center crotch line, obtaining points 9' and 11'.

Transfer lines 1 to 2, 2 to 3, 3 to 4, 4 to 5, 5 to 6, 6 to 7', 7' to 8, and 7' to 8' from Figure 1 to the base line of the true-length triangle in Figure 2. Transfer lines 8' to 9', 9' to 10, 10 to 11', 11' to 12, and 12 to 13' from Figure 1 to line C–D in Figure 3 from point C toward D.

To lay out the pattern as in Figure 4, use the slant true-length lines in Figure 2 to develop the pattern 7' to 7' and 8' to 8'. Use the slant true-length lines at the crotch in Figure 3 to develop the remaining pattern 7' to 13' and 8' to 14 in the same manner as in Plate 77.

Lay out the rivet holes at the crotch, and complete the pattern as shown.

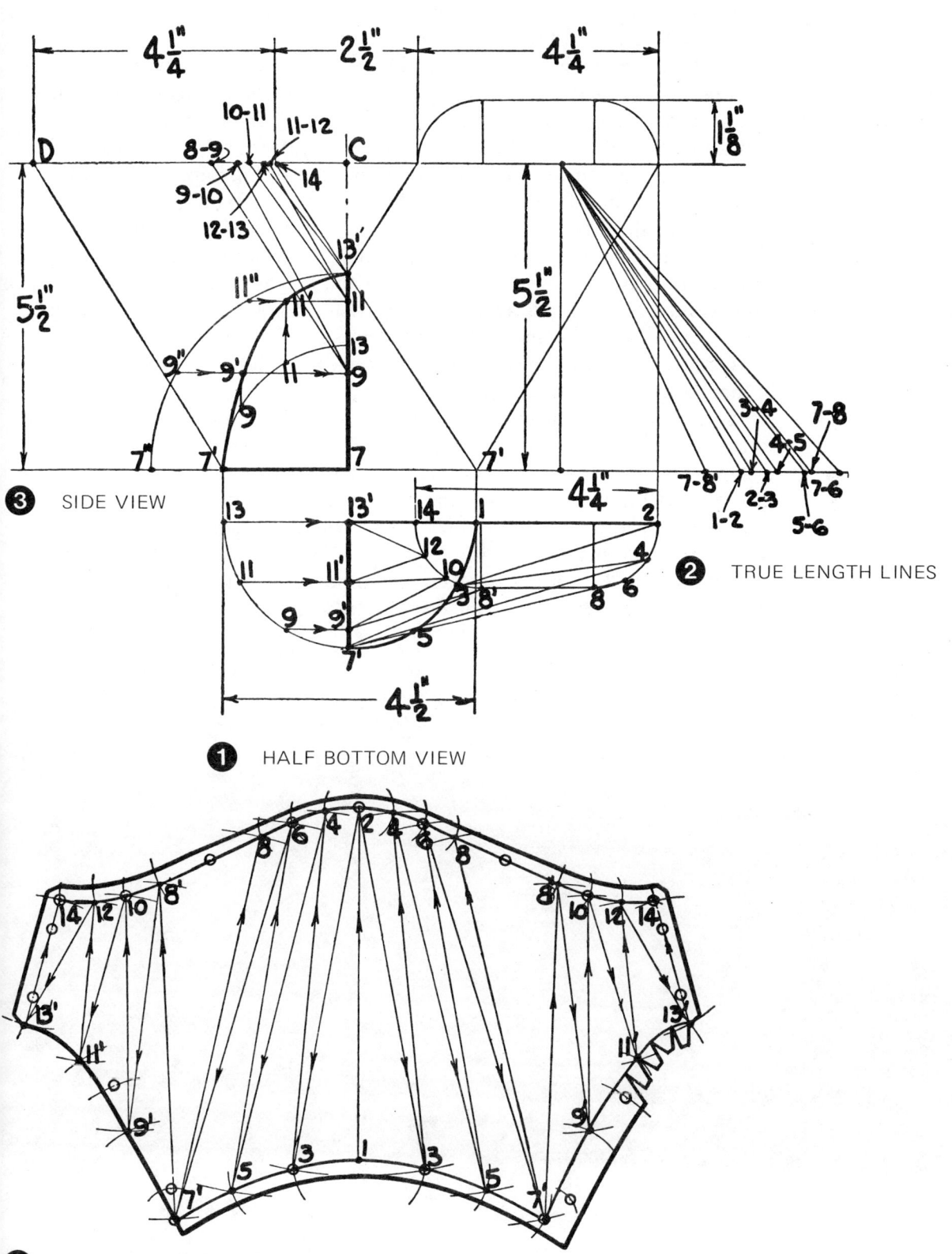

PLATE 83 Y BRANCH WITH ROUND MAIN AND OBLONG SECONDARY BRANCHES

The procedure for this plate is the same as for Plate 81.

Transfer the height 2 to 2′ from the oblong to the half circle at the base from point 3 to 2–3. Transfer the height of line 4 (the distance from point 4 to the center line 2′–14′) to the half circle at the base from point 3 to 3–4, and point 5 to 4–5. Transfer the distance from point 6 to the center line 2′–14′ to the half circle at the base from point 5 to 5–6, and point 7 to 6–7. Transfer the height of line 8 to line 7 from point 7 to 7–8.

Transfer the height of line 8 to the crotch line from point 9′ to 8–9. Transfer the height of line 10 to the crotch from point 9′ to 9–10, and to line 11 from point 11′ to 10–11. Transfer the height of line 12 to the crotch from point 11′ to 11–12, and to line 13 from point 13′ to 12–13. Transfer the distance 14–14′ to the crotch from point 13′ to 13–14.

Transfer the slant lines from the side view to the base line of the main branch, and to the crotch center line as shown.

To lay out the pattern as in Figure 2, draw the center line 1 to 2′ equal to the slant line 1 to 2′ in the side view. Draw the line 2′ to 2 equal to the distance 2′ to 2 on the oblong. Use the slant true-length lines at the base of the main branch to develop the pattern from points 2 to 8 and 1 to 7′; then use the slant true-length lines at the crotch to complete the remaining pattern from points 8 to 14 and 7′ to 13′. The spaces 7′ to 13′ are equal to the spaces 7′ to 13′ on the freehand crotch curve in Figure 1.

Lay out the rivet holes at the crotch, and complete the pattern as shown.

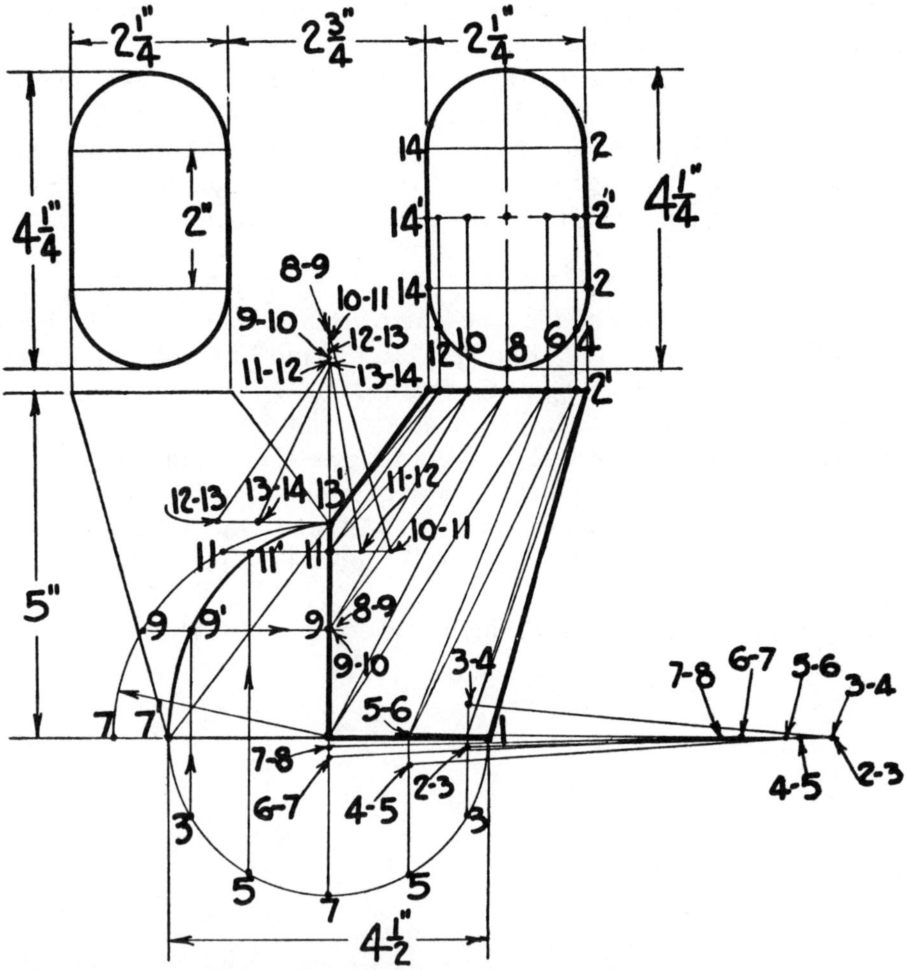

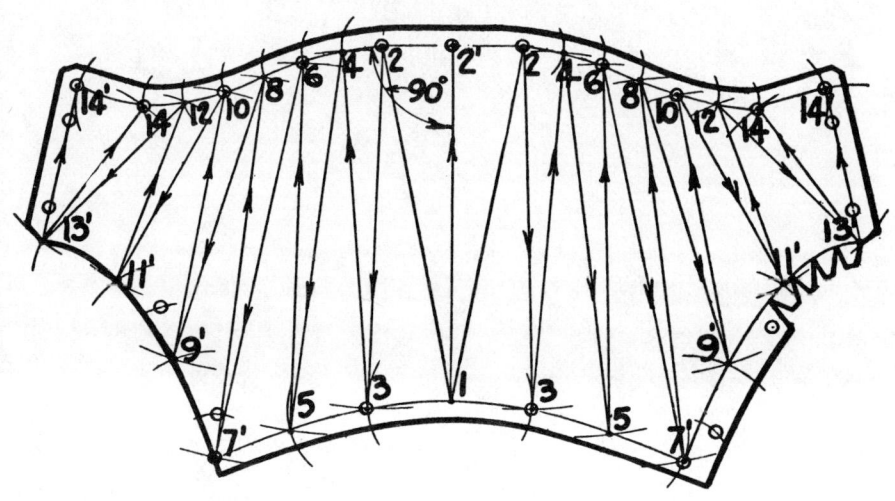

1 SIDE VIEW & TRUE LENGTH LINES

2 FULL PATTERN — TWO REQUIRED

PLATE 83

PLATE 84   SHORT METHOD: ROUND-TO-OBLONG Y BRANCH

This fitting is the same as the Y branch in Plate 83, but this method of laying out the pattern is shorter and simpler.

Draw only a portion of the side view as in Figure 3. Develop the freehand curved crotch 13′ to 7′ in the same manner as in Plate 76.

Draw the half bottom view as in Figure 1. Divide both the half circle on the oblong and the large half circle 1 to 13 into equal spaces. Draw a line from points 9 and 11 to intersect the center crotch line obtaining points 9′ and 11′.

Transfer lines 1 to 2′, 2 to 3, 3 to 4, 4 to 5, 5 to 6, 6 to 7′, and 7′ to 8 from Figure 1 to the base line of the true-length triangle in Figure 2. Transfer lines 8 to 9′, 9′ to 10, 10 to 11′, 11′ to 12, 12 to 13′, and 13′ to 14 from Figure 1 to line C–D in Figure 3 from point C toward D.

To lay out the pattern as in Figure 4, draw the center line 1 to 2′ equal to the slant true-length line 1–2′ in Figure 2. Use the remaining slant true-length lines in Figure 2 to lay out the pattern from points 2 to 8 and 1 to 7′. To lay out the remaining pattern from 8 to 14′ and 7′ to 13′, use the slant true-length lines at the crotch in Figure 3, and follow the same procedure as in Plate 77.

Lay out the rivet holes at the crotch, and complete the pattern as shown.

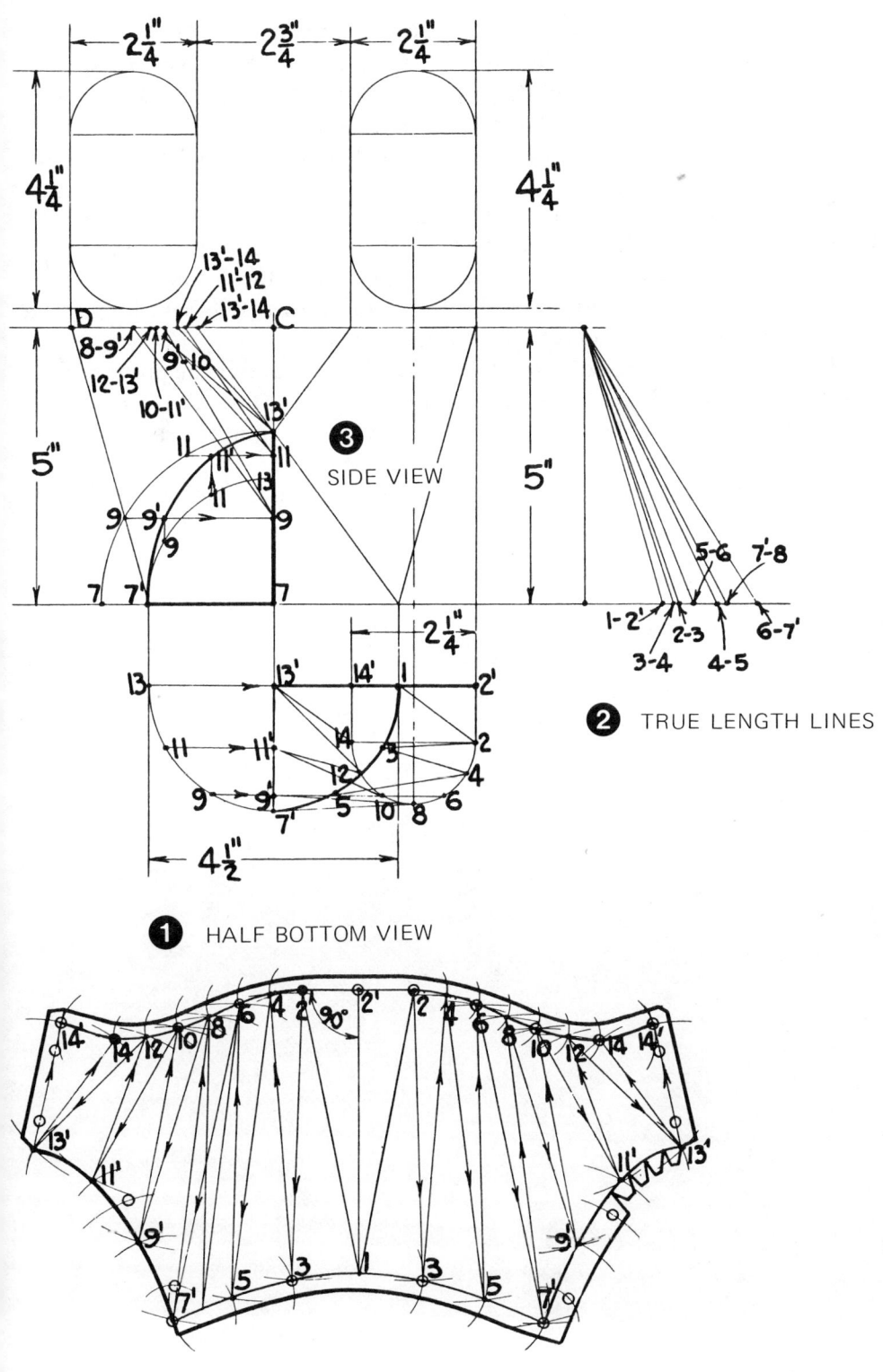

PLATE 85   Y BRANCH WITH SECONDARY BRANCHES AT 30 DEG.

When this type of Y branch is used for a blower exhaust system, the angle of the secondary branches should be about 30 deg. and never more than 45 deg. The length of the center crotch line should be at least equal to three quarters of the main diameter, or greater if possible.

Draw half of the side view as in Figure 1, and construct the freehand crotch curve from 13' to 7'.

Transfer the heights of lines 4, 6, and 8 from the small half circle to lines 3, 5, and 7 on the large half circle at the main branch. Transfer the height of lines 8, 10, and 12 to lines 9', 11', and 13' at the crotch.

Transfer the slant length lines from the side view to the base line of the main branch and to the center crotch line as shown.

Use the slant true-length lines at the base line to lay out the pattern from 7' to 7' and 8 to 8, Figure 2. Use the slant true-length lines at the crotch to lay out the remaining pattern from 7' to 13' and 8 to 14 in the same manner as in Plate 76.

Lay out the rivet holes at the crotch, and complete the pattern as shown.

PLATE 85

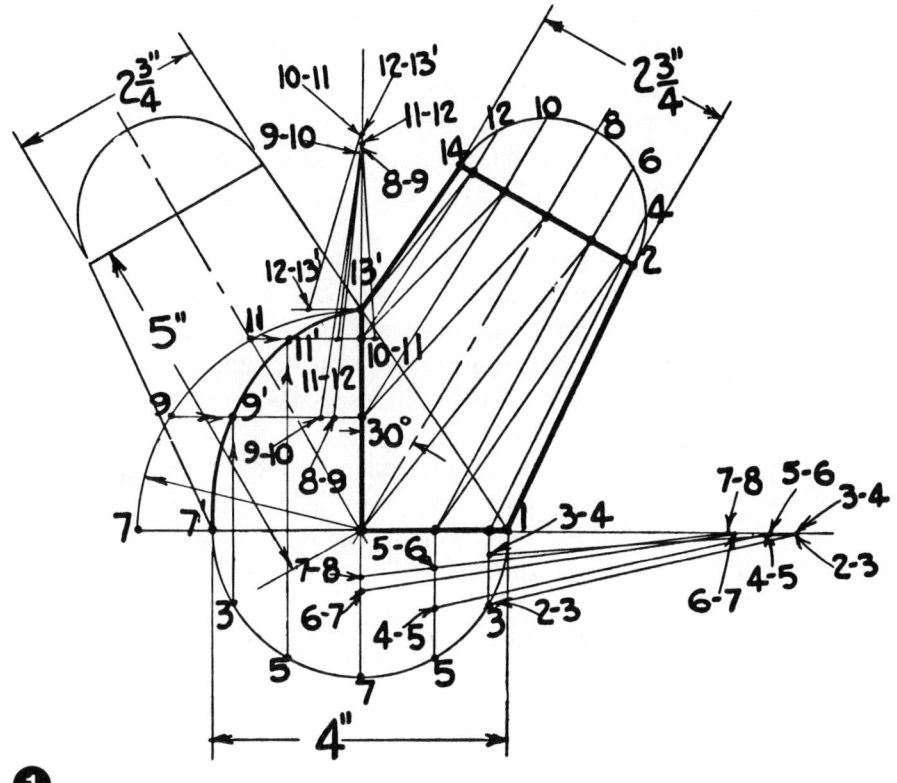

1 SIDE VIEW & TRUE REQUIRED

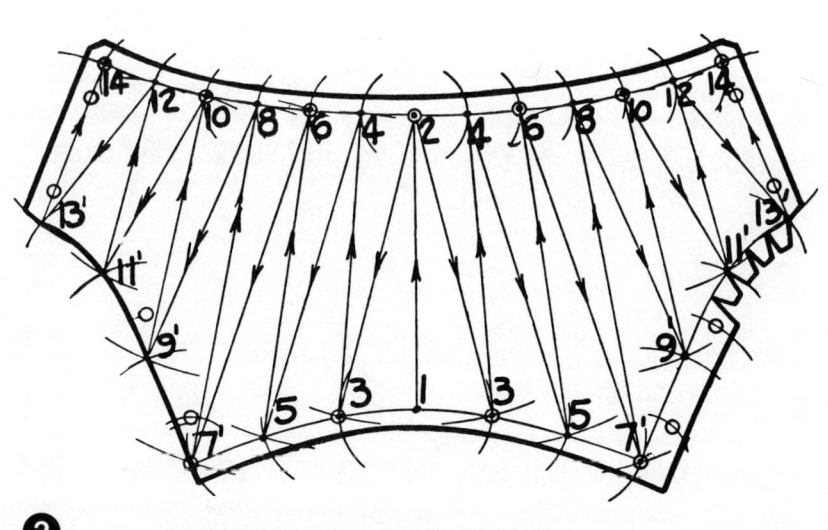

2 FULL PATTERN — TWO LENGTH LINES

PLATE 86   Y BRANCH WITH EQUAL SPREAD

To draw the half side view as in Figure 1, use point 14′ as a center to draw an arc tangent to line 2. Draw a line from point 1 tangent to the arc to intersect line 2 at point 2′. Draw a line from point 2′ to point 14′.

Construct the freehand crotch curve 13′ to 7′. Transfer the heights of lines 4, 6, and 8 from the small diameter to lines 3, 5, and 7 at the large diameter. Transfer the heights of lines 8, 10, and 12 to the crotch lines 9′, 11′, and 13′.

Transfer the slant length lines from the side view to the base line of the main branch and to the center crotch line as shown.

To lay out the pattern as in Figure 2, draw line 8 to 8 equal to the circumference of a $2\frac{3}{4}$-in. diameter, and divide it into equal spaces. Transfer the heights of the lines in section A to their respective lines on the pattern in Figure 2.

To lay out pattern B as in Figure 3, the spaces 14′ to 14′ are equal to the spaces 2′ to 14′ on the freehand curve in Figure 2. The spaces 7′ to 7′ are equal to the spaces 1 to 7 on the half circle at the main branch in Figure 1. The spaces 7′ to 13′ are equal to the spaces 7′ to 13′ on the freehand crotch curve in Figure 1.

Lay out the rivet holes at the crotch, and complete the pattern.

PLATE 86

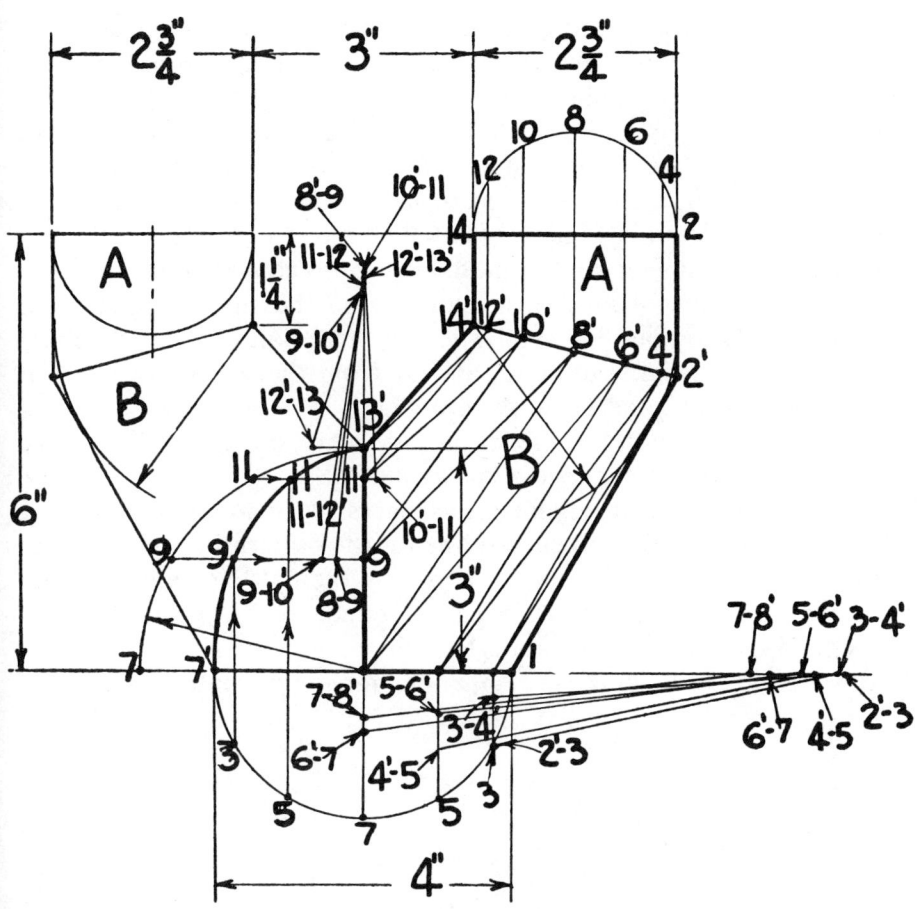

① SIDE VIEW & TRUE LENGTH LINES

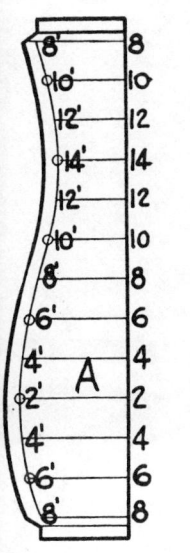

② PATTERN A

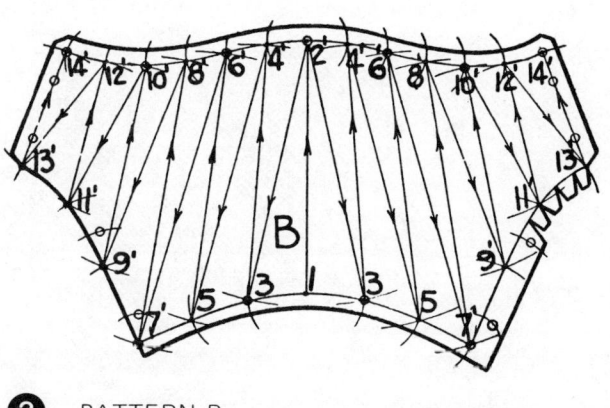

③ PATTERN B

PLATE 87   Y BRANCH WITH HORIZONTAL BRANCHES

This type of branch is not recommended for blower exhaust systems due to the sharp angles of the secondary branches, but this fitting may be used for discharging smoke and fumes.

Draw the side view as in Figure 1. Use point 2' as a center to draw an arc tangent to line 14. Draw a line from point 13' tangent to this arc to intersect line 14 at point 14'; then draw a line from 2' to 14'.

To lay out the pattern in Figure 2, draw line 8 to 8 equal to the circumference of a $2\frac{3}{4}$-in.-diameter circle, and divide it into equal spaces.

Transfer the heights of lines 4, 6, and 8 from the small half circle at section A, Figure 1, to lines 3, 5, and 7 on the large half circle at the main branch. Transfer the heights of 8, 10, and 12 from the small half circle at section A to lines 9', 11', and 13' at the crotch.

Transfer the slant length lines from the side view to the base line of the main branch and to the center crotch line as shown.

To lay out pattern B as in Figure 3, the spaces 14' to 14' are equal to the spaces 2' to 14' on the freehand curved line on the pattern in Figure 2. The spaces 7' to 7' are equal to the spaces 1 to 7 on the half circle at the main branch in Figure 1. The spaces 7' to 13' are equal to the spaces 7' to 13' on the freehand crotch curve in Figure 1.

Lay out the rivet holes at the crotch, and complete the pattern as shown.

PLATE 87

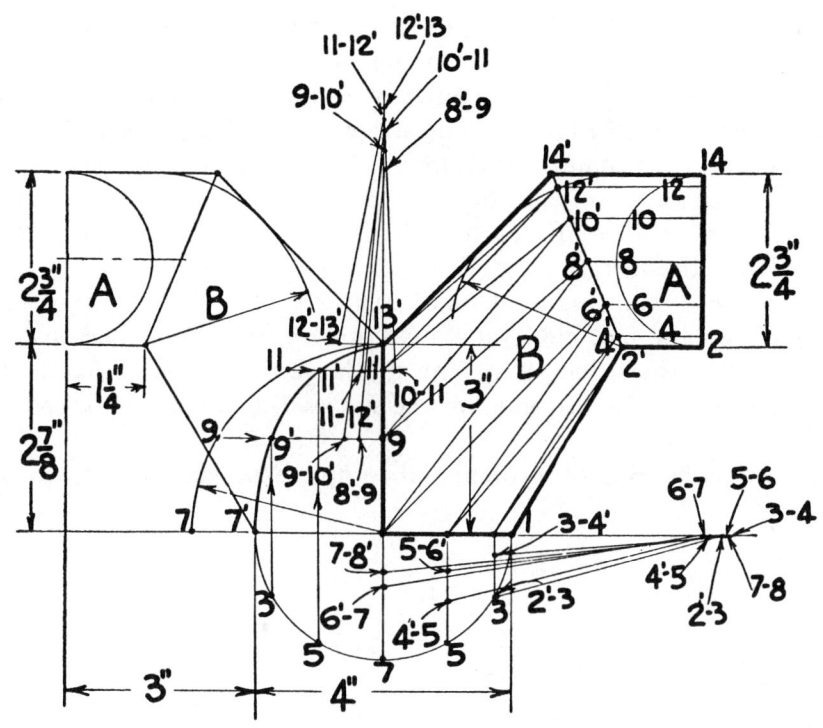

① SIDE VIEW & TRUE LENGTH LINES

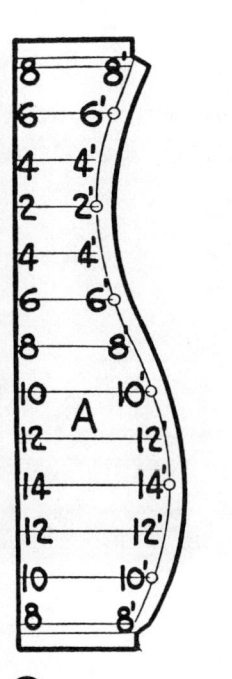

② PATTERN A

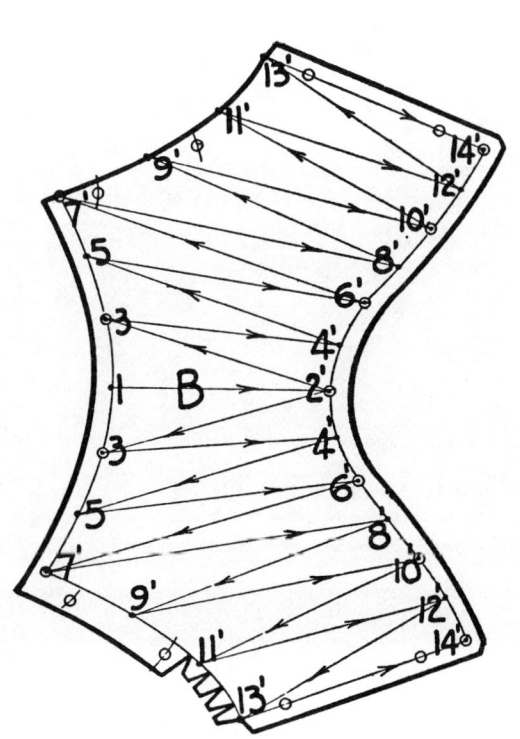

③ PATTERN B

PLATE 88 Y BRANCH WITH HORIZONTAL AND VERTICAL BRANCHES

Draw the side view as in Figure 1, and erect the true-length triangles at the base of the main branch and at the crotch as shown.

Lay out pattern A as in Figure 2. This may be a full pattern.

Lay out pattern B as in Figure 3. The spaces 14′ to 14′ are equal to the spaces 2′ to 14′ on the freehand curved line in Figure 2′. The spaces 7′ to 7′ are equal to the spaces 1 to 7 on the large half circle at the main branch in Figure 1. The spaces 7′ to 13′ are equal to spaces 7′ to 13′ on the freehand curved line at the crotch.

Lay out pattern C as in Figure 4. The spaces 14″ to 14″ are equal to the spaces 2″ to 14″ on the half circle at the secondary branch C. The spaces 7′ to 7′ are equal to the spaces 1 to 7 on the large half circle at the main branch. The spaces 7′ to 13′ are equal to the spaces 7′ to 13′ at the freehand crotch curve.

Lay out the rivet holes at the crotch, and complete the patterns as shown.

PLATE 88

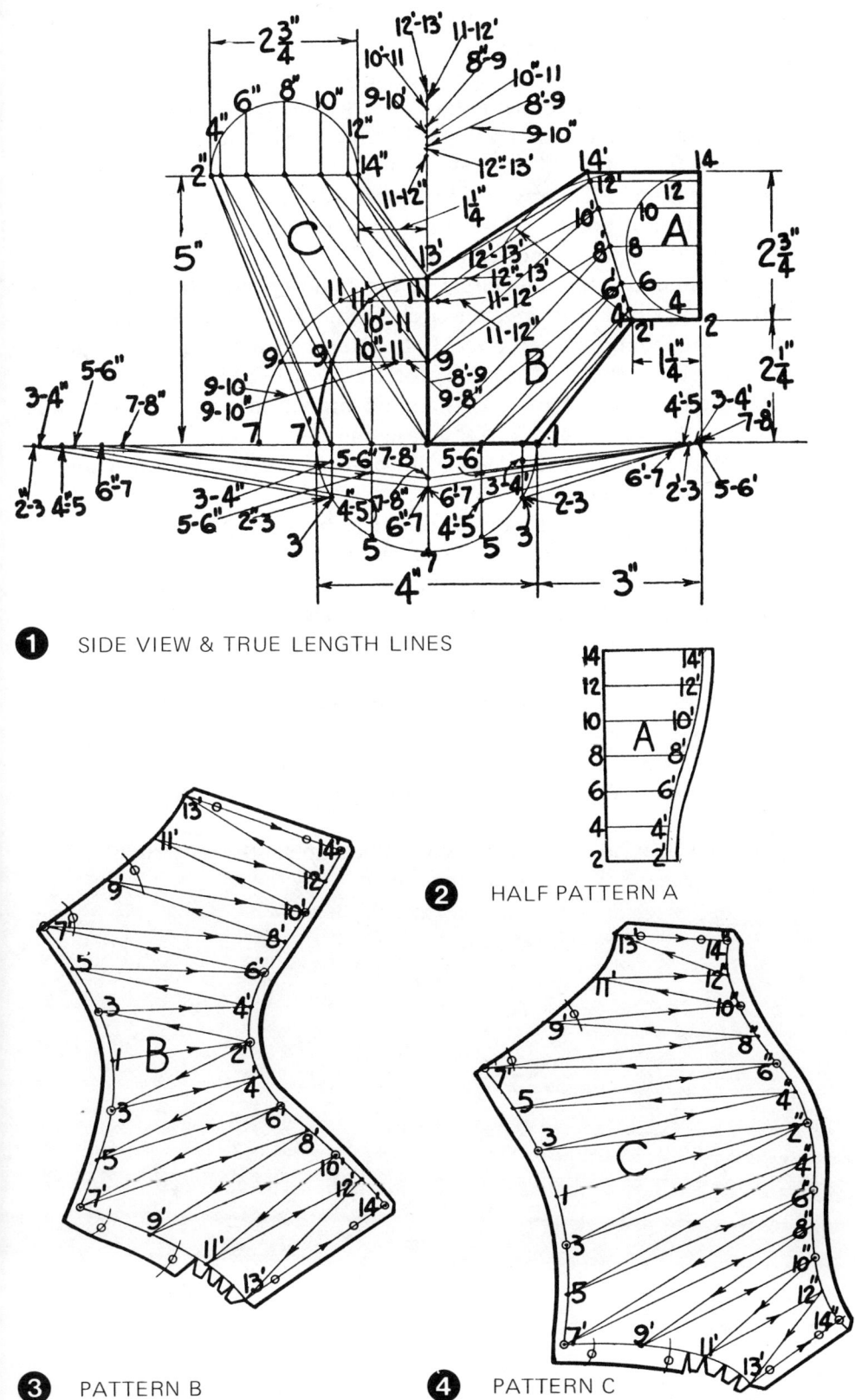

① SIDE VIEW & TRUE LENGTH LINES

② HALF PATTERN A

③ PATTERN B

④ PATTERN C

PLATE 89 Y BRANCH WITH HORIZONTAL AND VERTICAL BRANCHES OF DIFFERENT DIAMETERS

This plate is similar to Plate 88, and the procedures are identical except the method of obtaining the freehand crotch curve.

Draw the center crotch line at a 15-deg. angle. This is to direct the greater portion of air to enter or return from the large diameter branch B without retarding the flow of air and creating additional resistance.

Draw a line from point 14″ to point 1′, Figure 1, crossing the center crotch line, obtaining point 13′. Draw a line squaring from the center crotch line from point 7 toward point 7′. Use point 7 as a center to draw the quarter circle from point 13′ to 7′, and from point 13″ (the intersecting point of the center crotch line and the half circle at the main branch) to intersect line 7–7′ at point 7″. Divide this quarter circle from 13″ to 7″ into equal spaces, and draw a line from points 9″ and 11″ to intersect lines 9 and 11 at points 9′ and 11′, thus obtaining the freehand crotch curve from 13′ to 7″.

Divide the half circle 1 to 1′ at the main branch into equal spaces, and draw a line from points 3 and 5 to the base line 1–1′.

Erect the true-length triangles at the base of the main branch and at the center crotch as shown.

Lay out pattern A as in Figure 2; this may be a full pattern.

Lay out pattern C as in Figure 3. The spaces 7 to 7 are equal to the spaces 1 to 7 on the half circle at the main branch, Figure 1. The spaces 14″ to 14″ are equal to the spaces 2″ to 14″ on the half circle at branch C in Figure 1. The spaces 7 to 13′ are equal to the spaces 7″ to 13″ on the freehand crotch curve.

Lay out pattern B as in Figure 4. The spaces 14′ to 14′ are equal to the spaces 2′ to 14′ on the freehand curved line on the pattern in Figure 2. The spaces 7 to 7 are equal to the spaces 1 to 7 on the large half circle at the main branch. The spaces 7 to 13′ are equal to the spaces 7″ to 13′ on the freehand crotch curve.

Lay out the rivet holes at the crotch, and complete the pattern as shown.

PLATE 89

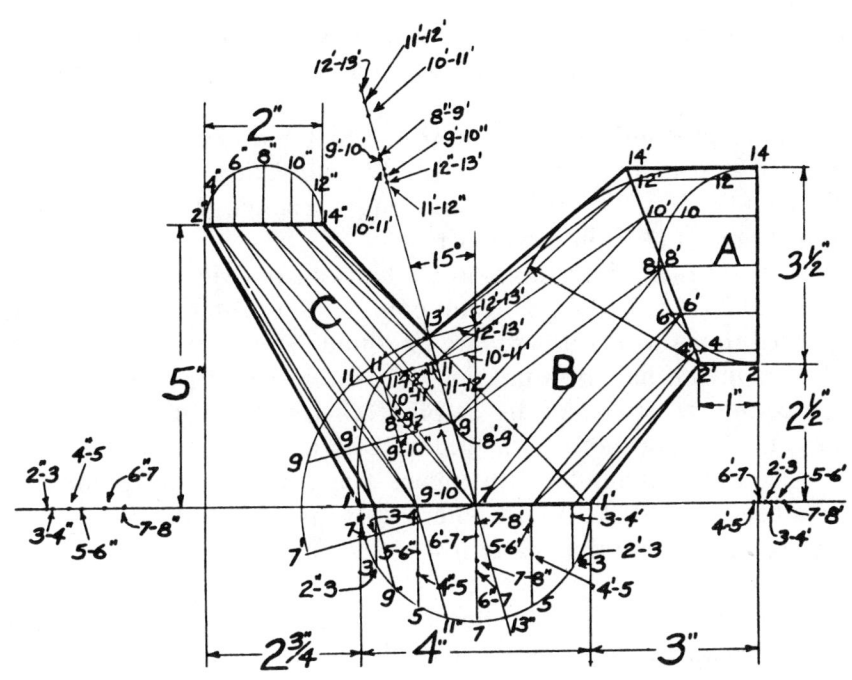

① SIDE VIEW & TRUE LENGTH LINES

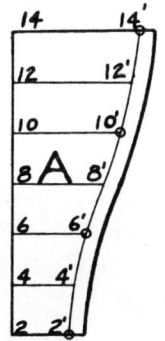

② PATTERN A

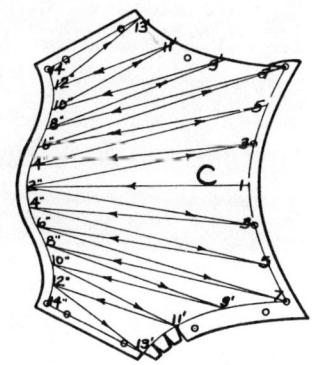

③ PATTERN C

④ PATTERN B

167

PLATE 90 Y BRANCH WITH ROUND MAIN AND RECTANGULAR BRANCHES

Draw a half portion of the side view, and a half bottom view as in Figure 1. The freehand crotch curve 4′ to 7′ may be obtained in the same manner as in Plate 76.

To erect the true-length triangles as in the half bottom view, use point A as a center to draw arcs from points 1, 2, 3, and 4 to intersect line $A-E$ which represents the base line for the true-length triangle. Transfer the distance C to 4 to line $A-E$ from point A to $C-4$.

Use the slant true-length triangles in the half bottom view to lay out the pattern in Figure 2 from 1 to 4′, by using point A as a center, and following the same procedure as used for a square-to-round. Then use the slant true-length line $C-4$ in the half bottom view as a radius, and point 4′ on the pattern as a center, to strike an arc at point C. Use the width A to C in Figure 1 as a radius, and point A in Figure 2 as a center, to strike an arc at point C. Use the width A to C in Figure 1 as a radius, and point A in Figure 2 as a center, to strike an arc to cross the arc at point C.

To complete the remaining pattern 4′ to 7′, use the slant true-length lines at the crotch, and the spaces 4′ to 7′ on the freehand crotch curve in the manner as shown. Use the slant line 7′ to $C-D$ in Figure 1 as a radius, and point 7′ in Figure 2 as a center, to strike an arc at point D. Use the distance C to D in Figure 1 as a radius, and point C in Figure 2 as a center, to strike an arc crossing the arc at point D.

Lay out the rivets at the crotch, and complete the pattern as shown.

PLATE 90

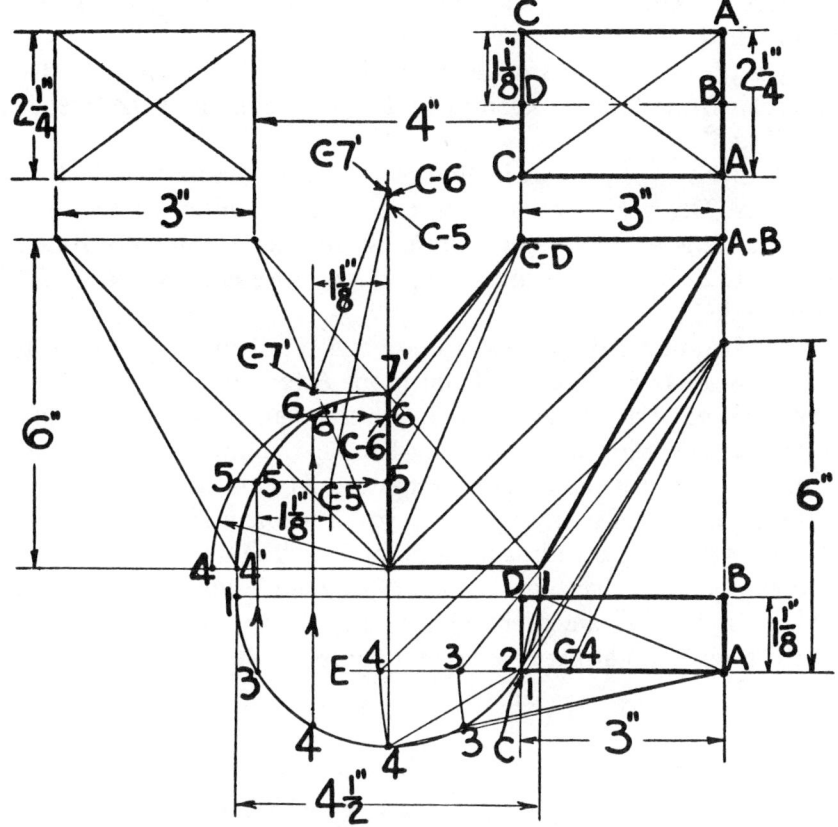

1 SIDE VIEW, HALF BOTTOM VIEW & TRUE LENGTHS

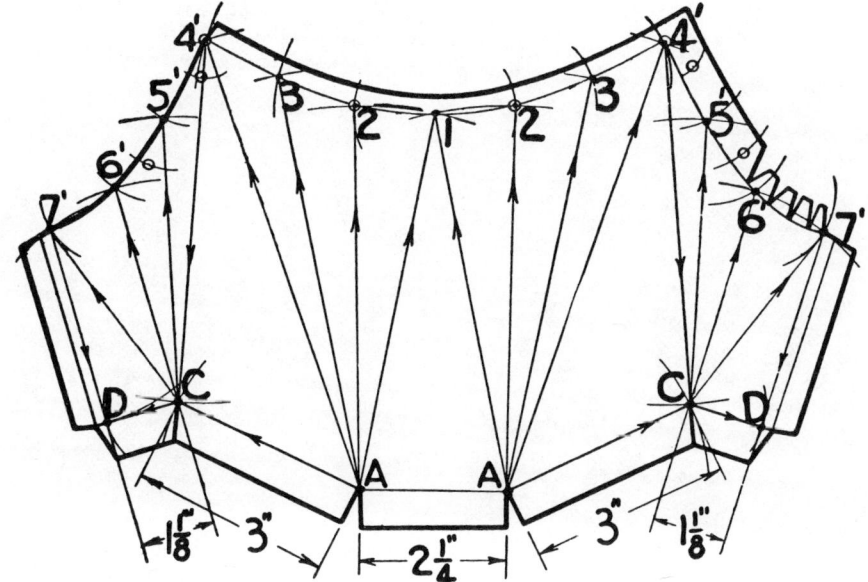

2 FULL PATTERN — TWO REQUIRED

169

PLATE 91 SHORT METHOD: Y BRANCH WITH ROUND MAIN AND RECTANGULAR BRANCHES

This Y branch is the same fitting as in Plate 90, but the procedure is shorter and more simplified.

Draw only a portion of the side view in Figure 2. Use point 4″ to draw the quarter circles from 7′ to 4 and from 4′ to 7; then construct the freehand crotch curve 7′ to 4′.

Draw the half bottom view as in Figure 1. Divide the half circle 1 to 7 into equal spaces. Draw a line from points 5 and 6 to intersect the center crotch line, obtaining points 5′ and 6′. Use point A as a center to draw an arc from points 1, 2, 3, and 4 to line A–E, which represents the base of the true-length triangle.

Transfer the distance from point C to points 4, 5′, 6′, and 7′ at the crotch in Figure 1 to line E–D in Figure 2, from point E toward D.

To lay out the pattern as in Figure 3, use the slant true-length lines in Figure 1 to lay out the pattern from points 1 to 4′. Use the slant true-length lines in Figure 2 to lay out the remaining pattern from 4′ to 7′ in a manner similar to the procedure for laying out a square-to-round. Use the slant line 7′ to C in Figure 2 as a radius, and point 7′ in Figure 3 as a center to strike an arc at point D. Use the distance C to D in Figure 1 as a radius, and point C in Figure 3 as a center to strike an arc at point D.

Lay out the rivet holes at the crotch, and complete the pattern as shown.

PLATE 91

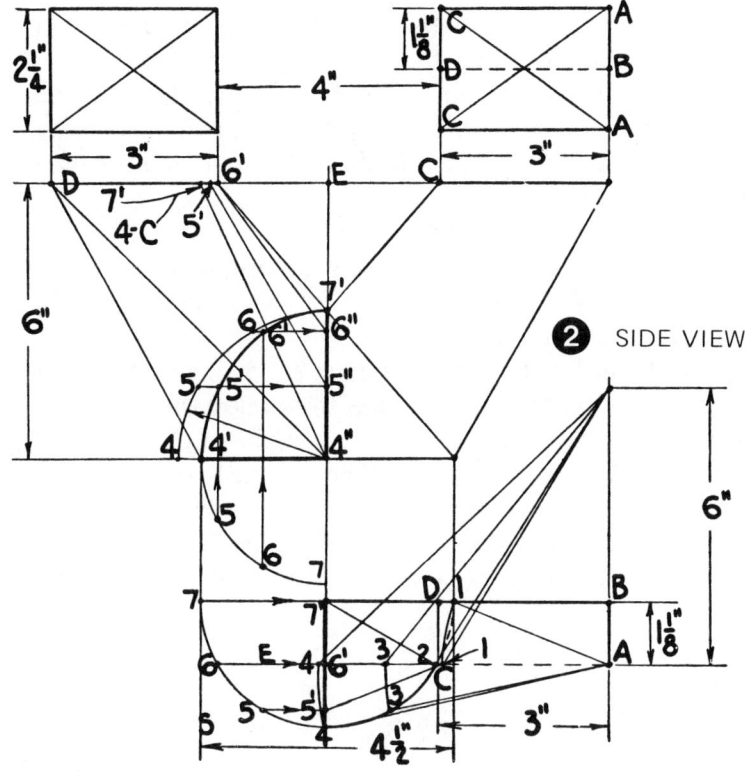

① HALF BOTTOM VIEW & TRUE LENGTH LINES

② SIDE VIEW

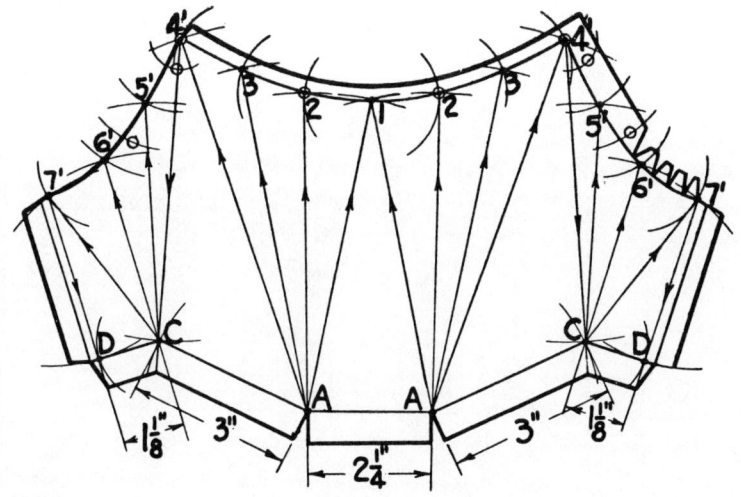

③ FULL PATTERN – TWO REQUIRED

PLATE 92 Y BRANCH WITH RECTANGULAR MAIN AND ROUND BRANCHES

The procedure for this plate is similar to that for Plate 90. Draw the half side view, Figure 2, and the half bottom view as in Figure 1. Use point *B* as a center to draw the quarter circle from 9′ to *B″*, Figure 2. Use half the distance *A* to *A* in Figure 1 as a radius, and point *B* in Figure 2 as a center, to draw the quarter circle from *B′* to *B″*. Divide this quarter circle into equal spaces and develop the freehand crotch curve 9′ to *B′*.

Use point *A*, Figure 1 as a center to draw arcs from points 1, 2, 3, and 4 from the quarter circle to line *A–C*. Use these slant true-length lines to develop the pattern, Figure 3, from point 1 to 4, following the same procedure as for a square-to-round.

Transfer the heights of lines 4, 6, and 8 to lines *B′*, 5′, 7′, and 9′, thus obtaining the heights for the slant true-length lines to develop the remaining pattern 4 to 10 and *B′* to 9′. The spaces *B′* to 9′ are equal to the spaces *B′* to 9′ on the freehand crotch curve in Figure 2. The spaces 1 to 4 are equal to the spaces 1 to 4 in the quarter circle on the half bottom view in Figure 1. The spaces 4 to 10 are equal to the spaces 4 to 10 in Figure 2.

Lay out the rivet holes at the crotch, and complete the pattern as shown.

PLATE 92

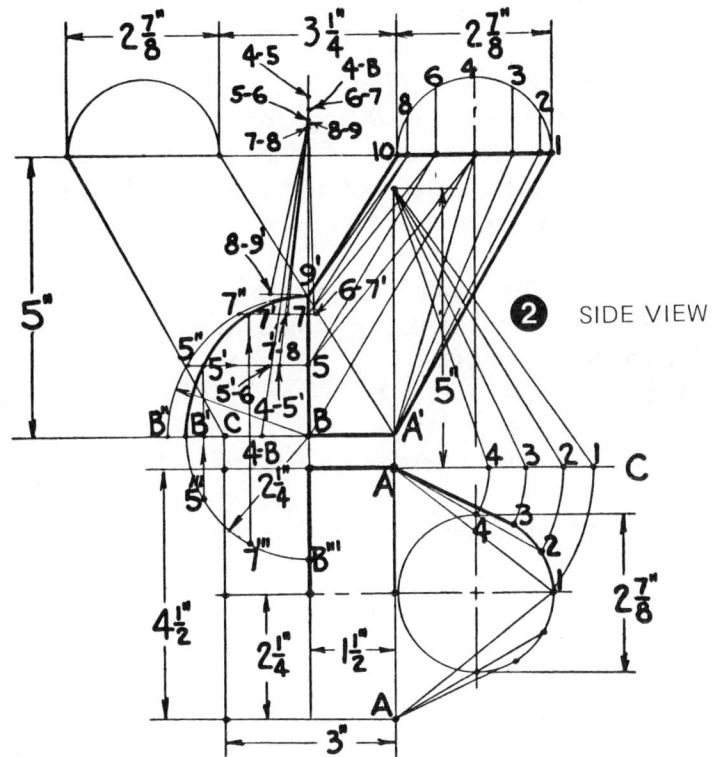

① HALF BOTTOM VIEW — TRUE LENGTH LINES

② SIDE VIEW

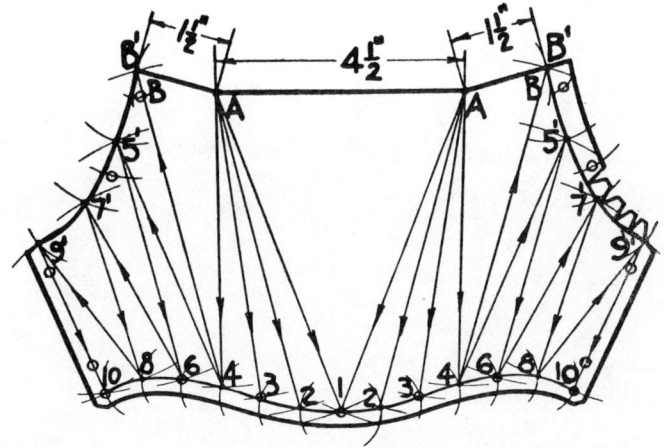

③ FULL PATTERN — TWO REQUIRED

PLATE 93 SHORT METHOD: Y BRANCH WITH RECTANGULAR MAIN AND ROUND BRANCHES

This Y branch is the same fitting as in Plate 92, but this method is shorter and more simplified.

Draw a portion of the side view as in Figure 2. Use point B'' as a center to draw the quarter circle from point $9'$ to B, and from point 9 to B'.

Draw the half bottom view as in Figure 1. Divide the half circle 1 to 10 into equal spaces, and divide the quarter circle 9 to B into equal spaces. Draw a line from points 5 and 7 to intersect the center crotch line, obtaining points $5'$ and $7'$.

Use point A as a center to draw arcs from points 1, 2, 3, and 4 on the half circle to line A–C. Use these slant true-length lines to develop the pattern in Figure 3 from point 1 to 4.

Transfer the distance 4 to B, 4 to $5'$, $5'$ to 6, 6 to $7'$, $7'$ to 8, and 8 to $9'$ from Figure 1 to line C–D in Figure 2, from point C toward D. Use these slant true-length lines to develop the remaining pattern in Figure 3 from points 4 to 10 and B' to $9'$.

Lay out the rivet holes at the crotch, and complete the pattern as shown.

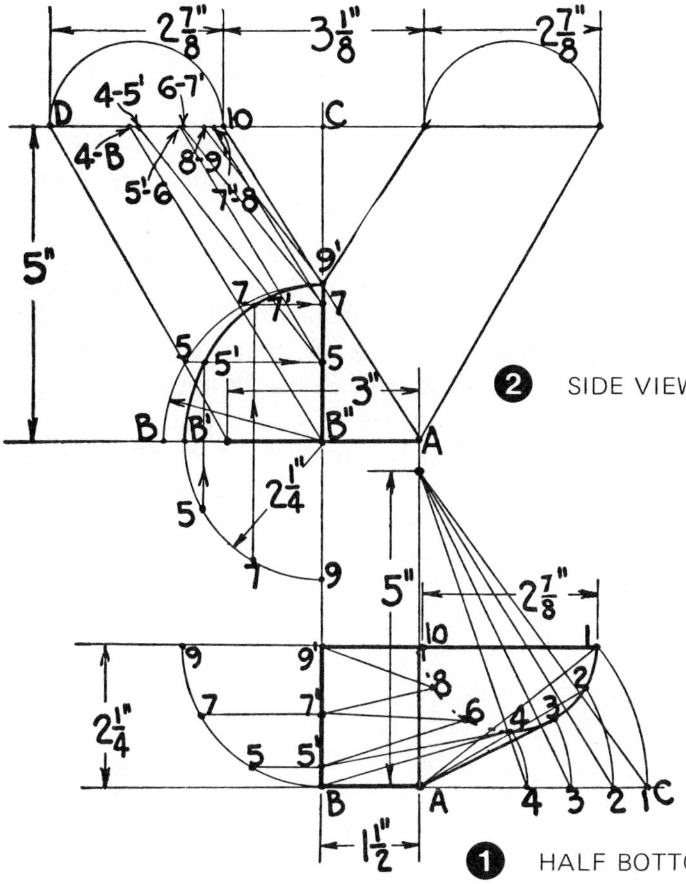

PLATE 93

② SIDE VIEW

① HALF BOTTOM VIEW & TRUE LENGTH LINES

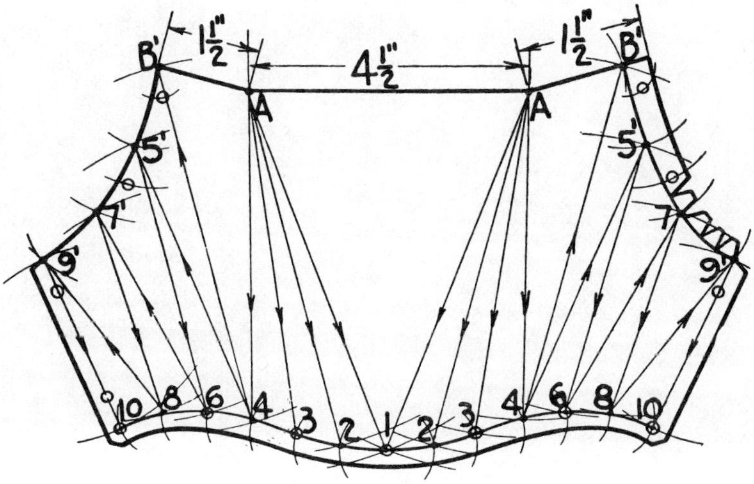

③ FULL PATTERN — TWO REQUIRED

PLATE 94 Y BRANCH WITH A FLAT CROTCH

This type of crotch is used when a splitter is required.

It is not necessary to draw any portion of the side view in Figure 2 except to obtain the height of E' to C' at the crotch, and the slant length line A to $C-D$.

The procedure for this plate is identical to the procedure in Plate 16. Draw the half top view as in Figure 1. Use point A as a center to draw the arcs from points 1, 2, 3, and 4 to line $A-E$. Use point C as a center to draw the arcs from points 4, 5, 6, and 7 to line $A-E$.

To lay out the pattern as in Figure 3, use the long slant true-length lines in Figure 1 to develop the portion of the pattern from point 1 to 4. To obtain point C in Figure 3, use the distance A to $C-D$ in Figure 2 as a radius to draw an arc from point A to C. Use the short true-length lines in Figure 1 to develop the remaining pattern from points 4 to 7. The distance C to D is equal to the distance C to D' in Figure 1. The distance 7 to D is equal to the distance 7 to $C-D$ in Figure 2. The distance C to B is equal to the distance B to $C-D$ in Figure 2.

Complete the pattern with the rivet holes laid out as shown.

PLATE 94

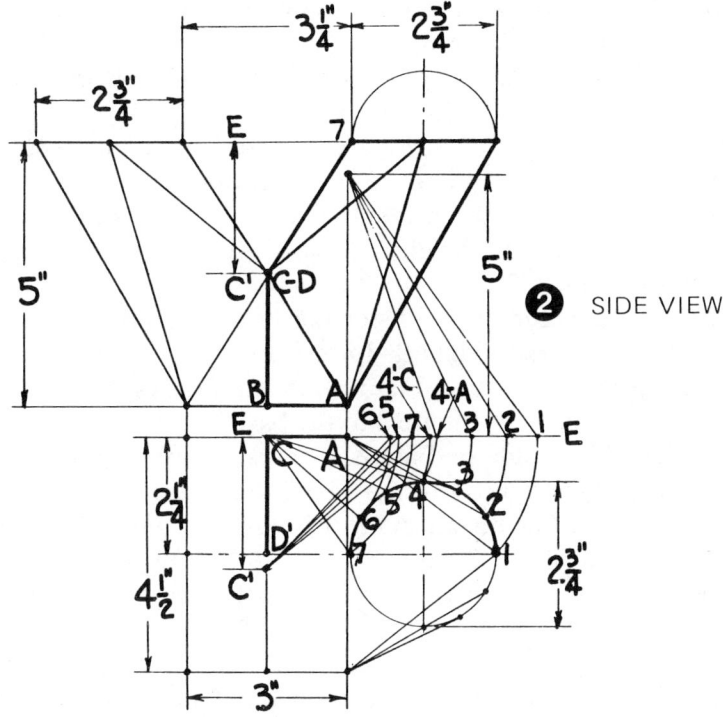

① HALF BOTTOM VIEW — TRUE LENGTH LINES

② SIDE VIEW

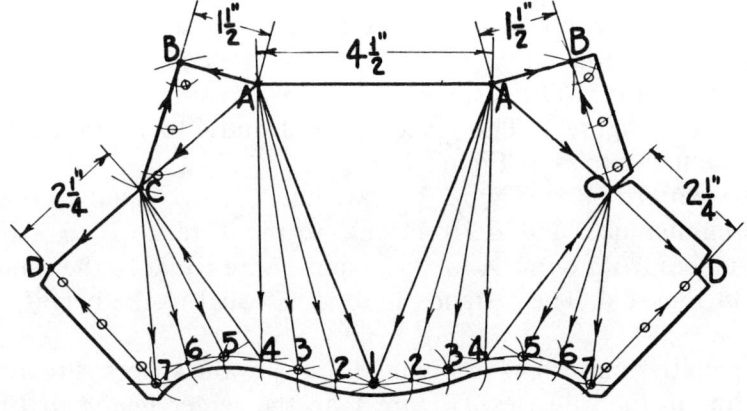

③ FULL PATTERN — TWO REQUIRED

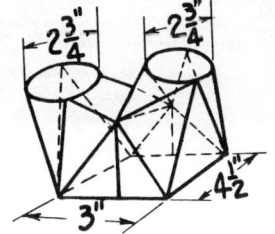

④ ISOMETRIC VIEW

PLATE 95 ROUND Y BRANCH WITH CENTER SPLITTER

This type of Y branch is made with all branches of equal area with a center splitter damper as shown from point A to $10'$.

Draw the side view as in Figure 1, and a portion of the bottom view, obtaining point A and the quarter circle $1''$ to $5''$.

To erect the true-length triangle as in Figure 3, transfer the lengths 5 to 6, 7 to 8, and 9 to 10, from the side view, Figure 1, to the base line of the triangle. Transfer the difference of the heights of lines 5 and 6 on the half circles in Figure 1 to represent the height 5–6 in Figure 3. Transfer the difference of the heights of lines 7 and 8 on the half circles in Figure 1 to represent the height 7–8 in Figure 3. Transfer the height of line 9 on the half circle in Figure 1 to represent height 9–10 in Figure 3.

To erect the true-length triangles as in Figure 4, transfer the distance from point A to points $1''$, $2''$, $3''$, and $5''$ in Figure 2 to the base line of the triangle.

To erect the true-length triangle in Figure 5, transfer the lengths from point A to points 1, 2, and 3 in Figure 1 to the base line in Figure 5 as represented by points $1'$, $2'$, and $3'$. The distance 5 to 1 is equal to one half of the diameter. Transfer the height of line 2 from Figure 1 to Figure 5 from point 1 to 2. Transfer the height of line 3 from Figure 1 to Figure 5 from point 1 to 3.

To lay out the pattern as in Figure 6, draw line C to B equal to one quarter of the circumference, and B to A equal to the distance B to A in Figure 1. The distances 4 to 5 and A to 5 are equal to the length of lines 4 to 5 and A to 5 in Figure 1. The distances 5 to 6, 7 to 8, and 9 to 10 are equal to the slant true-length lines in Figure 3. The distances 6 to 7, 8 to 9, and 10 to 11 are the same length as the lines 6 to 7, 8 to 9, and 10 to 11 in the side view, Figure 1. The distances A to $3'$, $2'$, and $1'$ are equal to the slant true-lengths lines in Figure 5. The distance $1'$ to A' is equal to the distance A to 1 in the side view, Figure 1. The spaces $1'$ to 11 and 4 to 10 are equal to the spaces on the half circles in Figure 1.

To lay out the pattern as in Figure 7, draw line C to B equal to one quarter of the circumference, and B to A equal to the distance B to A in Figure 1. The distances from point A to $3''$, $2''$, and $1''$ are equal to the slant true-length lines in Figure 4. The distance $1''$ to A' is equal to the height A to $1'$ in Figure 1.

To lay out the splitter pattern as in Figure 8, transfer the spaces A to $10'$ on the diagonal line in the side view, Figure 1, to the center line A to $10'$ in Figure 8. Transfer the heights of lines 4 to 4, 6 to 6, and 8 to 8 on the half circle in Figure 1 to each side of the center line on their respective lines in Figure 8, obtaining the freehand curved line 4 to $10'$.

For practical work, the patterns in Figures 6 and 7 may be made as full pattern instead of halves.

PLATE 95

① SIDE VIEW

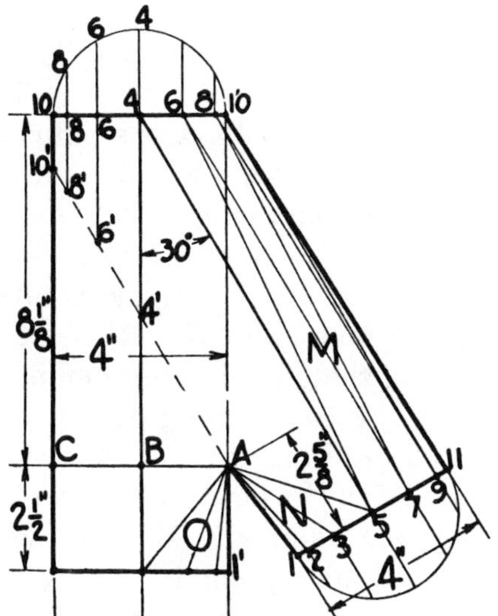

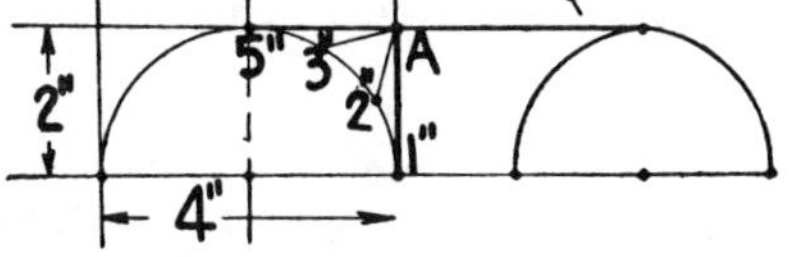

② HALF BOTTOM

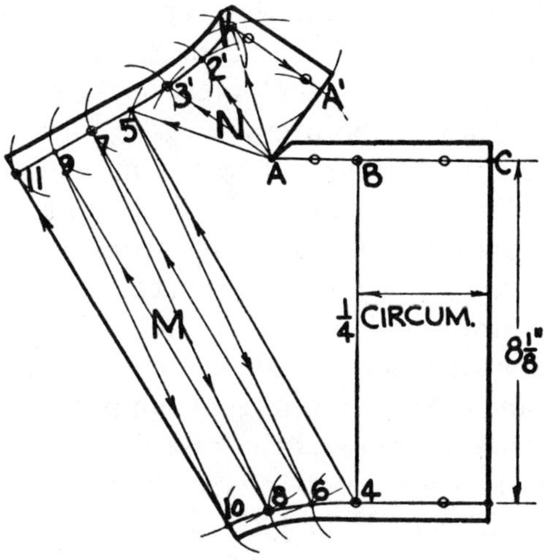

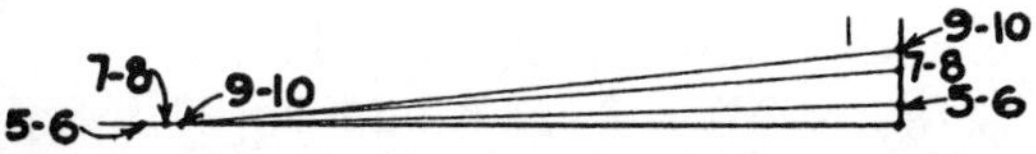

③ TRUE LENGTHS FOR SECTION M

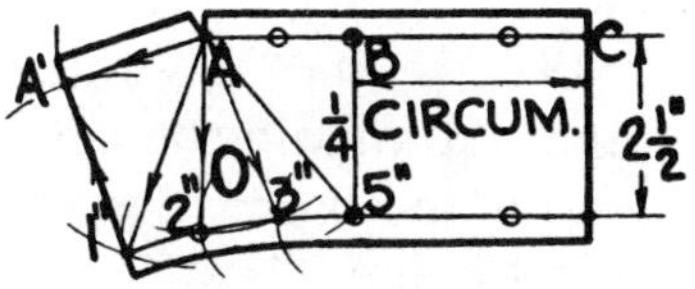

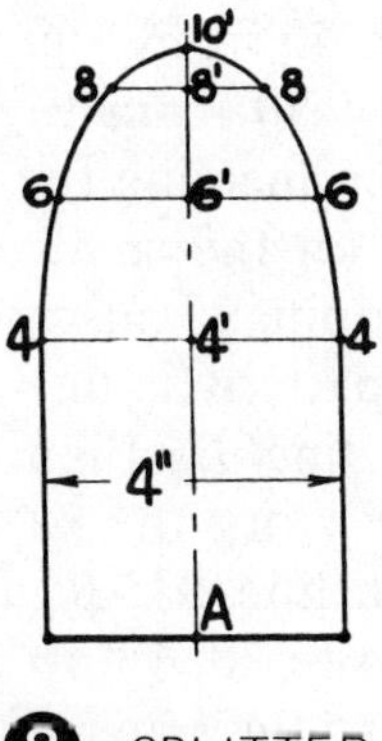

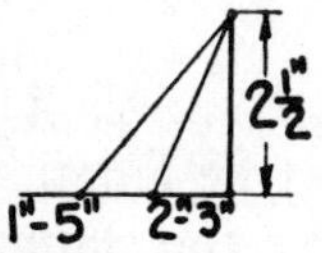

④ TRUE LENGTHS FOR SECTION O

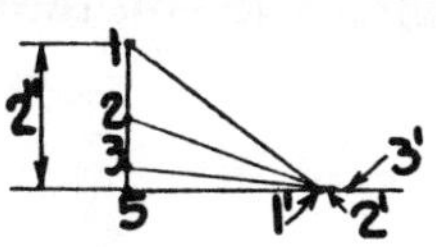

⑤ TRUE LENGTHS FOR SECTION N

⑥ HALF PATTERN

⑦ HALF PATTERN

⑧ SPLITTER PATTERN

PLATE 96 ROUND Y BRANCH FLAT ON ONE SIDE

The procedure for this type of Y branch is similar to the equal-tapered branch in Plate 77, except that a full bottom view of one branch must be drawn and the crotch must be developed as in Figure 2.

Draw the bottom view as in Figure 1. Divide the smaller circle as well as the large half circle 1 to 13, into equal spaces.

Draw a portion of the side view as in Figure 3. Use point 13–1 as a center to draw the quarter circle from point 19′ to 13″, and divide it into equal spaces. Draw a line from each division point to cross the center crotch line, obtaining points 15′–23′ and 17′–21′. Draw a line from point 24 to cross through point 15′–23′ at the crotch, to intersect the base line of the main branch. Also draw a line from point 18 to cross through point 17′–21′ at the crotch, to intersect the base line. Draw a line from these intersecting points on the base line to intersect the large circle in Figure 1, obtaining points 15 and 23, also 17 and 21. Draw a line from point 15 to point 16 on the small circle, crossing the center crotch line, obtaining point 15′. Draw a line from point 17 to point 18, crossing the crotch, obtaining point 17′. Draw a line from point 19 to point 20, crossing the crotch, obtaining point 19′. Draw a line from point 21 to 22, obtaining point 21′. Draw a line from point 23 to 24, obtaining point 23′.

To construct the freehand crotch curve as in Figure 2, draw a line from points 15′, 17′, 19′, 21′, and 23′ on the center crotch line in Figure 1, to the crotch in Figure 2. To obtain the heights for the crotch, transfer the heights from point 13–1 at the base, to point 15′–23′ on the center crotch line in Figure 3, to lines 15′ and 23′ in Figure 2. Transfer the height from the base line to point 17′–21′ in Figure 3, to lines 17′ and 21′ in Figure 2. Transfer the height from the base line to point 19′ on the crotch in Figure 3 to line 19′ in the Figure 2, obtaining the freehand crotch curve 1′ to 13′.

Transfer the lengths 14 to 15′, 15′ to 16, 24 to 23′, and 23′ to 22 from Figure 1 to line *A* in Figure 3. Transfer the lengths 16 to 17′, 17′ to 18, 22 to 21′, and 21′ to 20 from Figure 1 to line *B* in Figure 3. Transfer lengths 18 to 19′ and 19′ to 20 from Figure 1 to line *C* in Figure 3. Transfer the remaining lengths connecting the points 1 to 13 on the large half circle with the points 2 to 14 on the small circle in Figure 1, to line *D* in Figure 3, obtaining the remaining slant true-length lines.

To lay out the pattern as in Figure 4, draw line 7 to 8 equal to the slant true-length line at line *D*, Figure 3. Use the remaining slant true-length lines at line *D* to lay out the pattern 1′ to 13′ and 24 to 14. Use the slant true-length lines on lines *A*, *B*, and *C*, Figure 3, to lay out the remaining pattern 14 to 20, 24 to 20, 13′ to 19′, and 1′ to 19′ at the crotch. The spaces 20 to 20 are equal to the spaces 2 to 24 on the small circle in Figure 1. The spaces 1′ to 13′ are equal to the spaces 1 to 13 on the half circle in Figure 1. The spaces 13′ to 19′ and 1′ to 19′ are equal to the spaces 1′ to 13′ on the freehand crotch length curve in Figure 2.

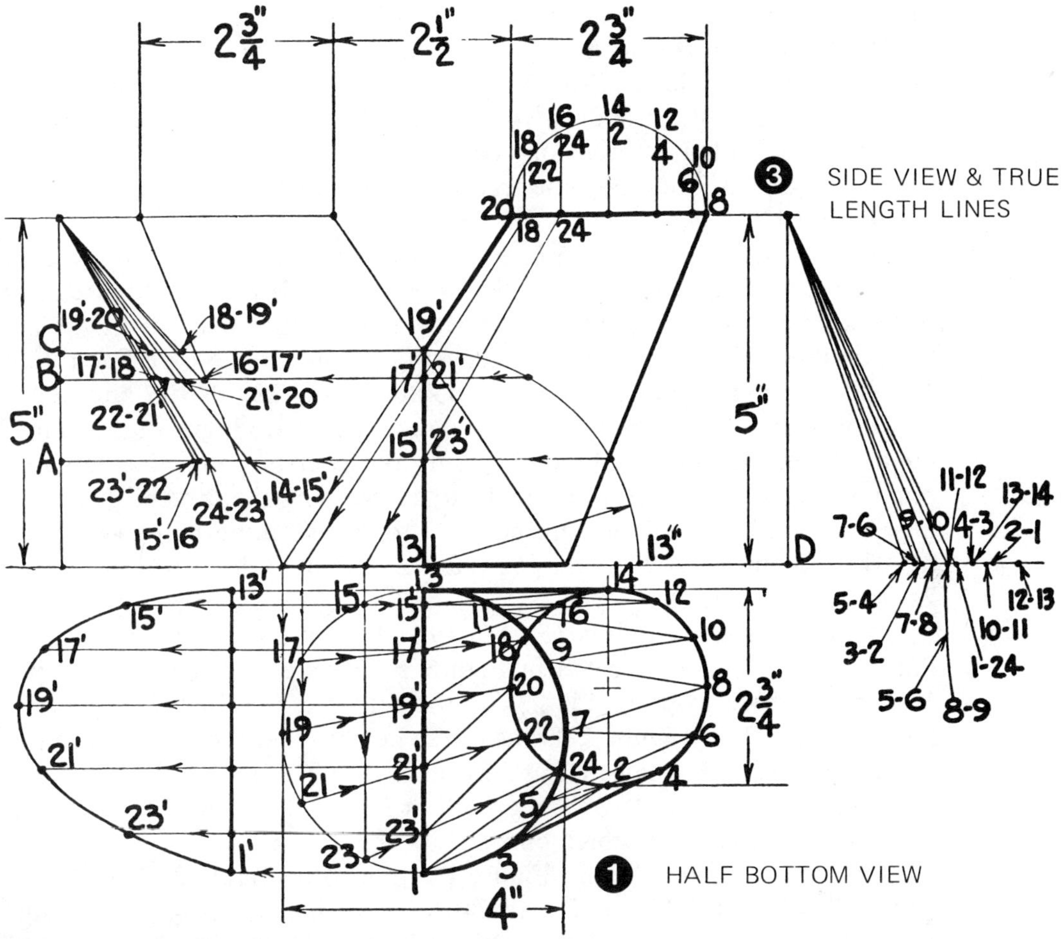

PLATE 96

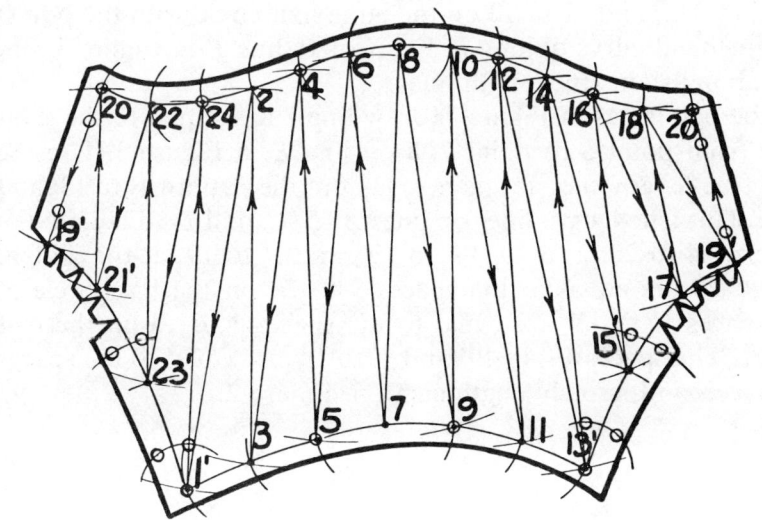

① HALF BOTTOM VIEW
② CROTCH LENGTHS
③ SIDE VIEW & TRUE LENGTH LINES
④ FULL PATTERN — TWO REQUIRED

PLATE 97   30-DEG. Y BRANCH FLAT ON ONE SIDE

The procedure for this plate is identical to Plate 96.

Draw the side view as in Figure 3, and the bottom view as in Figure 1. Draw the large circle, and divide the half circle 1 to 13 into equal spaces; draw the small half circle 2' to 14', and divide it into equal spaces. Draw a line from each of these division spaces to cross the lines drawn from line 8–20 in Figure 3, obtaining the freehand curved oblong 2 to 14 and 14 to 2 in Figure 1.

Draw a line from point 16 to cross through point 15'–23' on the crotch center line, to intersect the base line of the main branch in Figure 3, and from point 18 through point 17'–21', to intersect the base line of the main branch. Draw lines from these intersecting points on the base line to cross the half circle in Figure 1, obtaining points 15 and 23, also 17 and 21.

Draw a line from point 15 on the large half circle, Figure 1, to point 16 on the freehand curved oblong crossing the center crotch line, obtaining point 15'. Draw a line from point 17 to 18 crossing the crotch line, obtaining point 17'. Draw a line from point 19 to 20 obtaining point 19'. Draw a line from point 21 to 22 obtaining point 21'. Draw a line from point 23 to 24 obtaining point 23'.

Draw a line from points 15', 17', 19', 21', and 23' on the center crotch line in Figure 1 to the crotch in Figure 2. Transfer the height from point 13–1 at the base to point 15'–23' on the center crotch line in Figure 3, to lines 15' and 23' in Figure 2. Transfer the height from the base line to point 17'–21' in Figure 3 to lines 17' and 21' in Figure 2. Transfer the height from the base line to point 19' on the crotch in Figure 3, to line 19' in Figure 2, obtaining the freehand crotch curve 1' to 13'.

Transfer line 1 to 24 from Figure 1 to line A in Figure 3. Transfer lines 14 to 15', 15' to 16, 23' to 22, and 24 to 23' to line B in Figure 3. Transfer lines 16 to 17', 17' to 18, 21' to 20, and 22 to 21' to line C in Figure 3. Transfer lines 18 to 19' and 19' to 20 to line D. Transfer the remaining lengths connecting the points 1 to 13 on the large half circle with the points 2 to 14 on the freehand curve oblong in Figure 1 to line E in Figure 3, obtaining the remaining slant true-length lines.

To lay out the pattern as in Figure 4, draw line 7 to 8 equal to the slant true-length line from point 8 to point 7–8 on line E in Figure 3. Use the remaining slant true-length lines at line E to lay out the pattern 1' to 13' and 2 to 14. Use the slant true-length lines on lines A, B, C, and D to lay out the remaining pattern 14 to 20, 2 to 20, 13' to 19', and 1' to 19' at the crotch. The spaces 20 to 20 are equal to the spaces 8 to 20 on the half circle in Figure 3. The spaces 1' to 13' are equal to the spaces 1 to 13 on the half circle in Figure 1. The spaces 13' to 19' and 1' to 19' are equal to the spaces 1' to 13' on the freehand, crotch-length curve in Figure 2.

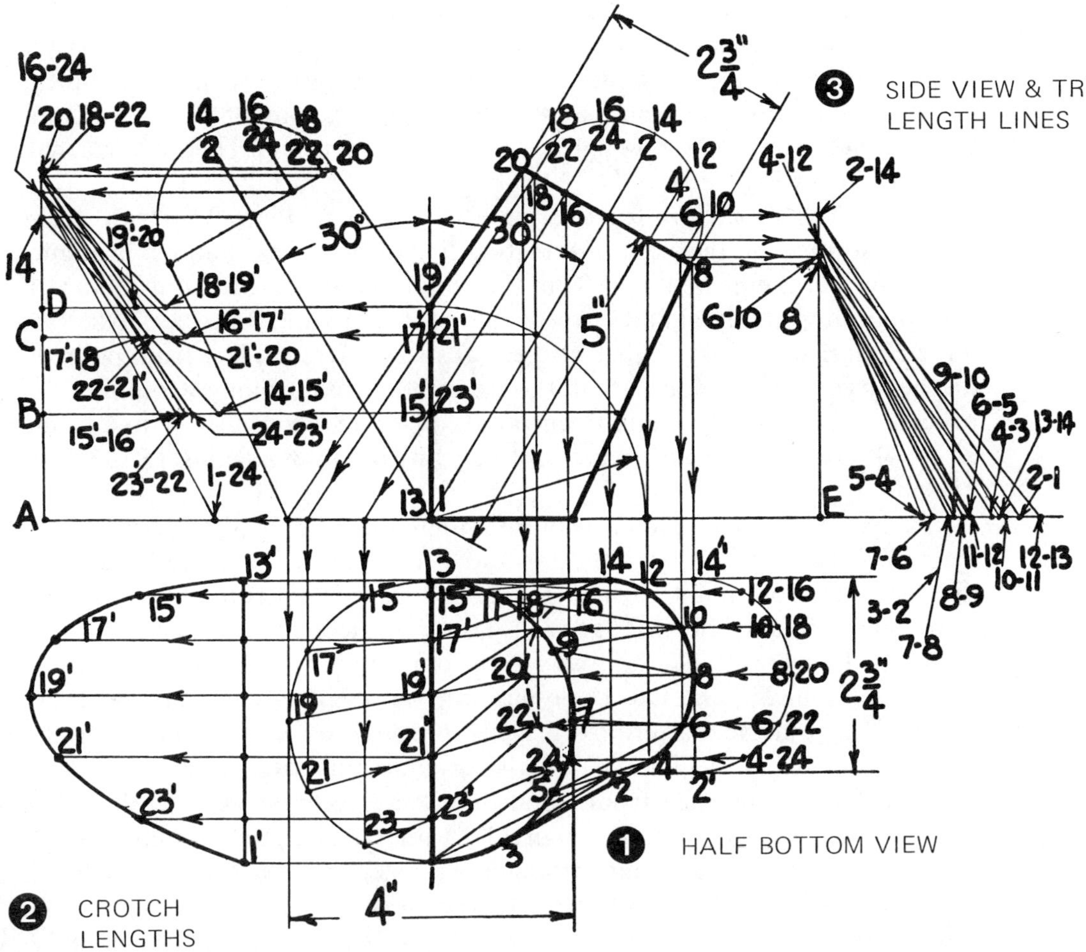

PLATE 97

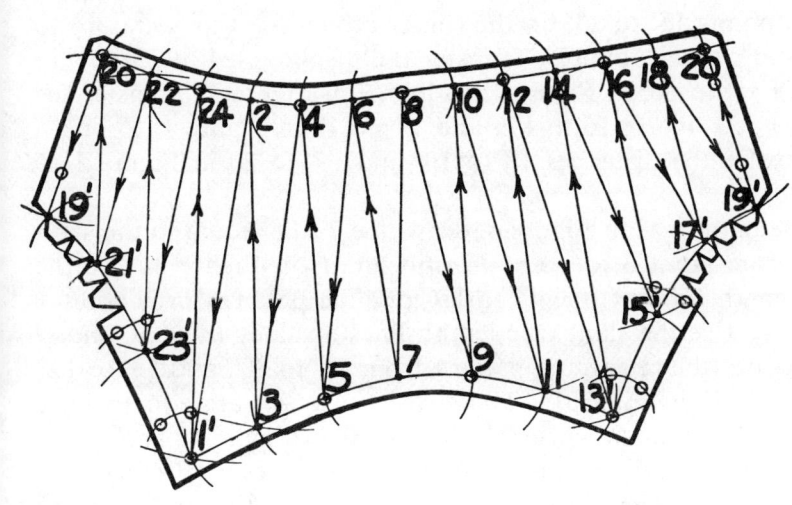

3 SIDE VIEW & TRUE LENGTH LINES

1 HALF BOTTOM VIEW

2 CROTCH LENGTHS

4 FULL PATTERN — TWO REQUIRED

PLATE 98 Y BRANCH, ONE SIDE FLAT WITH HORIZONTAL BRANCHES

To draw the side view as in Figure 3, draw the center line from point 13–1 in branch *A* on a 35-deg. angle to intersect with the center line of branch *B* at point 14′. Next bisect these two center lines and draw a line from the bisecting point crossing line 8 and 20, establishing points 8′ and 20′. Draw the distance from 8′ to 8 equal to the given dimension. Then draw a half side view as in Figure 3, and a half bottom view as in Figure 1.

NOTE: The procedure for this plate is similar to Plate 97. Draw a line from each point on the slant intersecting line of branches *A* and *B* in Figure 3, to intersect the lines drawn from the small half circle 2 to 14 in Figure 1, thus obtaining the freehand curved oblong 2 to 14 and 14 to 2. Divide the large circle 1 to 13 and 13 to 1 into equal spaces. Draw a line from point 15 on the large circle to point 16 on the freehand curved oblong. Continue this procedure by drawing a line from points 17 to 18, 19 to 20, 21 to 22, and 23 to 24, crossing the center crotch line, thus obtaining points 15′, 17′, 19′, 21′, and 23′.

To obtain the freehand curved crotch lengths as in Figure 2, draw a line from points 13, 15′, 17′, 19′, 21′, 23′, and 1 in Figure 1 to Figure 2. Transfer the distances from point 13–1 to 15′–23′ on the center crotch line in Figure 3, to lines 15′ and 23′ in Figure 2. Transfer the remaining heights from point 13–1 to points 17′–21′ and 19′ on the center crotch line in Figure 3, to lines 17′ and 21′, and line 19′, obtaining the freehand curve crotch length 13 to 1. Lay out the pattern for branch *B* equal to the circumference of a 2¾-in. diameter as in Figure 4, which may be a full pattern.

Transfer the distance from point 2 on the freehand curved oblong to point 1 on the large circle in Figure 1 to line *C* in Figure 3, representing the base line of the true-length triangle. Transfer the remaining length lines from the points on the large circle to the freehand curved oblong 3 to 2, 4 to 3, 5 to 4, 6 to 5, 7 to 6, 7 to 8, 8 to 9, 9 to 10, 10 to 11, 11 to 12, 12 to 13, and 13 to 14 in Figure 1 to line *C* in Figure 3.

The true lengths for the crotch may be obtained by transferring the distance 1 to 24 from Figure 1 to line *D* in Figure 3. (Transfer the remaining lengths from points 15′ to 23′ on the center crotch lines to points 16 to 24 on the freehand curved oblong.) Transfer the distances 24 to 23′, 23′ to 22, 14 to 15′, and 15′ to 16, in Figure 1, to line *E* in Figure 3. Transfer the distances 22 to 21′, 21′ to 20, 16 to 17′, and 17′ to 18 in Figure 1, to line *F* in Figure 3. Transfer the distances 18 to 19′ and 19′ to 20 in Figure 1 to line *G* in Figure 3.

To lay out the pattern as in Figure 5, draw line 7 to 8′ equal to the slant true-length line from point 8 to point 7–8 on line *C* in Figure 3. Use the remaining true-length lines at line *C* to lay out the pattern form point 1 to 13 and 2′ to 14′. Use the slant true-length lines on lines *D*, *E*, *F*, and *G* in Figure 3 to lay out the remaining pattern from 14′ to 20′, and 13′ to 19′, 2′ to 20′, and 1 to 19′, in Figure 5. The spaces 8′ to 20′ are equal to the spaces 8′ to 20′ on the freehand curved line 8′ to 20′ in Figure 4. The spaces 1 to 19′ and 13 to 19′ on the pattern are equal to the spaces on the freehand crotch curve in Figure 2. The spaces 1 to 13 are equal to the spaces on the large half circle 1 to 13 in Figure 1.

PLATE 98

3 SIDE VIEW & TRUE LENGTH LINES

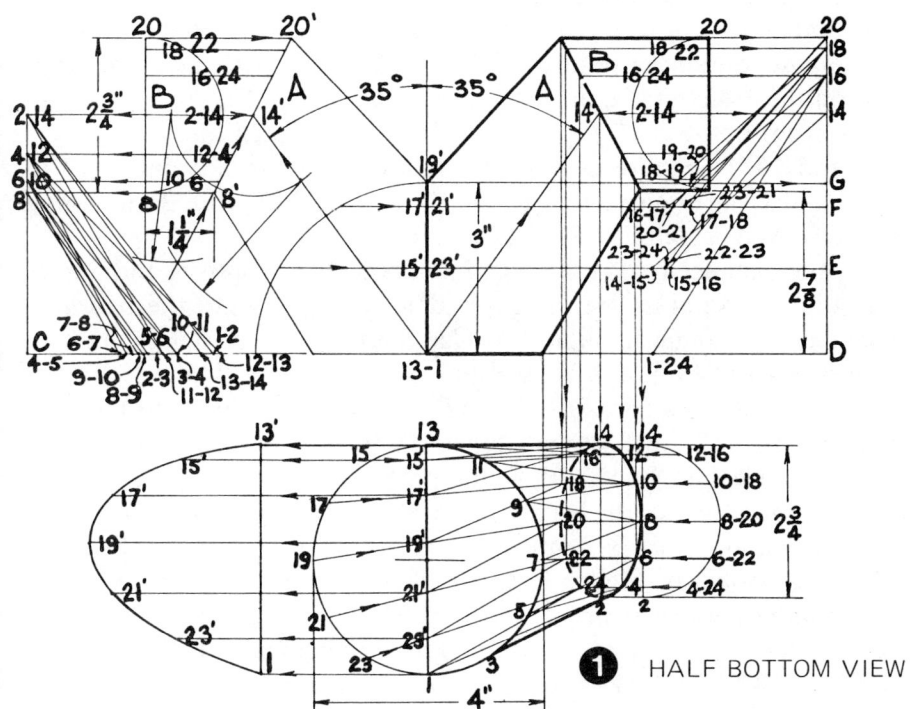

1 HALF BOTTOM VIEW

2 CROTCH LENGTHS

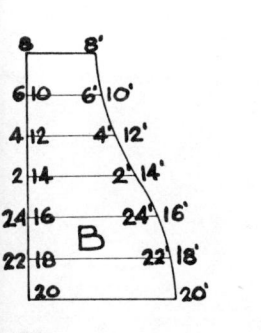

4 HALF PATTERN

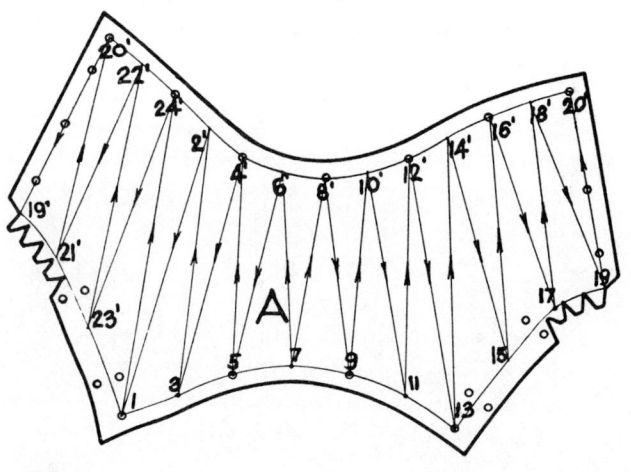

5 FULL PATTERN — TWO REQUIRED

185

PLATE 99 Y BRANCH, ONE SIDE FLAT WITH RECTANGULAR MAIN AND ROUND BRANCHES

Draw a portion of the side view as in Figure 3. Draw the half bottom view as in Figure 1. Divide the half circle 5 to 17 into equal spaces, and draw a line from points 6 to 7, 8 to 9, 10 to 11, 12 to 13, and 14 to 15 crossing the center crotch line, obtaining points 7', 9', 11', 13', and 15'. Draw a line from each point on the crotch in Figure 1 to cross line 5'–17 in Figure 2. Transfer the distance from point 5–17 to 7–15 on the center crotch line in Figure 3 to lines 7' and 15' in Figure 2. Transfer the distance from point 5–17 to points 9–13 and 11 in Figure 3 to lines 9', 13', and 11' in Figure 2, obtaining the freehand curved crotch line 5' to 17.

Transfer the distance from point A to points 1, 2, 3, 4, 4 to 5, and 5 to 6, and from point B to points 16, 18, 20, and 1, in Figure 1, to the base line of the true-length triangle in Figure 3. Use these slant true-length lines to lay out the pattern in Figure 4 from points 6 to 16 and B to 5'.

Transfer the distances 16 to 17' and 17' to 14 from Figure 1 to the base line 5–17 in Figure 3. Transfer the distances 14 to 15', 15' to 12, 6 to 7', and 7' to 8 from Figure 1 to line 7–15 at the crotch in Figure 3. Transfer the distances 12 to 13', 13' to 10, 8 to 9', and 9' to 10 from Figure 1 to line 9–13 at the crotch in Figure 3. Transfer the distance 10 to 11' from Figure 1 to line 11 at the crotch in Figure 3. Use these slant true-length lines to lay out the remaining pattern from 6 to 10, and 5' to 11', and from 16 to 10 and 17 to 11'. The spaces 1 to 10 are equal to the spaces on the circle in Figure 1. The spaces 5' to 11' and 17 to 11' at the crotch are equal to the spaces 5' to 17 on the freehand crotch curve in Figure 2.

NOTE: The rivet holes are placed on the lap and the pattern on both sides. This will represent the rivet holes for both patterns. On one pattern the lap allowance will be cut off.

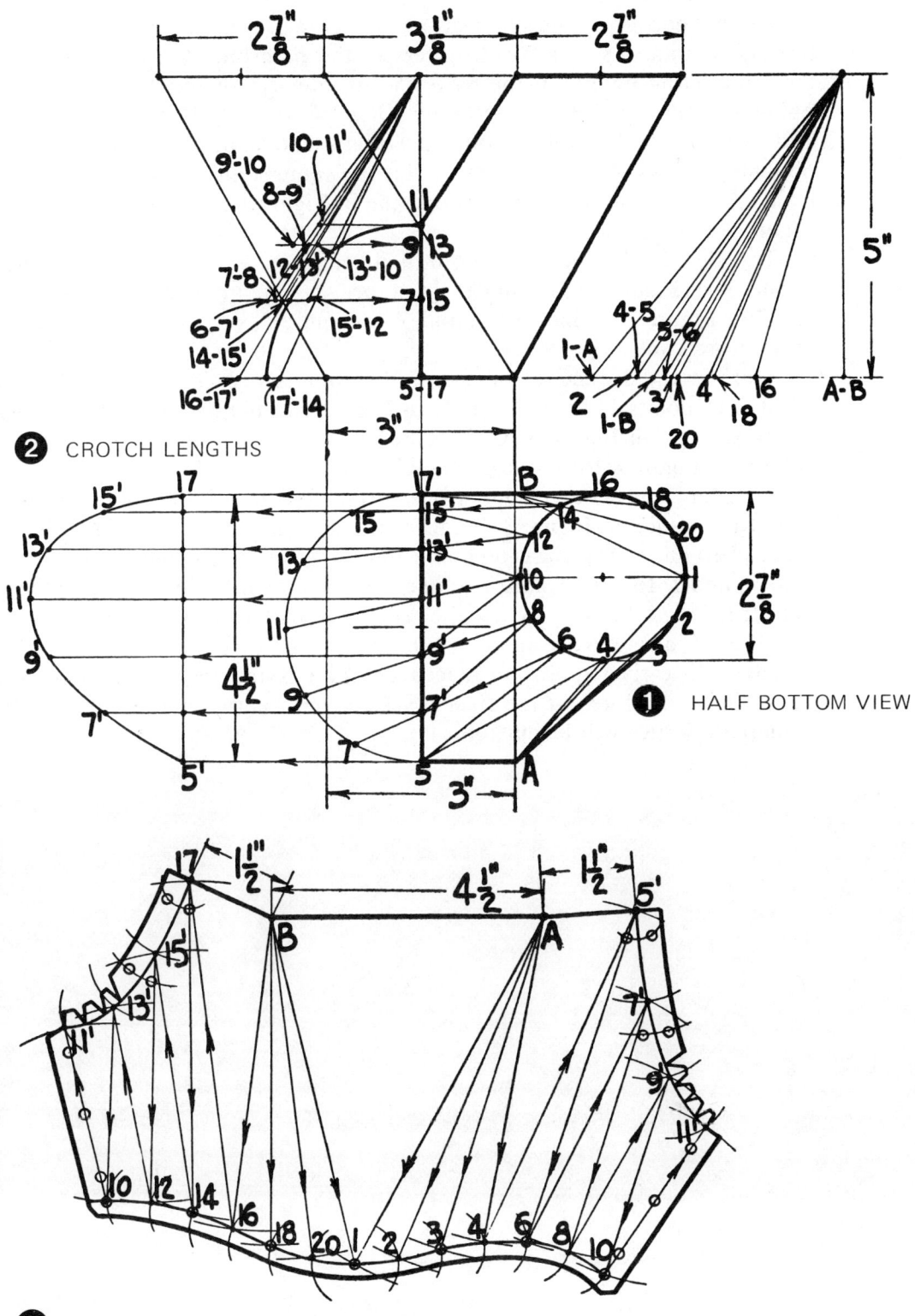

PLATE 99

PLATE 100 Y BRANCH, ONE SIDE FLAT WITH ROUND MAIN AND RECTANGULAR BRANCHES

Draw the half bottom view as in Figure 1. Divide the half circle 1 to 4′ and 1 to 10′ into equal spaces.

Draw a portion of the side view as in Figure 3. Draw lines from point C through points 6–8 and 5–9 to intersect the base line. Draw a line from the intersecting points on the base line to intersect the circle in Figure 1, obtaining points 5, 6, 8, and 9. Draw a line from points 5 and 6 to point C, and from points 8 and 9 to point E. Draw a line from point 7 to point D, obtaining points 5′, 6′, 7′, 8′, and 9′ on the center crotch line. Draw a line from each point on the center crotch line in Figure 1 to cross line 4′–10′ in Figure 2.

Transfer the distance from point 4–10 to point 5–9 in Figure 3 to lines 5′ and 9′ in Figure 2. Transfer the distance from point 4–10 to points 6–8 and 7 in Figure 3 to lines 6′, 8′, and 7′ in Figure 2, obtaining the freehand curve crotch line 4′ to 10′.

Transfer the distances from point B to points 1, 2, 3, and 4′, and from point A to points 1, 12, 11, and 10′ in Figure 1 to the base line of the true-length triangle in Figure 3. Use these slant true-length lines to lay out the pattern in Figure 4 from 4′ to 10′ and C to E.

Transfer the distances from points C to 5′ and E to 9′ in Figure 1 to line 5–9 at the crotch in Figure 3. Transfer the distances C to 6′ and E to 8′ in Figure 1 to line 6–8 at the crotch in Figure 3. Transfer the distances C to 7′, E to 7′, and D to 7′ in Figure 1 to line 7 at the crotch in Figure 3. Use these slant true-length lines to lay out the remaining pattern from points 10′ to 7′ and 4′ to 7′; also E to D and C to D.

NOTE: The rivet holes are placed on the lap and the pattern on both sides. This will represent the rivet holes for both patterns. On one pattern the lap allowance will be cut off.

PLATE 100

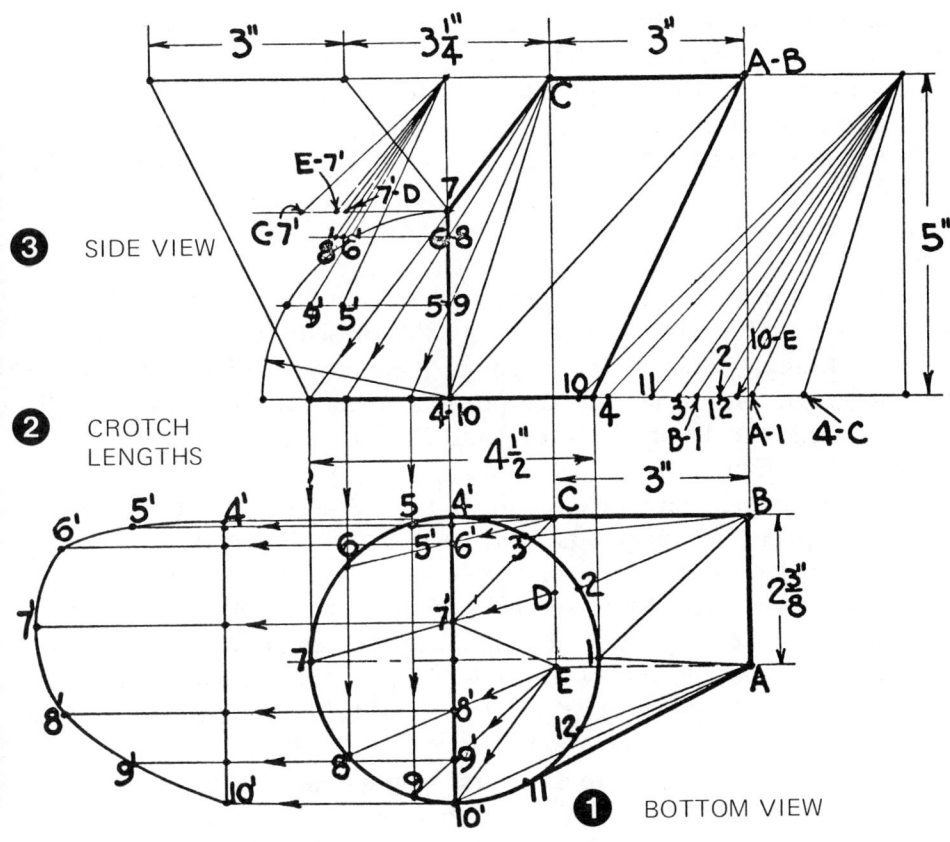

③ SIDE VIEW
② CROTCH LENGTHS
① BOTTOM VIEW

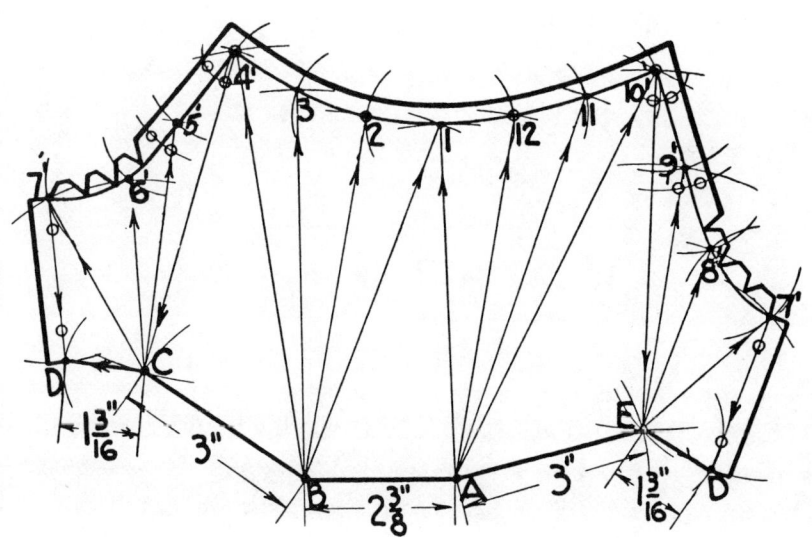

④ FULL PATTERN — TWO REQUIRED

PLATE 101 Y BRANCH, ONE SIDE STRAIGHT WITH FLAT CROTCH

The procedure for this plate is the same as for the rectangular-to-round in Plate 17 and the Y branch in Plate 94.

Draw the bottom view as in Figure 1, and divide the circle into equal spaces. Use point B as a center to draw arcs from points 1, 2, 3, and 4 to line B–K, as represented by $1'$–B, $2'$, $3'$, and $4'$–B. Draw line B to B' the height of the true-length triangle equal to the height of the side view (5 in.). Use the slant true-length lines to lay out the portion of the pattern from points $1'$ to $4'$.

Use point A in Figure 1 as a center to draw arcs from points 10, 11, 12, and 1 to the base line A–K. Draw line A–A' representing the height of the true-length triangle equal to the height of the side view (5 in.). Use these slant true-length lines to lay out the portion of the pattern from points $1'$ to $10'$. Note that the distance A' to D' and B' to C' is equal to the distance A–B to C–D in Figure 2.

Transfer the crotch height from point G to C–D in Figure 2 to the center crotch line represented by points C–F and E–D in Figure 1, from point E–D to C'–D'. Use point E–D as a center to draw arcs from points 7, 8, 9, and 10 to the base line A–K as represented by points $7'$, $8'$, $9'$, and $10'$–A. Then use these true-length lines to complete the remaining pattern from points $10'$ to $7'$. The distance $7'$ to H is equal to the distance from $7'$ to C–D in Figure 2; the distance D' to H is equal to the distance D–E to H in Figure 1.

Use point C–F as a center to draw arcs from points 4, 5, 6, and 7 to line B–K as represented by points $4'$–C, $5'$, $6'$, and $7'$. Use these true-length lines to complete the pattern from point $4'$ to $7'$. The distance C to H is equal to the distance C–F to H in Figure 1; the distance $7'$ to H is equal to the distance $7'$ to C–D in Figure 2.

Make lap allowances as desired.

PLATE 101

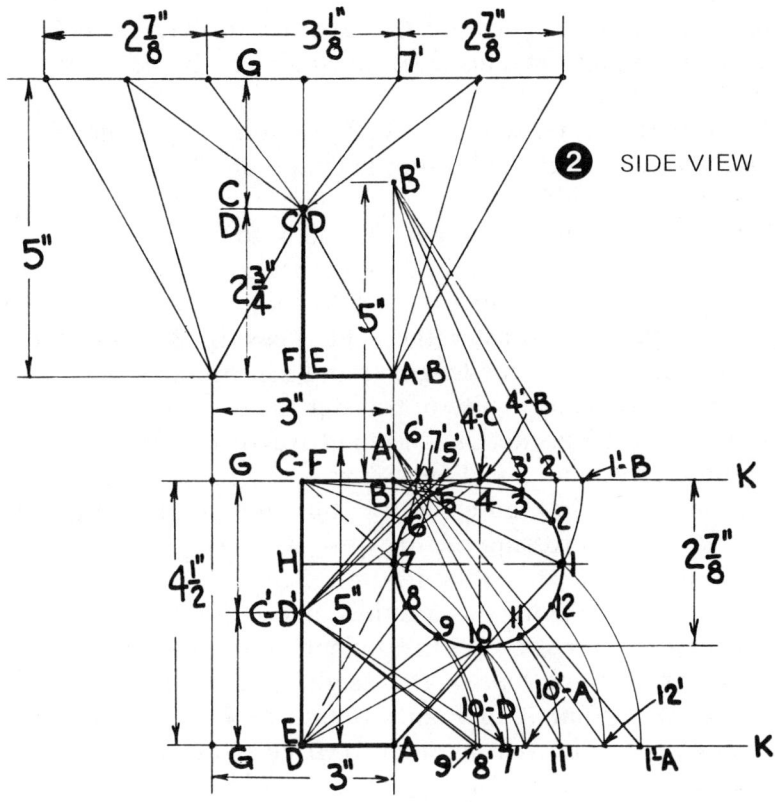

① BOTTOM VIEW & TRUE LENGTH LINES

② SIDE VIEW

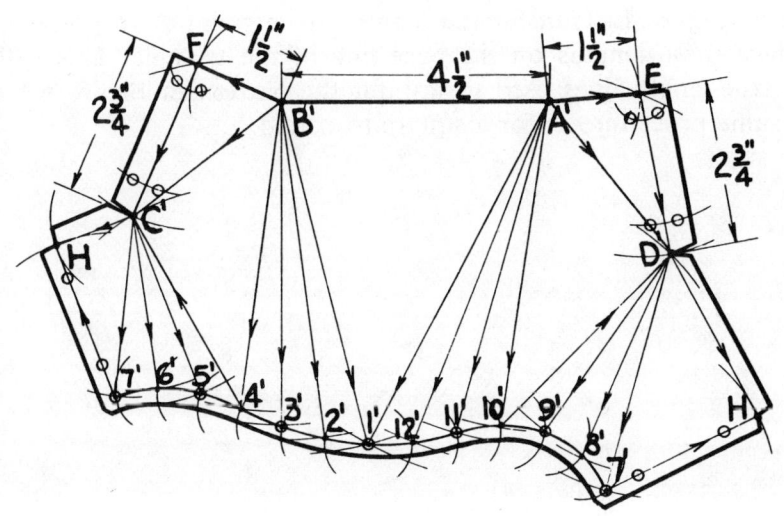

③ FULL PATTERN — TWO REQUIRED

PLATE 102 THREE-PRONGED, FLAT CROTCH BRANCH FITTING

This type of branch fitting is made with flat crotches so that dampers may be placed at each crotch point B. This will allow only one branch to operate at any one time.

To lay out the patterns, draw only a half portion of the side view in Figure 1, and only a quarter of the bottom view, Figure 2.

To simplify the procedure for obtaining the slant true-length lines, draw the height of the triangles $4'$ to $7'$ in Figure 3, and $4'$ to $1'$ in Figure 4 equal to one half of the diameters of the secondary branches. Transfer the length of the slant lines $5'$, $6'$, and $7'$ from Figure 1 to the base line of the true-length triangle in Figure 3. Transfer the slant lines $1'$, $2'$, and $3'$ from Figure 1 to the base line of the true-length triangle in Figure 4.

To lay out the pattern as in Figure 6, transfer the distances A to A and B to B from Figure 1 to Figure 6. Use the slant true-length line A to $4'$ in Figure 1 as a radius, and point A in Figure 6 as a center to strike an arc at $4'$. Use the slant line B to $4'$ in Figure 1 as a radius, and point B in Figure 6 as a center to strike an arc crossing the arc at $4'$. Use the slant true-length lines $5'$, $6'$, and $7'$ in Figure 3 to obtain points $5'$, $6'$, and $7'$ at Figure 6. Use the distance $7'$ to A in Figure 1 as a radius, and point $7'$ in Figure 6 as a center to strike an arc at point C. Set the dividers to span a distance equal to the dimensions representing half of the height of the main opening as shown in Figure 2. Use point A in Figure 6 as a center to strike an arc crossing the arc at C. Use point B as a center to strike an arc at point D. Use the slant true-length lines $3'$, $2'$, and $1'$ in Figure 4 to obtain the points $3'$, $2'$, and $1'$ in Figure 6. Use the distance $1'$ to B in Figure 1 as a radius, and point $1'$ in Figure 6 as a center to strike an arc crossing the arc at D, thus completing the half pattern for the main and the two side branches.

To erect the true-length triangle for the center branch pattern as in Figure 5, draw the height of the triangle in Figure 2 equal to the length of the center branch in Figure 1. Transfer the lengths from point B to points 2 and 3, and place these lengths on the base line of the triangle. This will give the slant true-length lines used to lay out the pattern in Figure 5 by following the same procedure as for a square-to-round.

PLATE 102

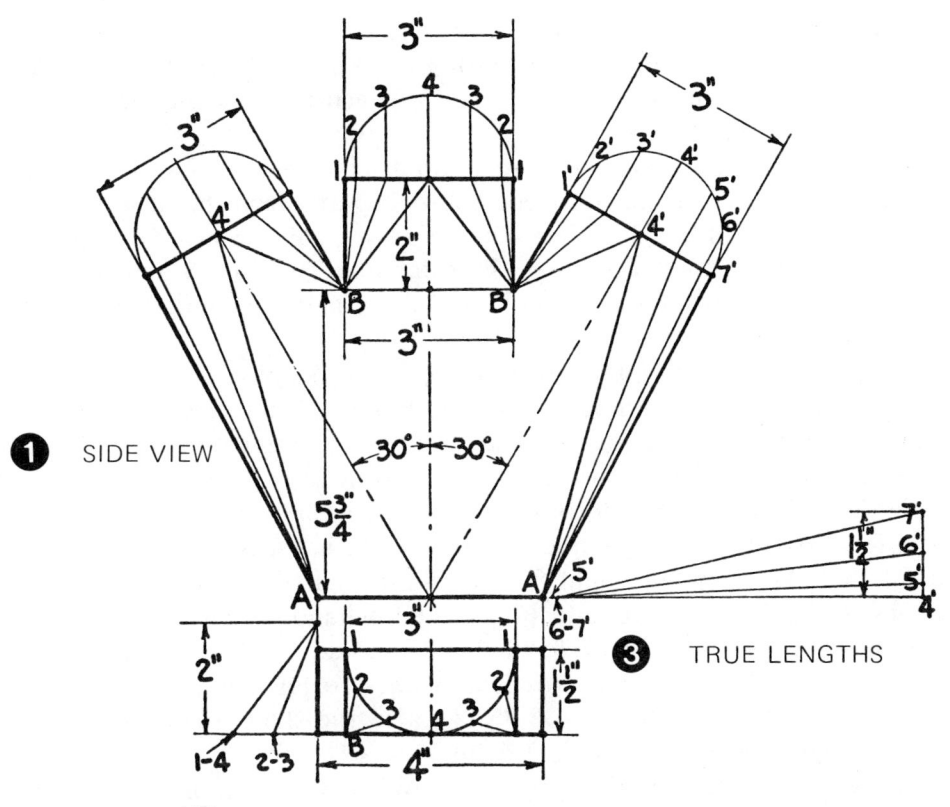

① SIDE VIEW

② HALF BOTTOM VIEW OF CENTER BRANCH

③ TRUE LENGTHS

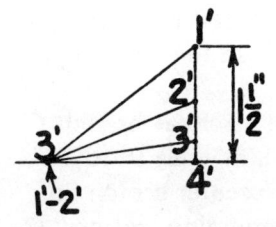

④ TRUE LENGTHS

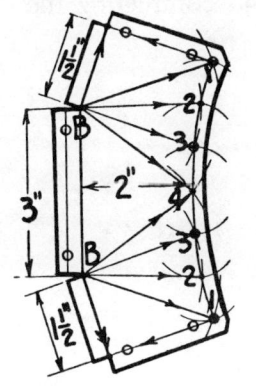

⑤ HALF PATTERN

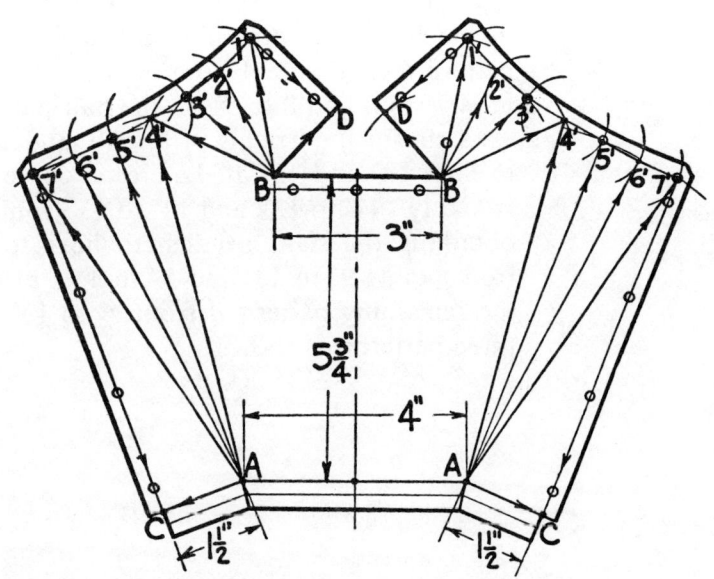

⑥ HALF PATTERN

193

PLATE 103 THREE-PRONGED BRANCH, 30 DEG. AND 45 DEG.

Draw the side view as in Figure 1. Draw the freehand crotch curves 8′ to 14′ and 8″ to 14″ by drawing a line squaring from line 8–14′ at point 8 to E, and from line 8–14″ at point 8 to F; then proceed in the same manner as in the previous Y branches.

Transfer the height of line 3 from the half circle on branch A to lines 2 and 4 at the main branch, from point 2 to 2–3 and from point 4 to 3–4. Transfer the height of line 5 to lines 4 and 6 at the main branch, from point 4 to 4–5 and from point 6 to 5–6. Transfer line 7 to line 6 and 8 from point 6 to 6–7 and from 8 to 7–8. Transfer the height of line 9 to line 8 from point 8 to 8–9.

Transfer the slant line 2 to 3 from branch A to the base line from point 2 to 2–3. Transfer the slant line 3 to 4 from branch A to the base line from line 4 to 3–4. Transfer line 4 to 5 to the base line from line 4 to 4–5. Transfer lines 59 to 6 and 6 to 7 to the base line from lines 6 to 5–6 and 6–7. Transfer lines 7 to 8 and 8 to 9 to the base line from lines 8 to 7–8 and 8–9, obtaining the slant true-length lines to lay out the half pattern A from 2 to 8′ and 1 to 9. Transfer the height of lines 9 and 11 from the half circles on branches A and B to lines 10′ and 12′ on the freehand crotch curve. Transfer the slant lines 9 to 10′, to 10′ to 11, 11 to 12′, and 12′ to 13 to the center crotch line obtaining the slant true-length lines to lay out the remaining pattern A from points 9 to 13 and 8′ 14′; also the slant true-length lines to lay out the pattern B in Figure 3 from 9″ to 13″, and 10′ to 14′.

Transfer the heights of lines 3′, 5′, 7′, and 9′ from the half circle on branch C to lines 2′, 4′, 6′, and 8 on the half circle at the main. Transfer the slant lines 2′ to 3′, 3′ to 4′, 4′ to 5, 5′ to 6′, 6′ to 7′, 7′ to 8′, and 8′ to 9′ to the base line, obtaining the slant true-length lines to lay out the half pattern C in Figure 4 from points 1′ to 9′ and 2′ to 8″; also the true-length lines 7″ to 8 and 8 to 9″ on the half pattern B in Figure 3.

Transfer the heights of lines 9′ and 11′ from the half circle at branch C to lines 10″ and 12″ at the crotch curve. Transfer the slant lines 9′ to 10″, 10″ to 11′, 11′ to 12″, and 12″ to 13′ from branch C to the center crotch line, obtaining the slant true-length lines to lay out the remaining pattern C from points 9′ to 13′ and 8″ to 14″; also the true-length lines to complete the remaining pattern B from 9″ to 13″ and from 8 to 14″, completing the three patterns as shown.

PLATE 103

① SIDE VIEW & TRUE LENGTH LINES

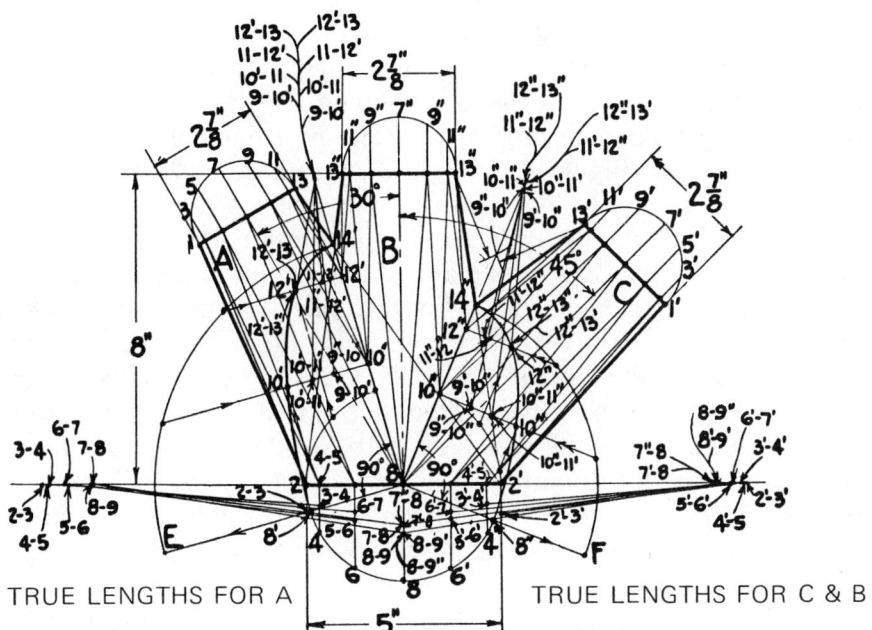

TRUE LENGTHS FOR A

TRUE LENGTHS FOR C & B

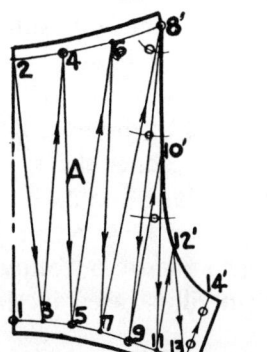

② HALF PATTERN A

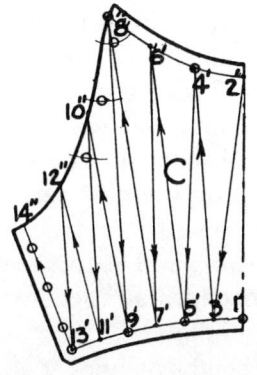

④ HALF PATTERN C

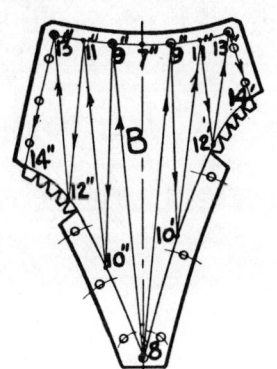

③ HALF PATTERN B

PLATE 104   30-DEG. TAPERING T ON A TAPER JOINT

This type of Y branch is efficient for exhaust systems because of its long crotch and small degrees of the angle at its branch, thereby minimizing the eddy currents and friction at the crotch.

Draw the side view as in Figure 1. The small half circle in Figure 2 represents the small diameter of the main branch.

To erect the true-length triangles, transfer the height of line 3 on the half circle in branch B, Figure 1, to Figure 2, from point 2 to point 2–3, and to line 4 from point 4 to 3–4. Transfer the height of line 5 in Figure 1 to lines 4 and 6 in Figure 2, from point 4 to 4–5 and from point 6 to 5–6. Transfer the height of line 7 to lines 6 and 8' from point 6 to 6–7, and from point 8' to 7–8. Transfer the height of line 9 to line 8' from point 8' to 8–9. Transfer the slant line 2 to 3 from branch B to line C in Figure 3, from point 2 on the half circle to point 2–3 at C. Transfer the slant lines 3 to 4 and 4 to 5 from branch B to Figure 3, from line 4 on the half circle to points 3–4 and 4–5 at C. Transfer the slant lines 5 to 6 and 6 to 7 from branch B to Figure 3, from line 5 to points 5–6 and 6–7 at C. Transfer lines 7 to 8 and 8 to 9 from branch B to Figure 3, from line 8' to points 7–8 and 8–9 at C, obtaining the slant true-length lines to lay out the half pattern B from points 2 to 8 and 1 to 9.

Transfer the heights of lines 9 and 11 from the half circle on branch B to lines 10' and 12' at the crotch. Transfer the slant lines 9 to 10, 10 to 11, 11 to 12, and 12 to 13 from branch B to the center crotch line, obtaining the slant true-length lines to lay out the remaining pattern B from points 8' to 14 and 9 to 13 in Figure 5.

Transfer the distances 1' to 2' and 2' to 3' from Figure 2 to the base line of the triangle in Figure 4. The slant line 1'–2' to D represents the true-length line 1' to 2' on the half pattern A in Figure 6. The slant line 2'–3' to D in Figure 4 represents the true-length lines 2' to 3', 1' to 4', 4' to 5', 3' to 6', 6' to 7', 5' to 8', and 8' to 9' on the half pattern A in Figure 6, following the same procedure as used for an equal taper joint illustrated in the beginning of the book.

Transfer the height of lines 9' and 11' from the half circle on branch A to lines 10' and 12' at the crotch. Transfer the slant lines 9' to 10, 10 to 11', 11' to 12, and 12 to 13' from branch A to the center crotch line, obtaining the slant true-length lines to lay out the remaining pattern A from points 8' to 14' and 9' to 13' in Figure 6. The same procedure is used as in Plate 89.

PLATE 104

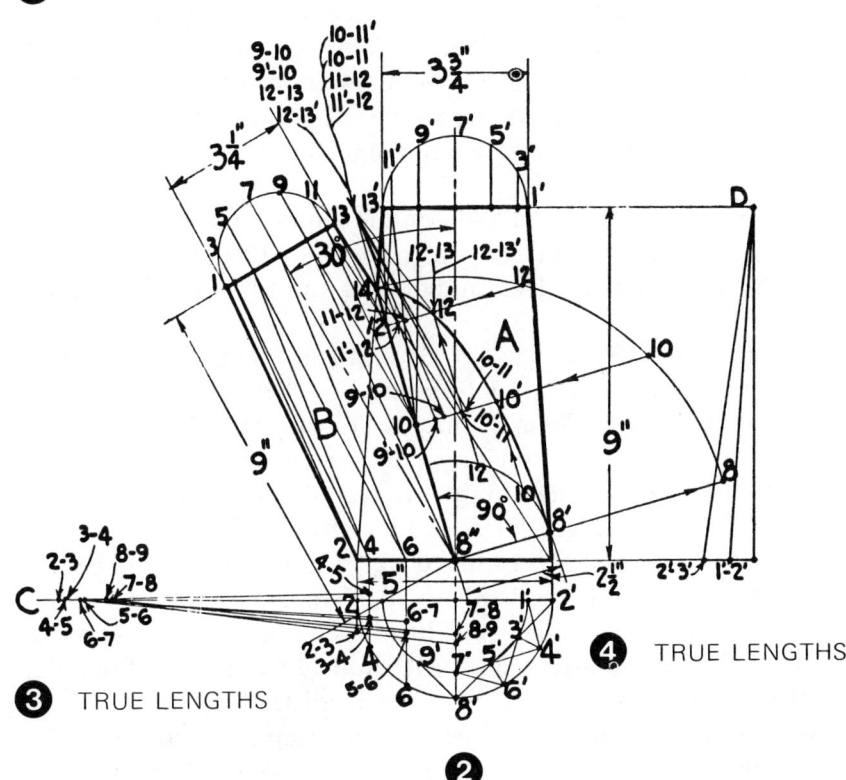

① SIDE VIEW

②

③ TRUE LENGTHS

④ TRUE LENGTHS

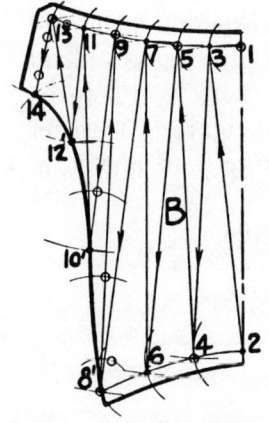

⑤ HALF PATTERN B

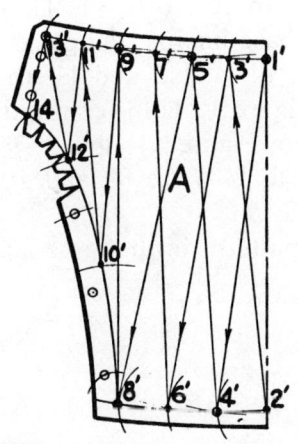

⑥ HALF PATTERN A

197

PLATE 105   40-DEG. TAPERING T ON A TAPER JOINT, FLAT ON ONE SIDE

Draw the side view as in Figure 1. To lay out the bottom view as in Figure 2, draw the large circle and the small circle equal to the dimensions. Draw lines from line 8′–8 on the T in Figure 1 to the bottom view in Figure 2. Draw the small half circle in Figure 2 equal to the diameter of the T, and divide it into equal spaces. Draw a line from each division point on the half circle to intersect their respective lines drawn from line 8′–8 in Figure 1, obtaining the freehand curved ellipse 2′ to 14′.

Draw lines from points 3′–11′ and 5′–9′ on the slant crotch line in Figure 1, to Figure 2, to intersect the lines drawn from 11 to 12, 9 to 10, 7 to 8, 5 to 6, and 3 to 4. This will be the freehand curved line 1 to 13 which represents the crotch.

Draw lines from points 13, 11′, 9′, 7′, 5′, 3′, and 1 in Figure 2 to Figure 3. Transfer the distance from point 1–13 to point 3′–11′ on the slant crotch line in Figure 1 to lines 3′ and 11′ in Figure 3. Transfer the distances from point 1–13 to points 5′–9′ and 7′ on the slant crotch line in Figure 1 to lines 5′, 7′, and 9′ in Figure 3, obtaining the freehand curve 1 to 13.

To obtain the slant true-length lines to lay out the taper pattern as in Figure 4, transfer the distances 13 to 14, 14 to 15, 15 to 16, 16 to 17, 17 to 18, 18 to 19, 19 to 20, 20 to 21, 21 to 22, 22 to 23, 23 to 24, and 24 to 1 from the bottom view in Figure 2 to the base line of triangle *A′* in Figure 1. Transfer the distances 2 to 3′, 3′ to 4, 11′ to 10, and 12 to 11′ from Figure 2 to line 3′–11′ at triangle *A′* in Figure 1. Transfer the distances 4 to 5′, 5′ to 6, 9′ to 8, and 10 to 9′ from Figure 2 to line 5′–9′ at triangle *A′* in Figure 1. Transfer the distances 6 to 7′ and 8 to 7′ from Figure 2 to line 7′ in Figure 1 at the true-length triangle *A′*.

To obtain the slant true-length lines to lay out the T pattern as in Figure 5, transfer the distances 1 to 2′, 3 to 4, 4 to 1, 5 to 6, 6 to 3, 7 to 8, 8 to 5, 8 to 9, 9 to 10, 10 to 11, 11 to 12, 12 to 13, and 13 to 14′ from Figure 2 to the base line of triangle *B′* in Figure 1. Transfer the distances 2′ to 3′, 3′ to 4′, 14′ to 11′, and 11′ to 12′ from Figure 2 to line 3′–11′ at triangle *B′* in Figure 1. Transfer the distances 4′ to 5′, 5′ to 6′, 12′ to 9′, and 9′ to 10′ from Figure 2 to line 5′–9′ at triangle *B′* in Figure 1. Transfer the distances 6′ to 7′, 7′ to 8′, and 10′ to 7′ from Figure 2 to line 7′ in Figure 1, triangle *B′*.

Lay out the patterns in the same manner as in Plate 104.

PLATE 105

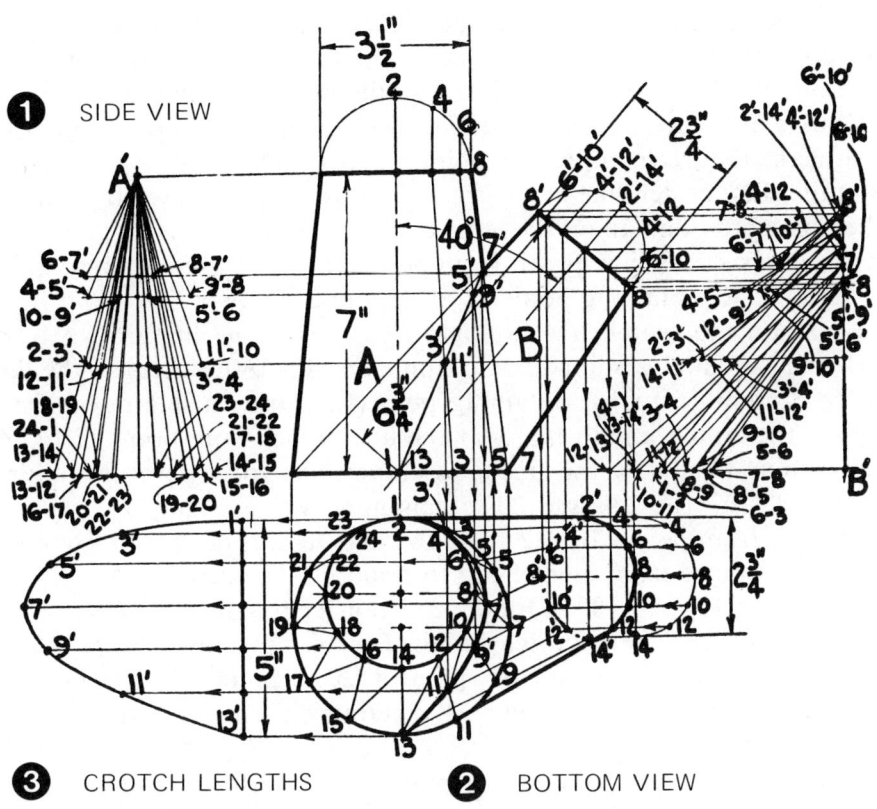

① SIDE VIEW

③ CROTCH LENGTHS ② BOTTOM VIEW

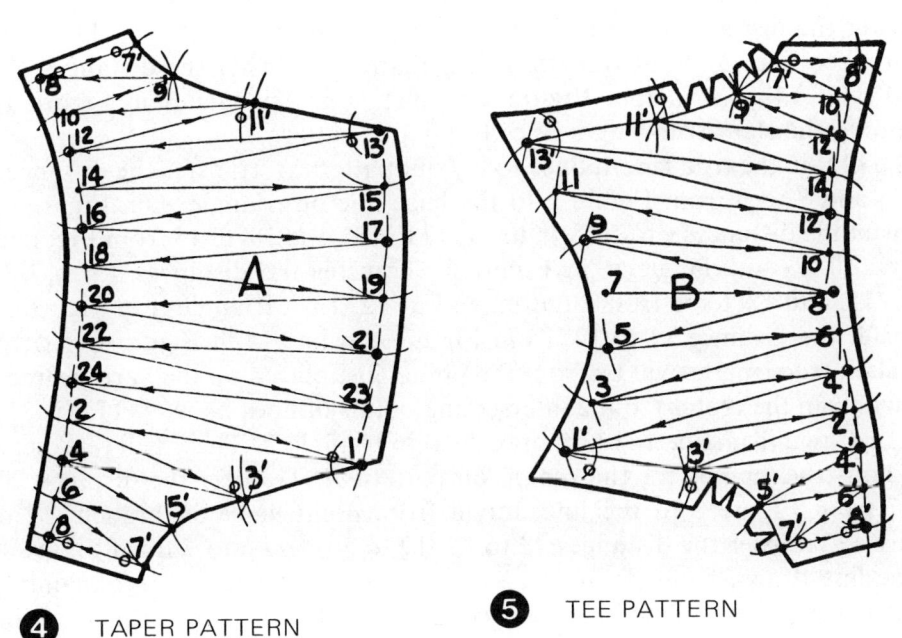

④ TAPER PATTERN ⑤ TEE PATTERN

PLATE 106 THREE-PRONGED BRANCH, TWO AT 35 DEG.

The procedure for this plate is identical to that for Plate 105. Draw the side view as in Figure 1. To obtain the half bottom, draw a line down from each point on line 1–13 at branch *A* to intersect the respective lines drawn from the half circle in Figure 2 (equal in diameter to the half circle on branch *A*). This will make the freehand curved oblong 7 to 19. Draw a line from point 10–18 at the slant crotch line in Figure 1 to cross the line drawn from point 9 on the oblong to point 10 on the large circle, obtaining point 10′; continue the line to intersect the line drawn from point 17 on the oblong to point 18 on the large circle, obtaining point 18′ in Figure 2. Draw a line down from point 12–16 on the crotch line in Figure 1 to cross the line drawn from point 11 to 12 obtaining point 12′. Continue line 12–16 to intersect the line drawn from point 15 to 16, obtaining point 16′. Draw a line down from point 14 at the crotch line to intersect the line drawn from point 13 to 14, obtaining point 14′. Draw the freehand curved line from point 8 crossing points 10′, 12′, 14′, 16′, 18′, and 20, obtaining curve in Figure 2.

Draw lines from points 10′, 12′, 14′, 16′, and 18′ in Figure 2 to Figure 3. Transfer the distance from point 8–20 to point 10–18 on the crotch line in Figure 1 to lines 10′ and 18′ in Figure 3. Transfer the distance from point 8–20 to point 12–16 on the crotch line in Figure 1 to lines 12′ and 16′ in Figure 3. Transfer the distance from point 8–20 to point 14 on the crotch line in Figure 1 to line 14′ in Figure 3, obtaining the true lengths for the crotch on patterns *A* and *B*.

To obtain the true-length lines for pattern *B*, transfer the distances 19′ to 20 and 8 to 9′ from Figure 2 to the base line on triangle *B* in Figure 1. Transfer the distances 9′ to 10′, 10′ to 11′, 17′ to 18′, and 18′ to 19′ from Figure 2 to the line drawn from point 10–18 to triangle *B* in Figure 1. Transfer the distances 11′ to 12′, 12′ to 13′, 15′ to 16′, and 16′ to 17′ from Figure 2 to line 12–16 at triangle *B* in Figure 1. Transfer the distances 13′ to 14′ and 14′ to 15′ from Figure 2 to line 14 at triangle *B* in Figure 1, obtaining the slant true-length lines to lay out pattern *B*.

To obtain the true-length lines to lay out pattern *A*, transfer the distances 7 to 8 and 8 to 9 from Figure 2 to the base line on triangle *A* in Figure 1. Transfer the distances 9 to 10′, 10′ to 11, 17 to 18′, and 18′ to 15 from Figure 2 to line 10–18 at triangle *A* in Figure 1. Transfer the distances 15 to 16′, 16′ to 13, and 12′ to 13 from Figure 2 to line 12–16 at triangle *A* in Figure 1. Transfer the distance 13 to 14′ from Figure 2 to line 14 in Figure 1. NOTE: The slant true lengths will be from the points just placed on the various lines (drawn from the crotch) to the intersecting points of lines 7–19, 9–17, 11–15, and 13 drawn from line 1–13 on branch *A*′ to triangle *A*. These will represent the slant true lengths for the crotch on pattern *A*. Transfer the distance 20 to 17 from Figure 2 to the line drawn from point 9–17 on branch *A*′ in Figure 1. Transfer the distances 22 to 19, 19 to 20, and 6 to 7 from Figure 1 to the line drawn from point 7–19 on branch *A*′. Transfer the remaining lengths from Figure 2 to their respective lines drawn from branch *A*′ in Figure 1. The distances from these points just placed on their respective lines to point *A* will represent the slant true-length lines to lay out the remaining pattern *A* from 8 to 20 and 9 to 17, thus completing the patterns as shown.

PLATE 106

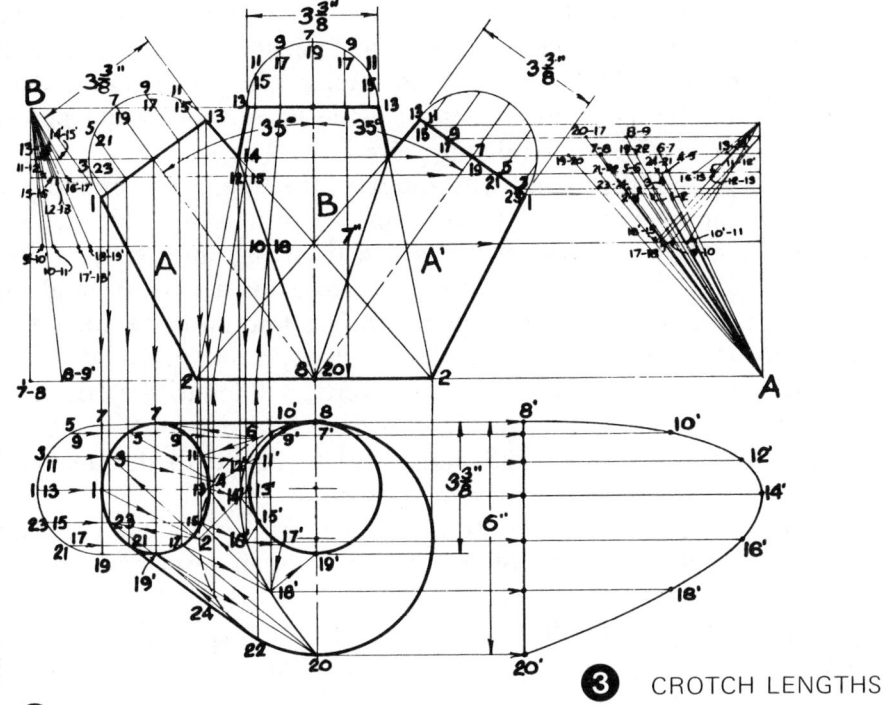

① SIDE VIEW & TRUE LENGTH LINES

② HALF BOTTOM PATTERN

③ CROTCH LENGTHS

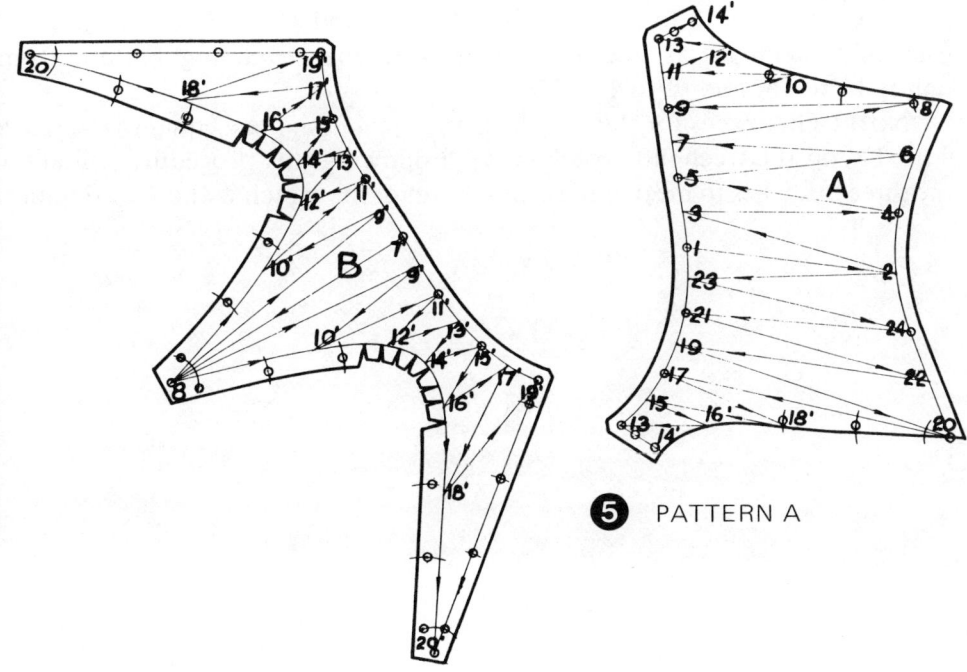

④ PATTERN B

⑤ PATTERN A

PLATE 107 THREE-WAY Y BRANCH

In drawing one third of the side view as in Figure 1, the distance 1 to 9 is equal to one third of the half circle; this will represent one half of the bottom of one branch. Draw the freehand crotch curve from point 9' to 17 in the same manner as for a two-pronged Y branch. Use point A as a center to draw arcs from points 11″, 13″, and 15″ to intersect line 9–A. Draw lines up from these intersecting points to intersect the lines drawn from points 11', 13', and 15', obtaining the freehand curved line 9 to 17.

Draw a 30-deg. line from point A to intersect the freehand crotch curve at point B. Use point A as a center to draw an arc from point 9' to intersect the slant line A–B. Draw a straight line down from this intersecting point to the base line 9'–1. Use point A as a center to draw an arc from B' to intersect the heavy slant line 9–A. Draw a straight line up from this intersecting point to intersect the line drawn across from point B. The distance from the intersection of the horizontal and vertical lines to the freehand curve line, which is represented by 17B, must be added to line 17–18 as represented by 17'. This same length must also be added to the freehand crotch length 15–17, and this is represented by point 17″.

To erect the true-length triangles, transfer the heights of lines 4, 6, 8, and 10 from the half circle at the top of the branch, to their respective lines 3, 5, 7, and 9 at the base. Transfer the slant lines 2 to 3, 3 to 4, 4 to 5, 5 to 6, 6 to 7, 7 to 8, 8 to 9, and 9 to 10 from the branch to the base line, obtaining the slant true-length lines to lay out the pattern from points 9' to 9' and 2 to 10.

To obtain the true lengths for the remaining pattern, transfer the heights of lines 11, 13, and 15 from the 60-deg. line 9–A to their respective lines 10, 12, 14, and 16 at the $3\frac{3}{8}$-in. half circle. Transfer the slant length lines 10 to 11, 11 to 12, 12 to 13, 13 to 14, 14 to 15, and 16 to 17' completing the true-length triangle at line 18–2. Complete the remaining pattern from points 10 to 18, and 9' to 17'–17″.

NOTE: The spaces for the crotch curve 9' to 17'–17″ are equal to the space 9' to 17″ on the freehand crotch curve. Following this procedure will allow the three branches to meet at the center point of the crotch without difficulties.

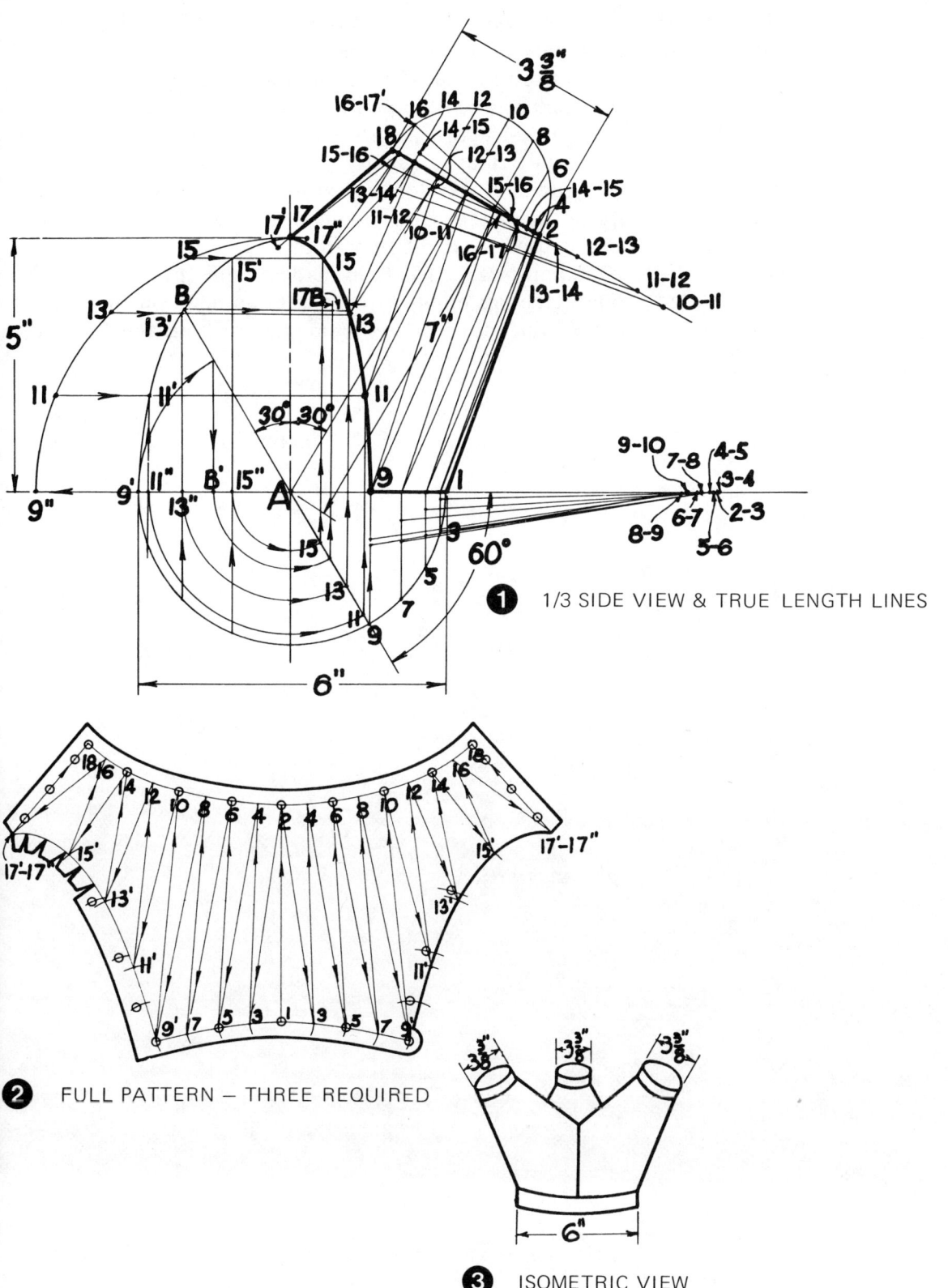

PLATE 108　FOUR-PRONGED BRANCH FITTING

Draw one quarter of the top view as in Figure 1. Divide the small circle and the large quarter circle into as many spaces as desired.

Transfer the lengths of the slant lines from points A and B to the small circle, and the difference between the points on the small and large circles to the base line of the true-length triangle in Figure 2.

To lay out the full pattern as in Figure 3, follow the same procedure as used for a square-to-round to develop the pattern from points 6 to 6 and B to B'. To complete the pattern from points 6 to 12 and B to 11, follow the same procedure as used for a round-to-round taper joint off center in Plate 25. NOTE: No lap allowance has been made on line A to B'. This has been omitted to allow the lap on A–B (bent out 180 deg.) on the mating branch to hook over this portion and eliminate riveting the branches together.

PLATE 108

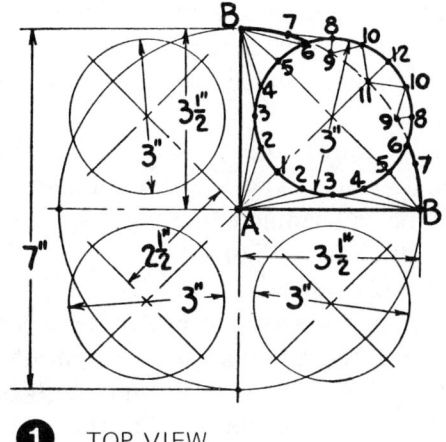

① TOP VIEW

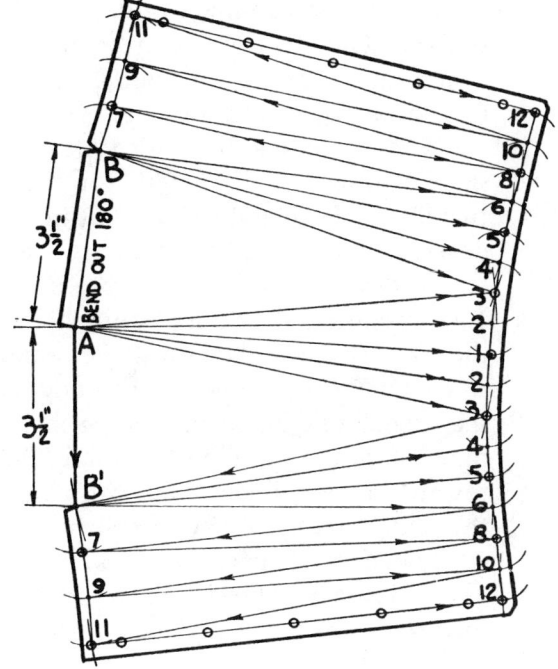

③ FULL PATTERN — FOUR REQUIRED

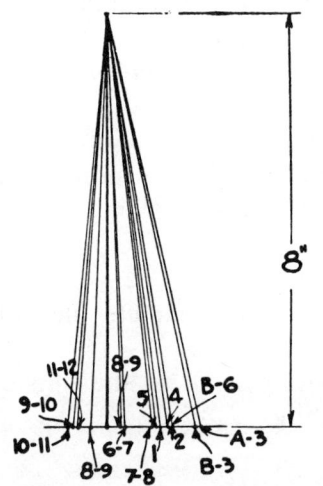

② TRUE LENGTH LINES

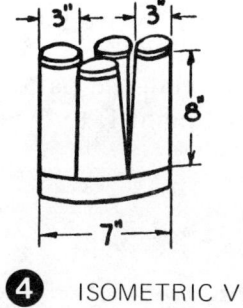

④ ISOMETRIC VIEW

PLATE 109 STREAMLINED TWO-WAY BRANCH

This type of streamlined fitting may be used as a T or a Y branch. Because of the extended angle of convergence the air may enter the main branch with a minimum amount of whirls and eddy currents which would retard the flow of air or create friction.

NOTE: The length of this type of fitting cannot be predetermined in the same manner as the length for the conventional-type T or Y branches; therefore, the length of this fitting will be determined in the process of developing the profile and the true-length lines as in Figure 1.

To develop the side view profile in Figure 1, draw the straight line 1 to 1 representing the ⅜-in. width of the flat portion of the crotch between the two branches, then draw the center line A and a line at point 10 parallel to center line A temporarily to any length desired. Now draw the width of the branch opening from point 1 to 11 on a 15-deg. angle equal to the given dimension, then draw a line on a 90-deg. angle from line 1–11 at point 11 toward the base line to intersect with line 10 (which was drawn parallel with the center line), thereby establishing the length of the fitting, and point 10.

Transfer the height of lines 3, 4, 5, 7, and 9 from the half circle at the 3½-in. branch to lines 2, 6, 8, and 10 on the half circle at the base. Transfer the slant lines from the branch profile to the base line and obtain the slant true-length lines.

To lay out the half pattern in Figure 2, draw line 1 to 1 the width of the flat crotch; then use as a radius the slant true-length line from point 2 on the large half circle in Figure 1 to point 1–2 on the base line. Next use each point 1 in Figure 2 as a center point to strike arcs to cross each other at point 2.

Use the remaining true-length lines in Figure 1 to complete the half pattern in Figure 2.

NOTE: When forming the patterns, place a sharp bend to the outside from point 2 to each point 1, to facilitate assembling.

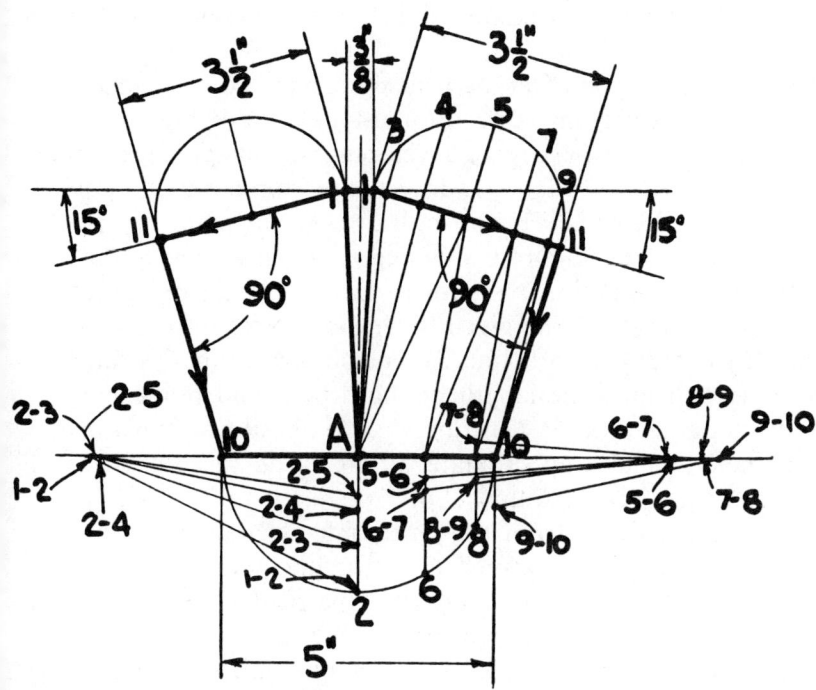

1 SIDE VIEW & TRUE LENGTH LINES

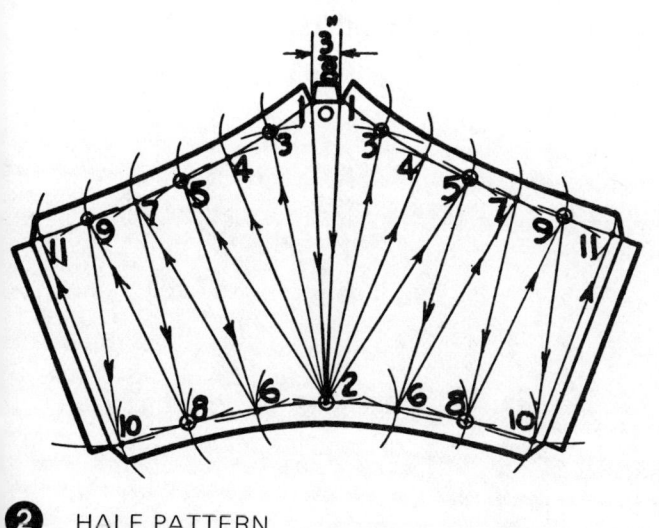

2 HALF PATTERN

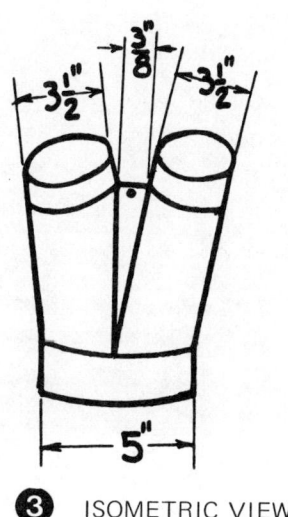

3 ISOMETRIC VIEW

PLATE 109

PLATE 110 STREAMLINED BRANCH FITTING

The procedure for this type of streamlined fitting is the same as that in Plate 109.

Since the length of this type of branch fitting cannot be predetermined in the same manner as the regular conventional-type T or Y branches, the length will be determined in the process of developing the side profile (Fig. 1).

To develop the side profile in Figure 1, draw a line from point 2 to point 1 temporarily to any desired length. Then also draw a line at point 21 parallel to line 2–1 to any desired length. Draw a 90-deg. angle line from point 1 to 12, and 12 to 13 to the dimensions as shown. Draw a line from point 13 to point 22 at a 15-deg. angle to the given dimension. Now draw a 90-deg. line from point 22 to intersect with line 21 (which was drawn parallel to line 2–1), thereby establishing the length of the fitting, and point 21.

NOTE: The distance from point 21 to point *A* is about one third of the base or main diameter, or a proportionate difference in the diameters of the two branch openings, such as *B* and *C*.

Transfer the heights 3 to 11 from the half circle at branch *B*, and the heights 14 to 20 at the half circle branch *C* to the lines on the larger half circle at the base. Transfer the slant lines from the profile of branches *B* and *C* to the base line and obtain the slant true-length lines.

To lay out the half pattern as in Figure 2, draw line 2 to 1 equal to the length of line 2 to 1 in the side profile in Figure 1. Use the remaining true-length lines to complete the half pattern. NOTE: The length of line 21 to 22 in Figure 2 is equal to the length of line 21 to 22 in Figure 1.

NOTE: When forming the patterns, place a sharp bend to the outside from point *A* to 12, and *A* to 13, to facilitate assembling.

PLATE 110

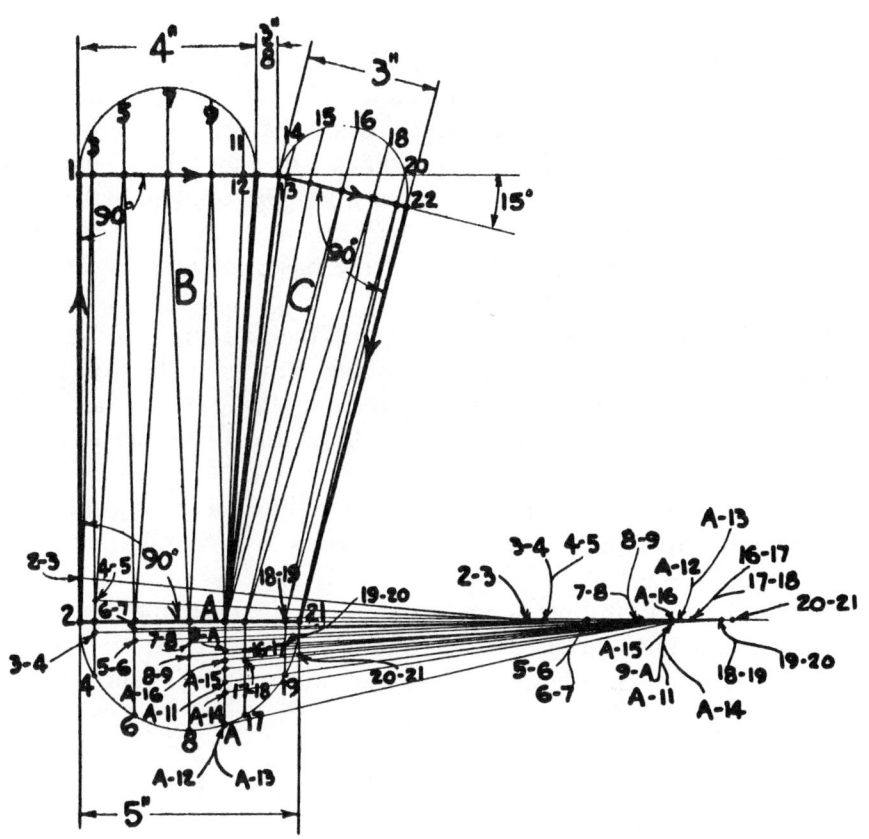

1 SIDE VIEW & TRUE LENGTH LINES

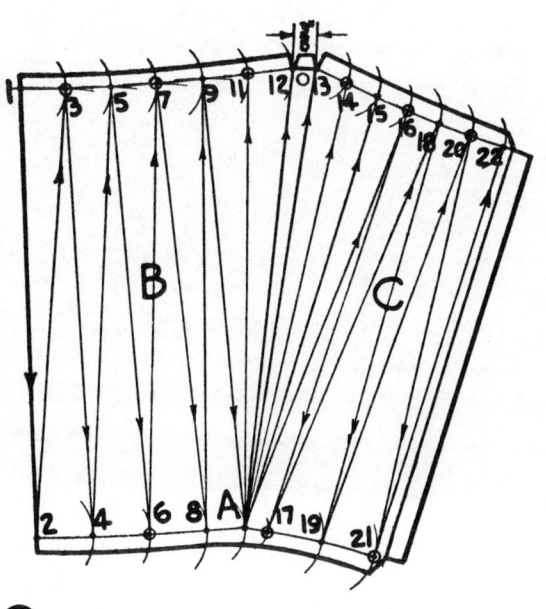

2 HALF PATTERN

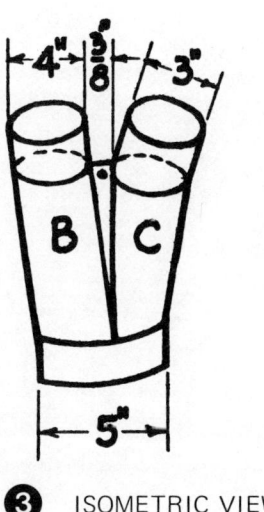

3 ISOMETRIC VIEW

PLATE 111 STREAMLINED THREE-WAY BRANCH FITTING

The procedure for this fitting is the same as the procedure followed in Plate 110, the two-way streamlined branch, except that this has one secondary branch on each side of the submain branch B.

To lay out the side view as in Figure 1, draw the center line 1–2 and line 14 parallel to any length desired. Draw lines 1 to 5 and 5 to 6 equal to the dimensions shown. Draw line 6 to 15 on a 15-deg. angle equal to the dimensions representing the diameter of the branch. Draw a 90-deg. line squaring from line 6–15 at point 15 toward the base, intersecting line 14, thus establishing the length of the fitting.

Transfer the height of line 1 and 3 to line 2 on the half circle at the base. Transfer the height of lines 3, 4, 7, 8, and 9 to line A on the half circle at the base. Transfer the height of lines 9 and 11 to line 10. Transfer the height of lines 11 and 13 to line 12. Transfer the height of line 13 to line 14 at the base.

Transfer the slant lines on branch B to the base line, obtaining the slant true-length lines to lay out the pattern from points 5 to 5 and A to A on section B in Figure 2.

Transfer the slant length lines from branch C to the base line, obtaining the slant true-length lines to lay out the remaining pattern from 5 to 15 and A to 14 on section C, in Figure 2, completing the pattern as shown.

PLATE 111

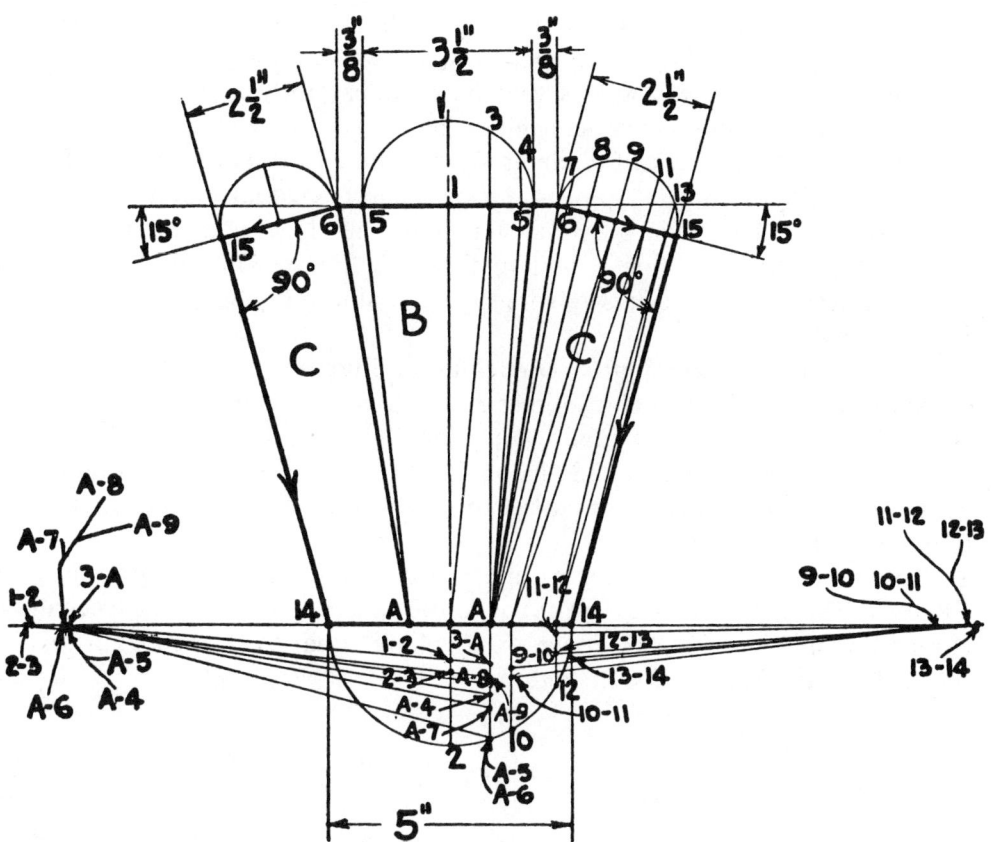

① SIDE VIEW & TRUE LENGTH LINES

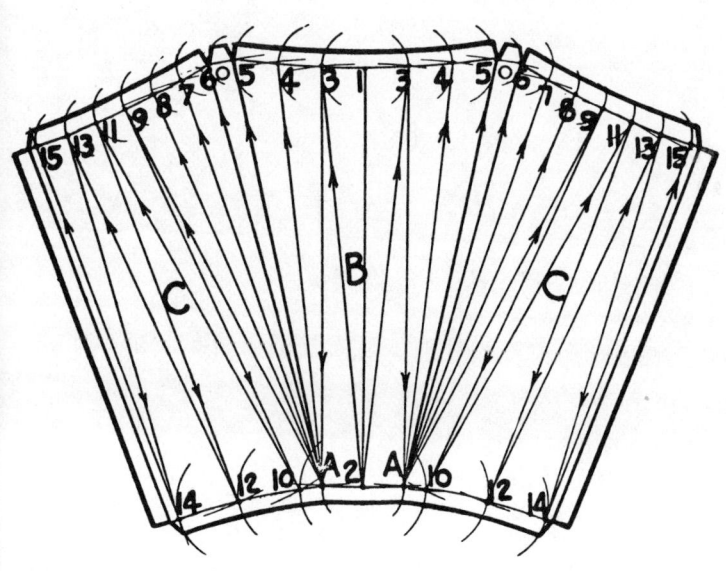

② HALF PATTERN — TWO REQUIRED

PLATE 112 RECTANGULAR 45 DEG. T, INTERSECTING A CYLINDER ON A TANGENT

To lay out the side view as in Figure 1, draw lines A to 1 and B to 1 to any angles desired or to the angles that may appear the most practical. Also draw lines 1 to 5 to any angles that may appear the most practical (the layout man in the shop often determines the angle).

Draw lines from the half circle at the top view to intersect lines 1–5 on the T, obtaining points 2, 3, and 4. Draw a line from each point 1, 2, 3, 4, and 5 in Figure 1 to Figure 2. Transfer the spaces 1 to 5 on the half circle in Figure 3 to the base line 1′ to 5 in Figure 2. Draw a line up from each point to intersect the lines drawn from Figure 1, obtaining the freehand curves 1′ to 5′ and 1″ to 5′.

To obtain the true-length triangles as in Figure 4, draw line B to C equal to the distance B to C in Figure 3. Transfer the heights of lines 2, 3, 4, and 5 on the half circle in Figure 3 to line B–C in Figure 4. Transfer the slant lines from point B to 1, 2, 3, 4, and 5 in Figure 1 to line C–E in Figure 4. Transfer the slant lines from point A to points 1, 2, 3, 4, and 5 in Figure 1 to line C–E in Figure 4.

To lay out the T pattern as in Figure 5, use the slant true-length lines in Figure 4. The spaces 1′ to 5′ are equal to the spaces 1′ to 5′ on the freehand curved line in Figure 2. The spaces 5′ to 1″ are equal to the spaces 5′ to 1″ on the freehand curved line in Figure 2.

PLATE 112

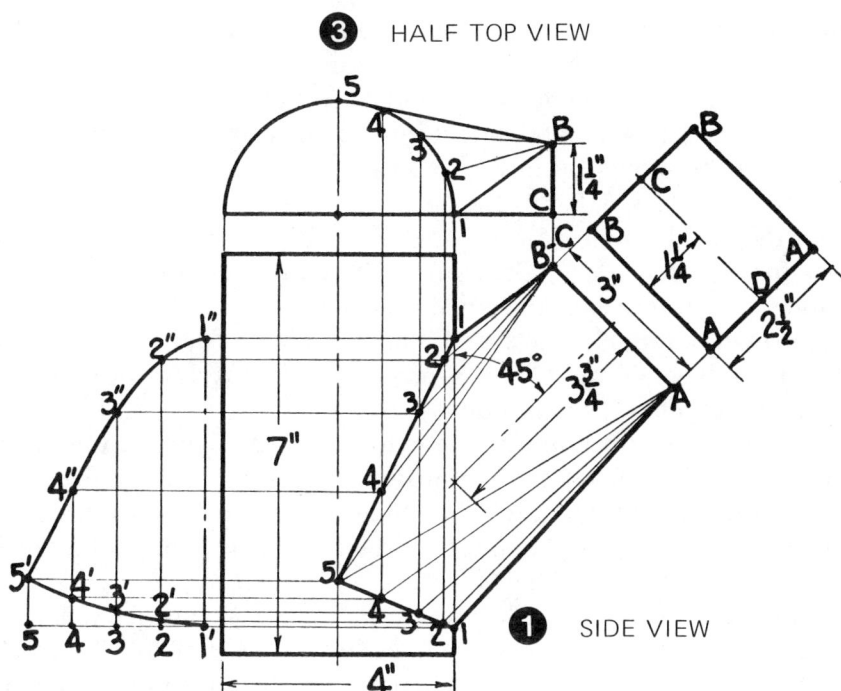

③ HALF TOP VIEW

① SIDE VIEW

② LENGTHS FOR MITER LINE ON TEE PATTERN & HALF CYLINDER CUT OUT

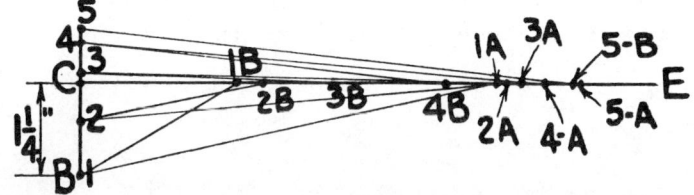

④ TRUE LENGTHS FOR A & B

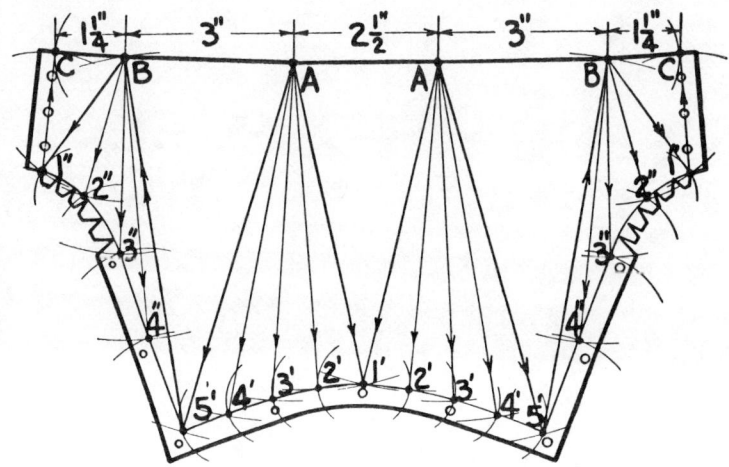

⑤ RECTANGULAR TO ROUND TEE PATTERN

213

PLATE 113   TAPERING T STRAIGHT ON ONE SIDE
INTERSECTING CYLINDER ON A TANGENT

To lay out the side view as in Figure 1, draw the diameter of the cylinder and the width of the T. Draw a line from point 13 tangent to the circle, obtaining point 14. Continue lines 1 to 2 and 13 to 14 to intersect at point A. Draw the half circle 1 to 13, and divide it into equal spaces. Draw lines from point A through the points on line 1 to 13 on the T to intersect the large circle, obtaining points 4, 6, 8, 10, and 12.

To draw the half bottom view of the T as in Figure 2, follow the same principle as for a scalene cone. Draw the half circle equal to the diameter of the T, and divide it into as many equal spaces as desired. Draw lines from point 1–2 through points 11, 9, 7, 5 and 3 on the half circle, intersecting the lines drawn down from points 12, 10, 8, 6, and 4 in Figure 1, thus obtaining the freehand curved lines through points 2 to 14.

To draw the cutout opening for the cylinder as in Figure 3, obtaining the lengths for the miter line on the T, transfer the spaces 2 to 14 on the circle in Figure 1 to the straight line 2 to 14 in Figure 3. Draw lines from points 4′, 6′, 8′, 10′, and 12′ in Figure 2 to intersect their respective lines in Figure 3, obtaining the freehand curved line 2 to 14.

To obtain the slant true-length lines as in Figure 4, transfer the slant lines 2 to 3, 3 to 4′, 4′ to 5, 5 to 6′, 6′ to 7, 7 to 8′, 8′ to 9, 9 to 10′, 10′ to 11, 11 to 12′, 12′ to 13, and 13 to 14 in Figure 2 to their respective lines in Figure 4.

To lay out the T pattern as in Figure 5, use the spaces 1 to 13 on the half circle in Figure 2 for the top diameter 1 to 13, and the spaces 2 to 14 on the freehand curved line in Figure 3 for the miter line 2 to 14 on the base.

The remaining procedure for laying out the T pattern is the same as in Plate 24.

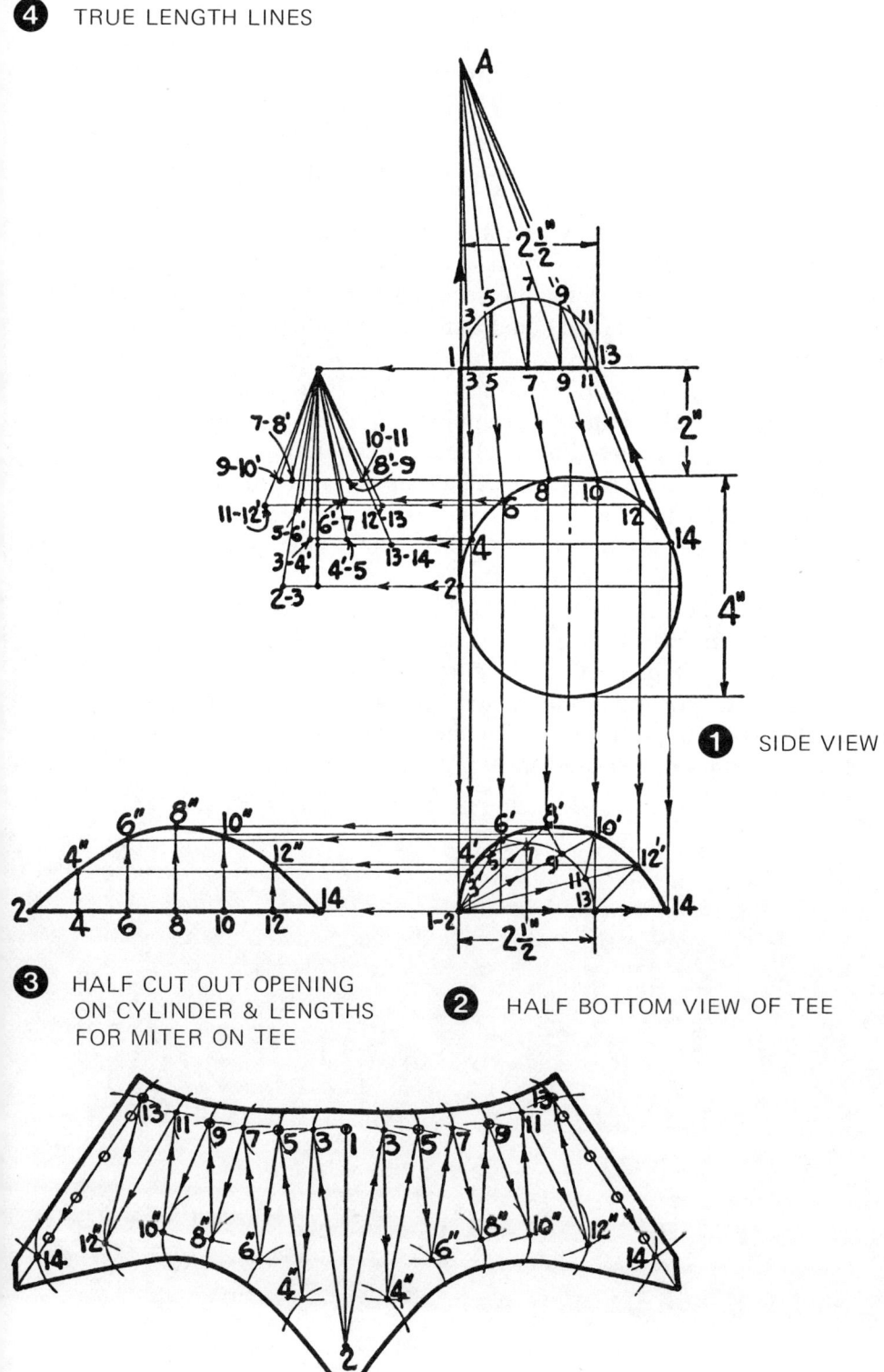

PLATE 113

PLATE 114 TAPERING OFFSET T INTERSECTING CYLINDER ON A TANGENT

To lay out the side view as in Figure 1, draw the height and center line of the T. Draw a line from point 2 tangent to the circle, obtaining point 1. Draw a line from point 14 tangent to the circle, obtaining point 13. Draw lines 13 to 14 and 1 to 2 to intersect at point A. Draw the half circle 2 to 14, and divide it into equal spaces. Draw lines from point A through the points on line 2 to 14 on the T to intersect the large circle, obtaining points 3, 5, 7, 9, and 11.

To lay out the half bottom view of the T as in Figure 2, draw the half circle 2 to 14, and divide it into equal spaces. Draw lines from point A through points 4, 6, 8, 10, and 12 to intersect their respective lines drawn down from points 3, 5, 7, 9, and 11 in Figure 1, obtaining the freehand curved line through points 1 to 13.

Transfer the spaces 1 to 13 on the circle in Figure 1 to the straight line 1 to 13 in Figure 3. Draw lines from points 3′, 5′, 7′, 9′, and 11′ in Figure 2 to intersect their respective lines in Figure 3 obtaining the freehand curved line 1 to 13. This represents the cutout opening on the cylinder and the length of the miter line on the T.

To obtain the slant true-length lines as in Figure 4, transfer the slant lines 1 to 2, 2 to 3′, 3′ to 4, 4 to 5′, 5′ to 6, 6 to 7′, 7′ to 8, 8 to 9′, 9′ to 10, 10 to 11′, 11′ to 12, 12 to 13, and 13 to 14 in Figure 2 to their respective lines in Figure 4.

To lay out the T pattern as in Figure 5, use the spaces 2 to 14 on the half circle in Figure 2 for the top diameter 2 to 14, and the spaces 1 to 13 on the freehand curved line in Figure 3 for the miter line 2 to 14 at the base.

The remaining procedure for laying out the T pattern is the same as in Plate 113.

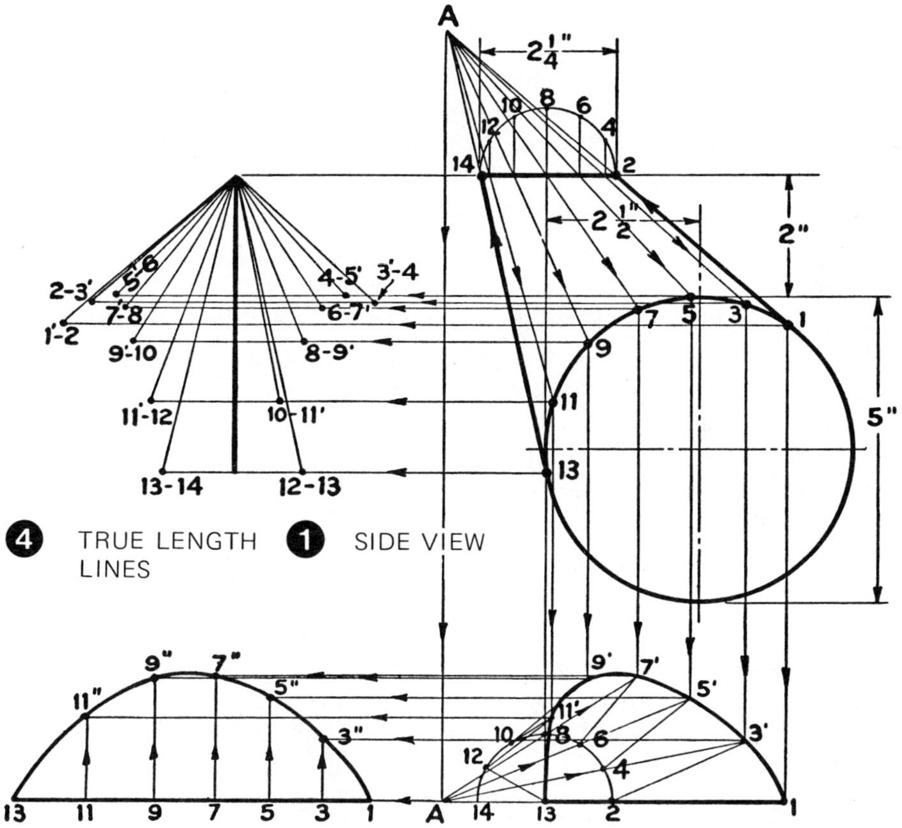

PLATE 114

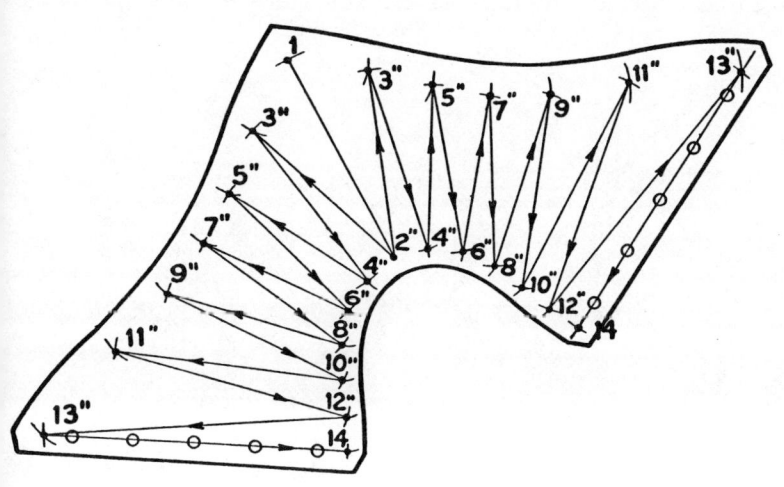

④ TRUE LENGTH LINES
① SIDE VIEW
③ HALF CUT OUT OPENING ON CYLINDER & LENGTHS FOR MITER ON TEE
② HALF BOTTOM VIEW OF TEE
⑤ TEE PATTERN

217

PLATE 115 TAPERING OFFSET T STRAIGHT ON ONE SIDE, INTERSECTING CYLINDER ON A TANGENT

To lay out the front view as in Figure 1, draw a line from point 1 tangent to the circle, obtaining point 1′. Draw a line from point 7 tangent to the circle obtaining point 7′. Draw lines 7–7′ and 1–1′ to intersect at point A. Draw the half circle, and divide it into equal spaces. Draw lines from point A through the points on line 1–7 on the T to intersect the large circle, obtaining points 6′–8′, 5′–9′, 4′–10′, 3′–11′, and 2′–12′.

Lay out the side view in Figure 2, by drawing lines from points 7′, 6′–8′, 5′–9′, 4′–10′, 3′–11′, 2′–12′, 1′ on the circle in Figure 1 to Figure 2. Draw a line from point A in Figure 1 to intersect the vertical line 10′–10 at point B in Figure 2, representing the straight side of the T. Draw lines from point B through points 9–11, 8–12, 7–1, 6–2, 5–3, and 4 to intersect their respective lines drawn from Figure 1, obtaining the freehand curved line through points 1′ to 7′.

Draw lines down from the points on the freehand curved line 1′ to 7′ in Figure 2 to Figure 3. Transfer the spaces 1′ to 7′ on the circle in Figure 1 to line C–D in Figure 3. Draw straight lines from each point on line C–D to intersect their respective lines drawn down from Figure 2, obtaining the freehand curved line 1″ to 7″. This represents the cutout opening on the cylinder and the lengths for the miter line of the T pattern.

Draw lines down from points 1′ to 7′ on the circle in Figure 1 to intersect line C–D in Figure 4. Transfer the distances from line C–D to the various points on the freehand curved line 1″ to 7″ in Figure 3 to their respective lines that intersected line C–D in Figure 4. This will give the freehand curved line through points 1″ to 7″ and complete the bottom view of the T.

To erect the true-length triangles as in Figure 5, the slant lines from the points on the freehand curve to the points on the circle in Figure 4 are transferred to their respective lines drawn from Figure 1 to line E–F in Figure 5.

To lay out the T pattern as in Figure 6, use the spaces 1 to 12 on the circle in Figure 4 for the top diameter 1 to 1, and the spaces 1″ to 7″ on the freehand curved line in Figure 3 for the miter line 1″ to 1″ at the base.

The remaining procedure for laying out the T pattern is the same as in Plate 113.

PLATE 115

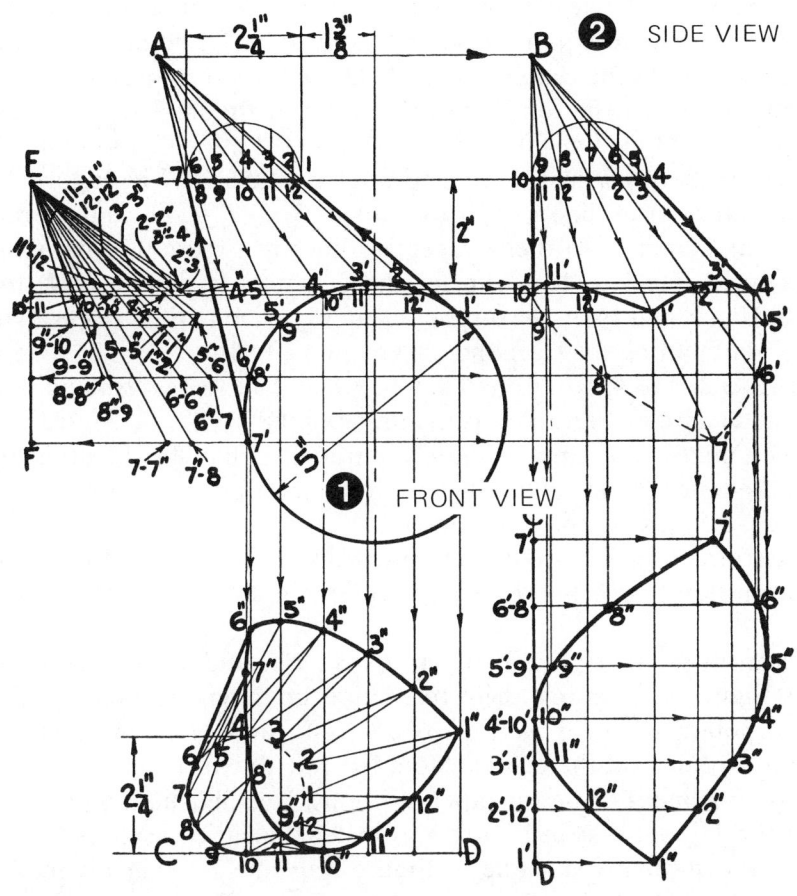

① FRONT VIEW
② SIDE VIEW
③ CYLINDER OPENING
④ BOTTOM VIEW OF TEE
⑤ TRUE LENGTH LINES

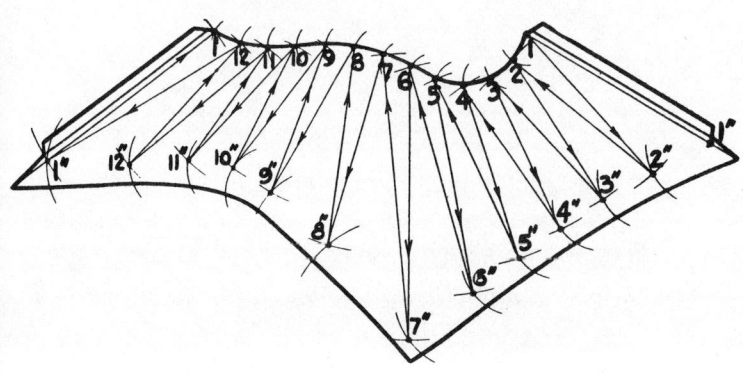

⑥ TEE PATTERN

219

PLATE 116 45-DEG. T ON A TAPER STRAIGHT ON ONE SIDE

Draw the top view as in Figure 1, and the side view as in Figure 2.

To lay out the auxiliary view as in Figure 3, draw the center line 1 from the T in Figure 2 to Figure 3 as represented by E to F. Draw line C–D in Figure 3 at right angles to the center line E–F. Draw lines from points 1–13, 3–11, 5–9, and point 7 on the top line of the taper in Figure 2 to intersect line C–D in Figure 3. Draw lines from points 2–14, 4–12, 6–10, and point 6 on the base line of the taper in Figure 2 to intersect line C–D in Figure 3. Transfer the distances from line C–D to points 1, 3, 5, 7, 9, 11, and 13 on the small circle in Figure 1, to their respective lines in Figure 3, obtaining the freehand curved line 1 to 13. Transfer the distances from line C–D to points 2, 4, 6, 8, 10, 12, and 14 on the large circle in Figure 1, to their respective lines in Figure 3, obtaining the freehand curved line 2 to 14.

Draw lines 1 to 2, 3 to 4, 5 to 6, 7 to 8, 9 to 10, and 11 to 12, crossing the $4\frac{1}{2}$-in., T-diameter circle, Figure 3, obtaining points 1′, 3′, 4′, 5′, 6′, 7′, 8′, 9′, and 10′. Draw a line approximately parallel to line 11–12 tangent to the $4\frac{1}{2}$-in.-diameter circle, obtaining point 11′ and points A and B on the freehand curves, thus completing Figure 3.

Draw a line from point B in Figure 3 to intersect the base line in Figure 2. Draw a line from point A in Figure 3 to intersect the top line of the taper in Figure 2.

Draw lines from points 5′, 9′, 3′, 1′, 4′, 6′, 8′, 10′, and 11′ in Figure 3 into the T in Figure 2, crossing their respective lines on the taper. This will obtain the points 1″, 3″, 5″, 7″, 4″, 6″, 8″, 10″, and 11″, representing the intersection of the T on the taper. (Note the line drawn from point 11′ in Figure 3, crossing line A–B on the taper in Figure 2, establishes point 11″.)

To lay out the T pattern as in Figure 4, transfer the various spaces (7′ to 7′) around the $4\frac{1}{2}$-in.-diameter circle in Figure 3 to line 7′–7′ in Figure 4. Transfer the lengths of the various lines on the T in Figure 2 to their respective lines in Figure 4, obtaining the freehand curved line 7″ to 1″.

The taper joint is laid out in the same manner as in Plate 24.

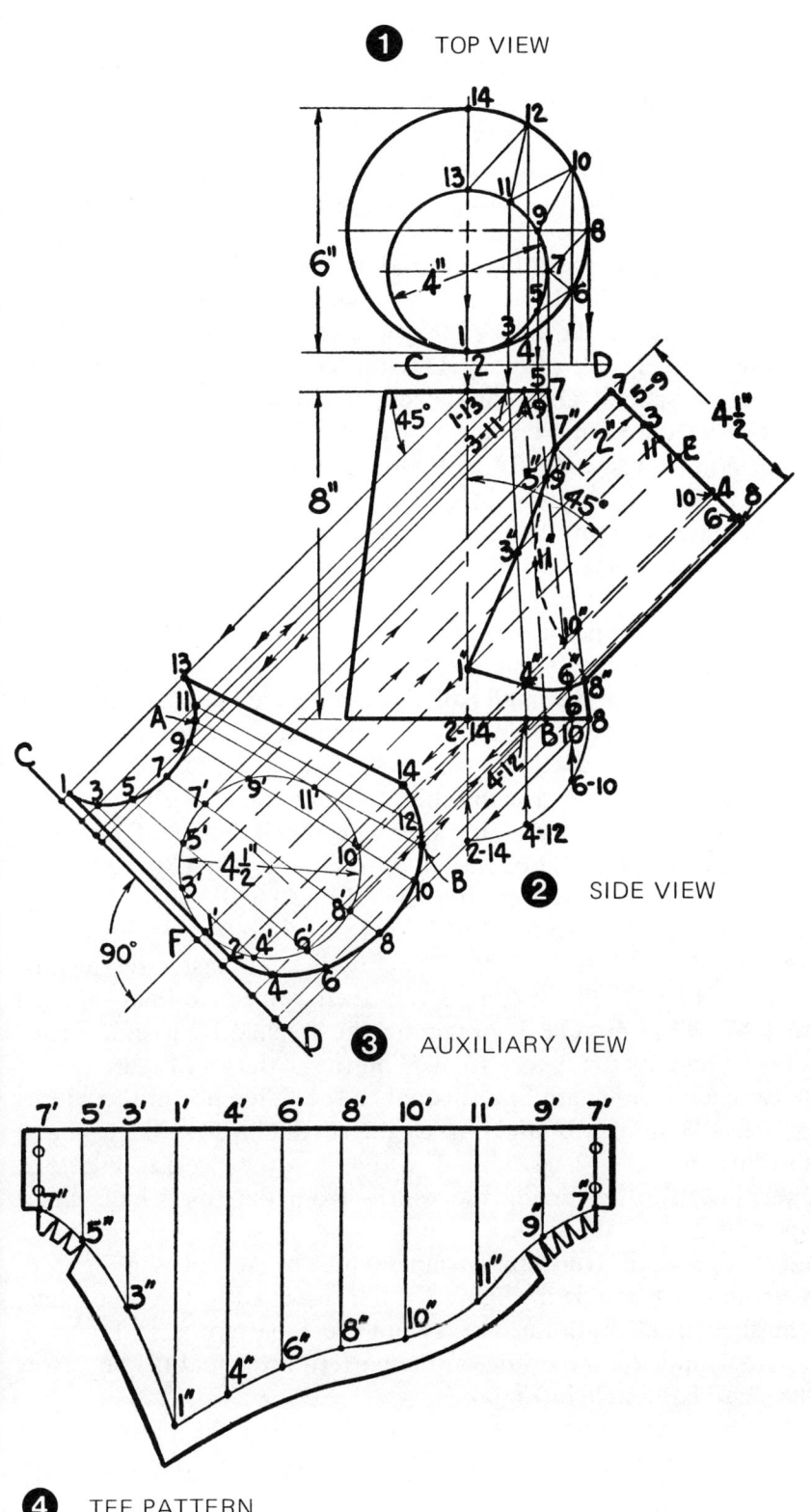

PLATE 117　TAPERING T CENTERED ON CYLINDER, AT 45 DEG.

Draw the center line D to C in Figure 1 equal to the given dimension. Draw a line from point C in Figure 1 to cross the center line $C-D$ in Figure 2. Mark points 4 on each side of the center line $C-D$ equal to the diameter of the T. Draw a line from each point 4 to intersect on a tangent the circle representing the cylinder, obtaining point BA'. Draw a line from point B in Figure 2 to intersect the center line $C-D$ in Figure 1 obtaining point $A4''$. Draw line B to B equal to the width B to B in Figure 2. Draw lines down from points 1 and 7 at the small diameter of the T, passing through points $B-B$ toward the base to establish line $1'''-7'''$ wherever desired (the farther down line $1'''-7'''$ may be the better; this will avoid confusion when the lines are drawn from Fig. 2 to Fig. 1). Draw a half circle at each end, and divide it into equal spaces. Draw lines from the points on line $1'''-7'''$ at the base to connect the points on line 1–7 at the small diameter. Draw a line from each point on line $1'''-7'''$ in Figure 1 to intersect the center line $C-D$ in Figure 2. Transfer the widths of lines 2–3, 4, 5, and 6 on the large half circle at the base in Figure 1 to their respective lines drawn to the center line $C-D$ in Figure 2, obtaining the freehand curved half ellipse 1 to 7 at the base.

Draw a line from each point on line 1–7 at the top diameter of the T in Figure 1 to intersect the center line $C-D$ in Figure 2. Transfer the widths of lines 2, 3, 4, 5, and 6 from the small half circle in Figure 1 to their respective lines drawn to the center line $C-D$ in Figure 2, obtaining the freehand curved ellipses 1 to 7 at the top. Draw a line from each point on the large freehand ellipse to connect their respective numbers on the small freehand ellipse at the top crossing the circle, obtaining points $6'$, $2'$, $5'$, and $3'$. Draw lines from points $6'$, $2'$, $5'$, and $3'$ on the circle in Figure 2 to intersect their respective lines in Figure 1, obtaining the freehand curved line $1''$ to $7''$.

Transfer the spaces $1'-7'$ to $6'$, $6'$ to $2'$, $2'$ to $5'$, $5'$ to $3'$, and $3'$ to $4'$ on the circle in Figure 2 to line $1'-7'$ to $4''$ in Figure 3. Transfer the lengths from each side of line $C-A4''$ in Figure 1 to their respective lines on each side of line $1'-7'-4''$ in Figure 3, obtaining the freehand curve $1''$ to $7''$. This will give the lengths for spaces $1''$ to $7''$ on the pattern in Figure 5.

The rise for each true-length line is equal to the difference in the widths of the lines 2–2, 3–3, 4–4, 5–5, and 6–6 on the small ellipse at the top, and the widths of lines $6'-6$, $2'-2$, $5'-5$, $3'-3$, and 4–4 at the circle in Figure 2. This may be done by transferring the widths from the small half ellipse to their respective lines at the circle.

To avoid confusion, the true-length triangles may be erected as in Figure 4. The lengths from the T profile in Figure 1 are transferred to the center line, obtaining the slant true-length lines to lay out the T pattern as in Figure 5. The spaces 1 to 7 on the curved line on the pattern are equal to the spaces 1 to 7 on the small half circle in Figure 1.

NOTE: A shorter and faster method for developing the pattern for a sharper tapering T may be found on Plate 94 in the book *Short Cuts for Round Layouts*

PLATE 117

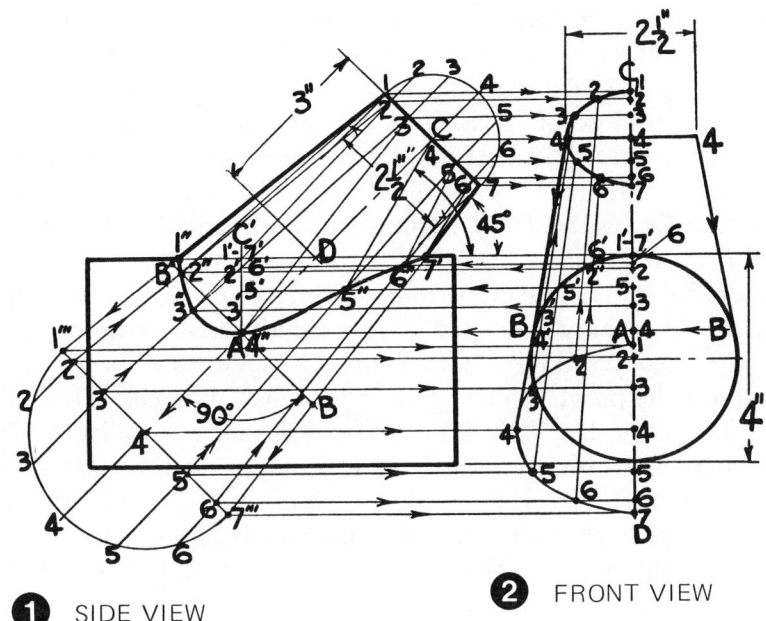

① SIDE VIEW

② FRONT VIEW

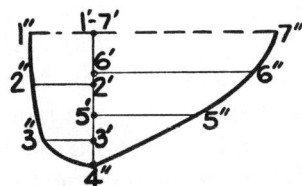

③ HALF CUT OUT OPENING FOR CYLINDER & LENGTHS FOR BASE OF TEE PATTERN

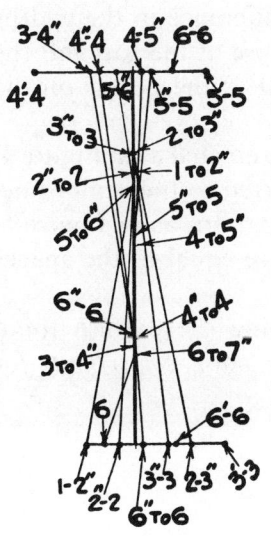

④ TRUE LENGTH LINES

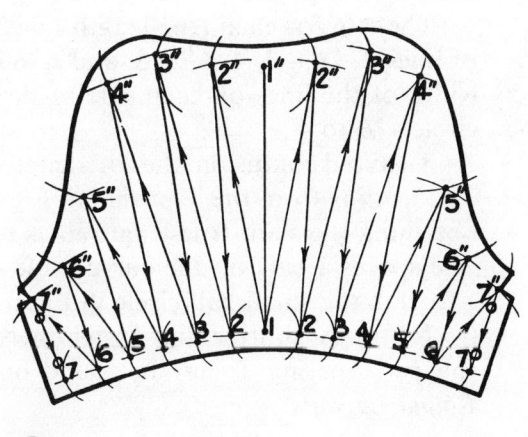

⑤ TEE PATTERN

223

PLATE 118 TAPERING T OFF CENTER, AT 45 DEG.

Draw the center line *F* to *G* in Figure 1 equal to the given dimension. Draw a line from point *G* in Figure 1 to cross the center line *C–D* in Figure 2. Mark points 4 on each side of the center line *C–D* equal to the diameter of the T. Draw a line from point 4*G* to intersect the circle on a tangent at 4″*B*. Draw a line from point 4″*B* in Figure 2 to intersect the center line, obtaining point 4″*A* in Figure 1. Draw line *B–B* to any length desired. Transfer the distance 4*A* to 4″*B* from Figure 2 to each side of point 4″*A*, obtaining points *B–B* in Figure 1. Draw lines down from points 1 and 7 at the small diameter of the T, passing through points *B–B* toward the base to establish line 1‴–7‴ wherever desired. Draw a half circle at each end, and divide into equal spaces. Draw lines from the points on line 1‴–7‴ at the base to connect the points on line 1–7 at the small diameter. Draw a line from each point on line 1‴–7‴ in Figure 1 to cross the center line *C–D* in Figure 2. Transfer the widths of lines 2, 3, 4, 5, and 6 on the large half circle at the base in Figure 1 to their respective lines on each side of the center line *C–D* in Figure 2, obtaining the freehand curved ellipse at the base.

Draw a line from each point on line 1–7 at the top diameter in Figure 1 to cross the center line *C–D* in Figure 2. Transfer the widths of lines 2, 3, 4, 5, and 6 from the small half circle in Figure 1 to their respective lines on each side of the center line *C–D* in Figure 2, obtaining the freehand curved ellipse at the top. Draw a line from each point on the large ellipse to connect their respective numbers on the small ellipse at the top crossing the circle, obtaining points 3″, 5″, 2″, 6″, 6′, 2′, 5′, 3′, and 4′. Draw a line from each point on the circle in Figure 2 to intersect their respective lines in Figure 1, obtaining the freehand curves 1″ to 7″ and 2′ to 6′. Transfer the spaces from point 4″*B* to 4′ on the circle in Figure 2 to the straight line 4″ to 4′ in Figure 3. Transfer the lengths from each side of line *F–H* in Figure 1 to their respective lines on each side of line 4″–4′ in Figure 3, obtaining the freehand curved line which represents the lengths for the spaces 1″ to 7″ and 1″ to 7′ on the pattern in Figure 5.

The rise for each true-length line is equal to the difference in the widths of lines 2–2, 3–3, 4–4, 5–5, and 6–6 on the small ellipse at the top and the width of the lines on each side of the center line *C–D* to the points on the circle 4″*B* to 4′.

To avoid confusion, the true-length triangles may be erected as in Figure 4. The lengths from the T profile in Figure 1 are transferred to the center line, obtaining the slant true-length lines to lay out the T pattern as in Figure 5. The spaces 1 to 7 on the curved line on the pattern are equal to the spaces 1 to 7 on the small half circle in Figure 1.

NOTE: A shorter and faster method for developing the pattern for a sharper tapering T may be found on Plate 96 in the book *Short Cuts for Round Layouts*.

PLATE 118

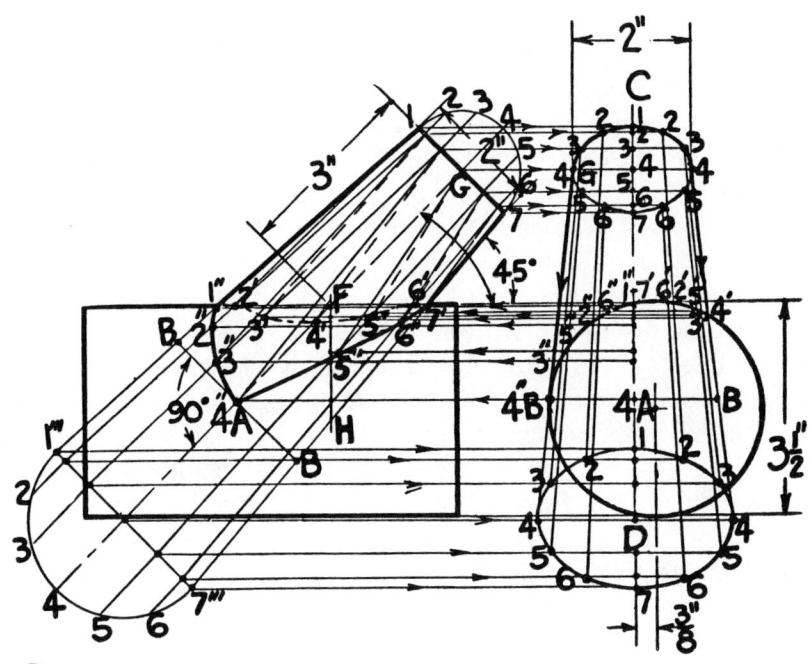

① SIDE VIEW

② FRONT VIEW

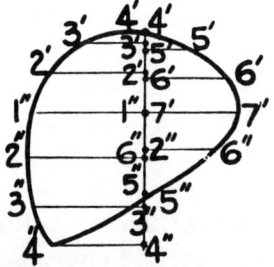

③ CUT OUT OPENING FOR CYLINDER & LENGTHS FOR BASE OF TEE PATTERN

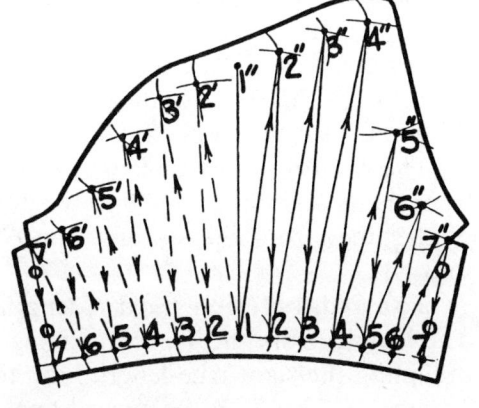

⑤ TEE PATTERN

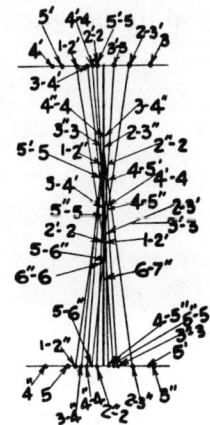

④ TRUE LENGTH LINES

225

PLATE 119 TAPERING T FLAT ON ONE SIDE, AT 45 DEG.

Draw the center line E to F in Figure 1 equal to the given dimension. Draw the small half circle, and divide it into equal spaces. Draw a line from each point on line 1–7 in Figure 1 to cross the straight line G–$4''B$ in Figure 2. Transfer the widths of lines 2, 3, 4, 5, and 6 from the small half circle in Figure 1 to each side of the center line 1–7 in Figure 2, obtaining the freehand curved ellipse 1 to 7. Draw a line from point $4H$ tangent to the circle to intersect the line drawn across from point $4''B$, obtaining point C. Divide in half the distance from $4''B$ to C, obtaining point A. Draw a line from $4''B$ in Figure 2 to intersect the center line in Figure 1, obtaining point $4''A$. Draw line B–C equal to the distance $4''B$ to C in Figure 2. Draw a line from the center point 4 on line $1''$–$7''$ in Figure 1 to intersect the slant line drawn from point $4H$ tangent to the circle in Figure 2, obtaining point 4. Divide the distance 4 to 4 in half, obtaining the center line 1–7 at the base. Transfer the widths of lines 2, 3, 4, 5, and 6 from the half circle in Figure 1 to each side of the center line 1–7 in Figure 2, obtaining the large freehand curved ellipse 1 to 7.

Draw a line from each point on the large ellipse to connect their respective numbers on the small ellipse crossing the circle, obtaining points $5''$, $3''$, $6''$, $2''$, $7''$, $1''$, $6'$, and $2'$–$5'$. Draw a line from each point on the circle in Figure 2 to intersect their respective lines in Figure 1, obtaining the freehand curved lines $1''$ to $7''$ and $2'$ to $6'$.

Transfer the spaces from point $4''B$ to $3'$–$4'$ on the circle in Figure 2 to the straight line $4''$ to $3'$–$4'$ in Figure 3. Transfer the lengths from each side of line $4''A$–D in Figure 1 to their respective lines on each side of line $4''$–$3'$–$4'$ in Figure 3, obtaining the freehand curved line which represents the lengths for the spaces $1''$ to $7''$ and $1''$ to $7'$ on the pattern in Figure 5.

The rise for each true-length line is equal to the difference in the widths from the straight line G–$4''B$ to the points on the small ellipse and the widths of the lines from the straight line G–$4''B$ to the points on the circle $4''B$ to $3'$–$4'$.

To avoid confusion, the true-length triangles may be erected as in Figure 4. The lengths from the T profile in Figure 1 are transferred to the center line, obtaining the slant true-length lines to lay out the T pattern as in Figure 5. The spaces 1 to 7 on the curved line on the pattern are equal to the spaces 1 to 7 on the small half circle in Figure 1.

PLATE 119

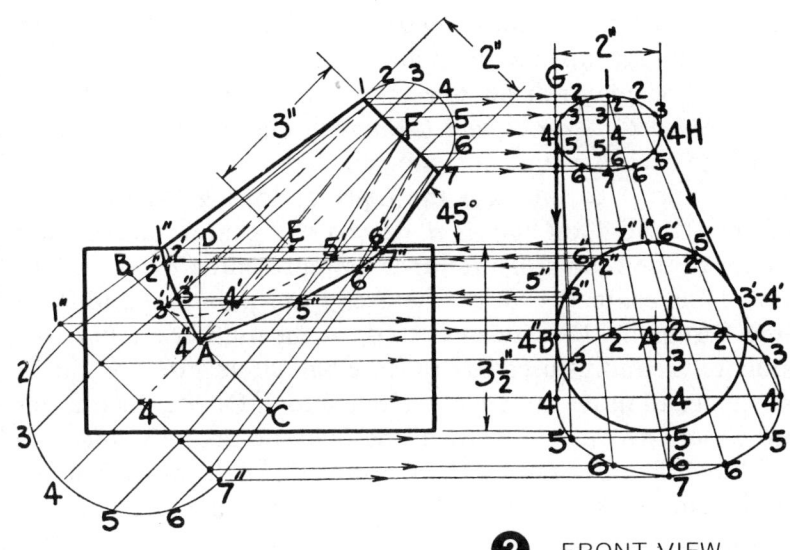

① SIDE VIEW

② FRONT VIEW

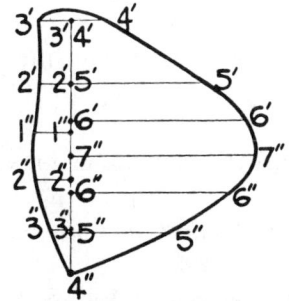

③ CUT OUT OPENING FOR CYLINDER & LENGTHS FOR BASE OF TEE PATTERN

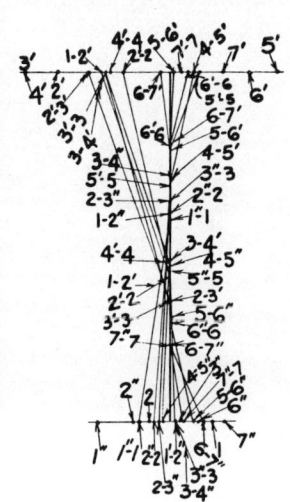

④ TRUE LENGTH LINES

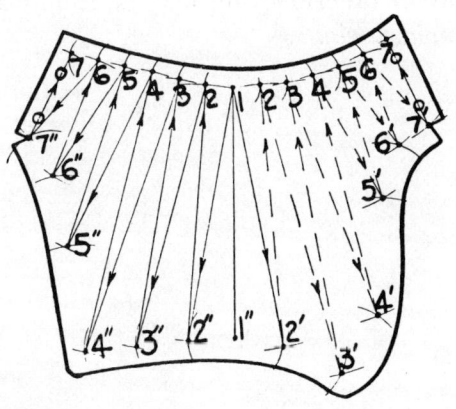

⑤ TEE PATTERN

227

PLATE 120   SHORT METHOD: ROUND T INTERSECTING A TAPER AT 45 DEG.

This plate illustrates a short, simplified method for laying out the pattern for a round T intersecting a taper joint at an angle.

Although this method may not be as accurate as some of the methods shown in the book, *Short Cuts for Round Layouts*, it may be used on many T intersections with an accuracy almost foolproof.

Draw the T in Figure 1 to intersect the taper at 45 deg. At the intersecting points of lines 1' and 7' on line $A-B$, draw a straight line across from each point to intersect the center of the taper as represented by C and D. Use points C and D as centers to draw the arcs from points 1' and 7' to any length desired. Use point 1' as a center to draw the quarter circle 1 to 4 with a radius equal to the radius used to draw the half circle 1 to 7 on the T. Use point 7' as a center to draw the quarter circle 4 to 7. Divide the quarter circle 1 to 4 into equal spaces, and draw a line from points 2, 3, and 4 to intersect the arc drawn from point 1'. Draw a line down from each intersecting point on the arc to intersect line $D-1$. Divide the quarter circle 4 to 7 into equal spaces, and draw a line from points 4, 5, and 6 to intersect the arc drawn from 7'. Draw a line up from each intersecting point on the arc to intersect line $C-7$. Draw a slant line from each intersecting point on line $D-1$ to connect to the intersecting points on line $C-7$. Draw lines from points 2, 3, 4, 5, and 6 on the half circle to intersect their respective lines in the taper to obtain the freehand curve 1' to 7'.

To lay out the T pattern as in Figure 3, draw the line 1 to 1 equal to the desired circumference, and divide it into any equal number of spaces as may be desired. Transfer the length of the lines from line 1–7, the top of the T to the freehand curved line 1'–7' in Figure 1, to their respective lines on line 1 to 1 in Figure 3.

To lay out the pattern for the taper, draw a quarter top view as in Figure 2; then follow the same procedure as for Plate 21.

NOTE: A shorter and faster method for developing the pattern for a sharper tapering joint may be found on Plate 71 in the book *Short Cuts for Round Layouts*.

PLATE 120

1 SIDE VIEW

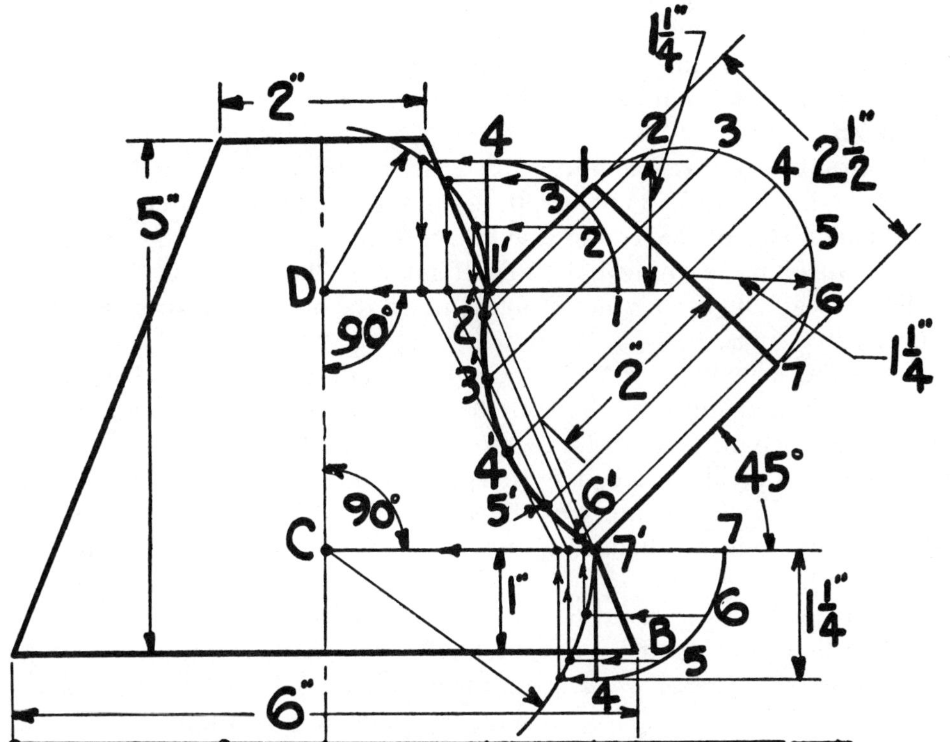

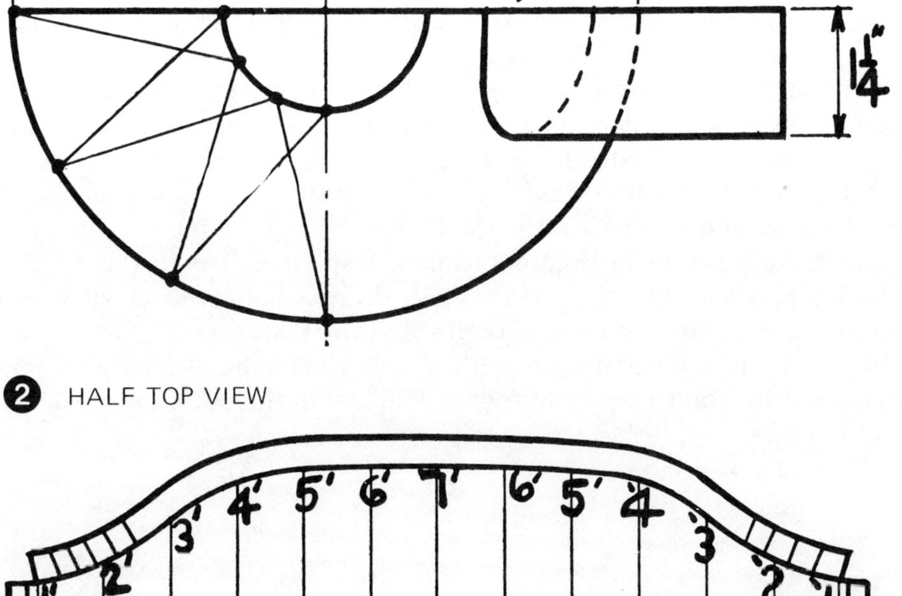

2 HALF TOP VIEW

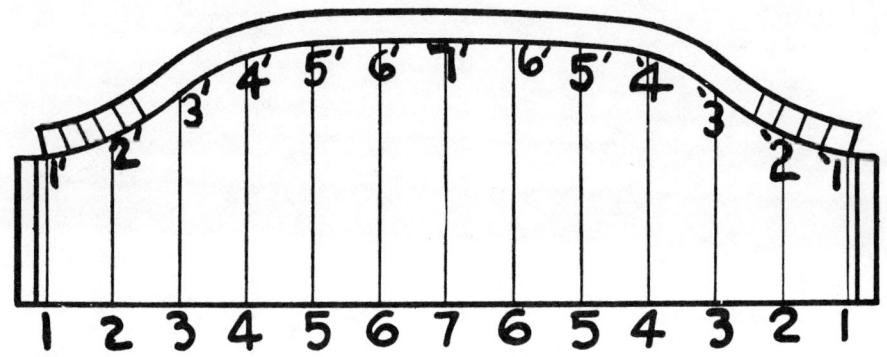

3 TEE PATTERN

229

PLATE 121 ROUND T AT 45 DEG. INTERSECTING A TAPER, ONE SIDE STRAIGHT

The procedures for obtaining the points of intersection of the T on the taper are the same as in Plate 120.

Draw the T to intersect the taper at 45 deg. as in Figure 2. At the intersecting points of line 4–4, represented by A and B, draw a straight line across from each point to intersect the center line of the taper as represented by points $7'$ and $7''$. To avoid confusion, use the distance from point $7'$ to A as a radius to draw the portion of a circle 7 to A at the top and draw the horizontal line 7 to 7 to any length desired. Draw the half circle 1 to 7 equal to the diameter of the T, and divide it into equal spaces. Draw a line from each division point on the half circle to intersect the arc drawn 7 to A. Draw a line down from each intersecting point on the arc to intersect line $A–7'$ as represented by $6', 1', 2', 5', 3'$, and $4'$.

Use the distance $7''$ to B (the lower intersecting point of the T) as a radius to draw the portion of a circle at the base, and draw the horizontal line 7 to 7 to any length desired. Draw the half circle 1 to 7 equal to the diameter of the T, and divide into equal spaces. Draw a line from each division point on the half circle to intersect the large arc. Draw a line up from each intersecting point on the arc to intersect line $B–7''$ as represented by $6'', 5'', 4''$, and $3''$.

Draw a line from point $6''$ on line $A–7'$ crossing point $6''$ on line $B–7''$. Draw a line from point $5'$ crossing line $5''$. Draw the remaining lines $4', 3', 2'$, and $1'$ from line $A–7'$ to cross through the points $4'', 3'', 2''$, and $1''$ on line $B–7''$ to intersect their respective lines drawn from the half circle 4 to 4 on the T, thus obtaining the freehand curved line which represents the intersecting points of the T on the taper.

NOTE: The above method may not be as accurate as the method in Plate 133 in the book *Short Cuts for Round Layouts*, but it will be accurate enough to be used on many T intersections.

To lay out the T pattern as in Figure 3, draw line 4 to 4 equal to the circumference, and divide it into equal spaces. Transfer the lengths of the lines from the T profile in Figure 2 to their respective lines in Figure 3.

To lay out the pattern for the taper, draw a half bottom view as in Figure 1; then follow the same procedure as for Plate 24.

NOTE: A shorter and faster method for developing the pattern for a sharper tapering joint may be found on Plate 89 in the book *Short Cuts for Round Layouts*.

PLATE 121

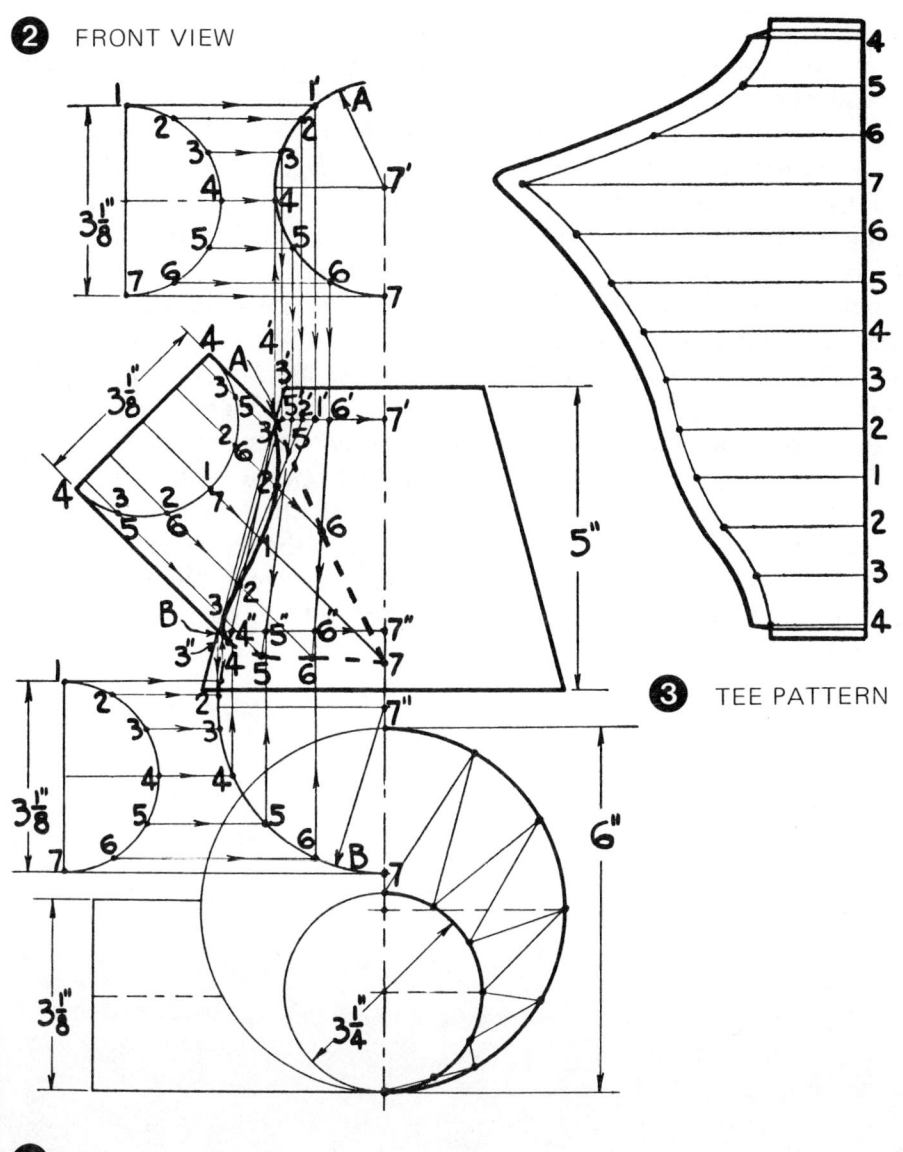

PLATE 122 TWO ELBOWS ON A TAPERING BASE

To draw the side view as in Figure 1, follow the regular procedure of constructing a four-piece elbow. Draw the tapering base to equal the height and diameter as shown. Draw a line from points 13 to 14 and 1 to 2, thereby adding the bottom segment of the elbow to the tapering base. Draw the half circle on the center line 1'–13' on the elbow segment A, and divide it into equal spaces. Draw a line from each division point to intersect line 1–13.

To lay out the end segment C in Figure 2, draw line 7 to 7 equal to the circumference. Transfer the length of the lines from line 1'–13' to the slant miter line 1–13 on the elbow in Figure 1, to their respective lines in Figure 2, obtaining the freehand curved line 7' to 7'.

The center-segment pattern in Figure 3 may be laid out by doubling the end segment in Figure 2. The notched portion between 3' and 3' will be cut off on one A pattern to facilitate connecting the two A segments at point 1.

Transfer the height of lines 3, 5, 7, 9, and 11 from the half circle on the elbow, Figure 1, to lines 2, 4, 6, 8, 10, and 12 on the half circle at the base.

Transfer the slant lines from the tapering base to the base line, obtaining the slant true-length lines to lay out the pattern in Figure 4. Note the spaces 1' to 13' are equal to the spaces 1' to 13' on the freehand curved line on the pattern in Figure 3.

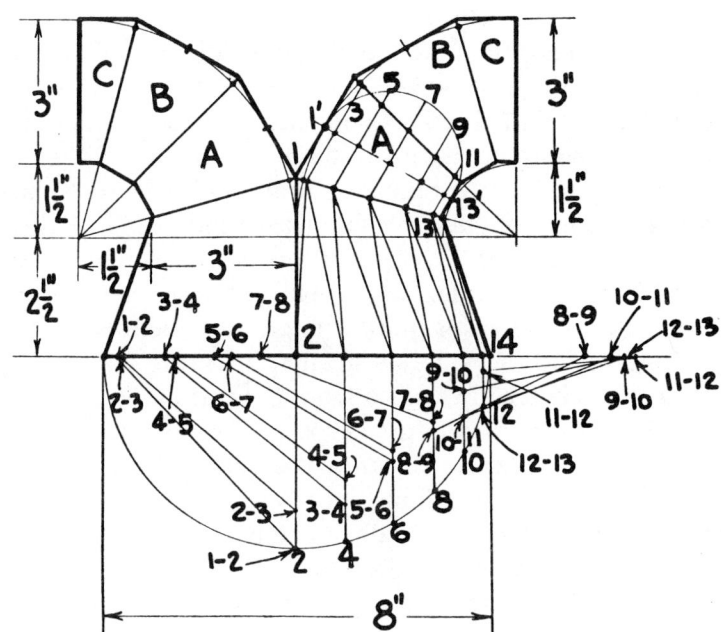

1 SIDE VIEW & TRUE LENGTH LINES

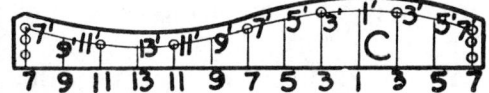

2 END SEGMENT

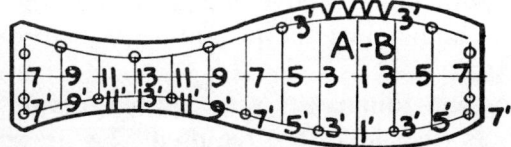

3 CENTER SEGMENT

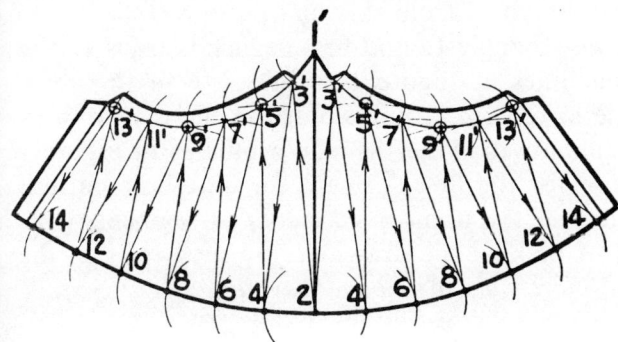

4 ROUND TAPERING BASE PATTERN

PLATE 123   QUARTER PATTERN FOR ROUND ELBOWS

To obtain the degrees in the first segment, divide the degrees in the elbow or angle by the number of spaces that the heel curve would be divided. Thus, in a 4-piece, 90-deg. elbow, $4 \times 2 = 8$, $8 - 2 = 6$ spaces, $90 \div 6 = 15$ deg. In Figure 1, draw the center line C–D then set half the diameter from point R to A, and draw a straight line up crossing the slant line at B.

To develop the quarter pattern as in Figure 2, draw line F–E to equal one quarter of the circumference, or half the diameter times 1.57; in this case, $2.5 \times 1.57 = 3.925$ or about $3\tfrac{15}{16}$. Then transfer the height A–B from Figure 1 to line F–5 in Figure 2. Draw the quarter circle using F as a center and F to 5 as a radius. Divide this quarter circle 1 to 5 and the base line F–E into the same number of spaces. Draw straight lines through points 2, 3, and 4; then transfer the heights 2, 3, and 4 from the end curve to lines 2, 3, and 4 on line F–E. Draw a freehand curve crossing points 1 to 5, and notch the pattern as in Figure 3. The $\tfrac{3}{16}$-in. allowance at the bottom and one end should be used as a guide in developing the full pattern.

To develop the full pattern as in Figure 4, draw the base line C to equal the full circumference. Transfer the height C–D from Figure 1 to obtain the height C–D in Figure 4. Divide the base line C into four equal spaces, thus representing four quarters. Draw straight lines up, crossing line D. Place the quarter pattern with line E–F resting on line D. Beginning at point D–1, draw the curve up from $1E$ to 5. Turn the quarter pattern over end for end, and draw the curve down from 5 to $1E$. Repeat this procedure but with the curved edge of the quarter pattern facing down toward the base line C. Draw the curved line down from 1 to 5, and again turn the pattern end for end and draw the curved line up from 5 to 1. This completes the pattern with the exception of the allowances for seaming and assembling. If a peening edge is required, draw a parallel line above line D to equal the height of the peening edge allowance, and repeat the procedure as before with the line E–F resting on the new line.

If riveting is preferred and more than four rivets are required, divide the base line C into as many spaces as rivet holes are required, and draw straight lines up crossing the curve as shown at points 1, 2, 3, and 4. Use the same procedure to draw the riveting edge allowance as for peening.

The remaining patterns are obtained from this first pattern.

NOTE: To guarantee the exact angle of the first segment, transfer the heights of the heel and throat lines in Figure 1 to line 5–F in Figure 4, from points F at the base line to represent points 5 at the curve.

NOTE: It is possible to eliminate drawing the side profile as in Figure 1 to obtain the angle or the rise for the first segment for an elbow pattern or any angle pattern regardless of the number of pieces or segments contained in the fitting. Instead of making the drawing, you may refer to the elbow chart and rise table in *Short Cuts for Round Layouts*.

PLATE 123

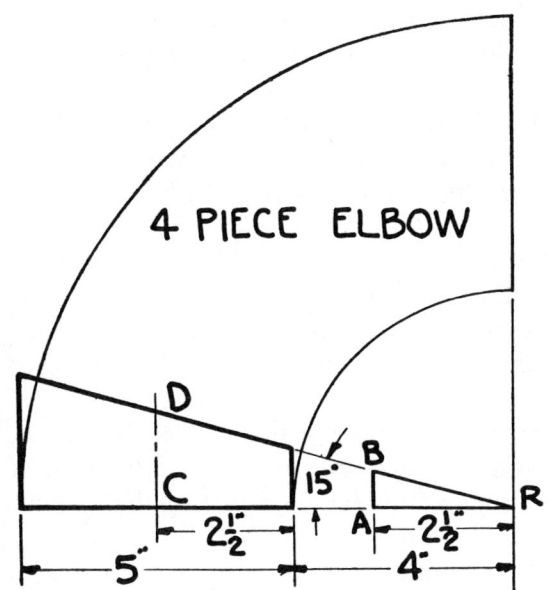

1 SIDE PROFILE

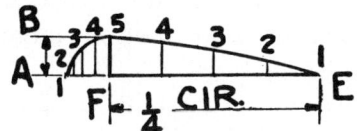

2 ¼ PATTERN

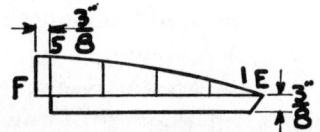

3 NOTCHED PATTERN

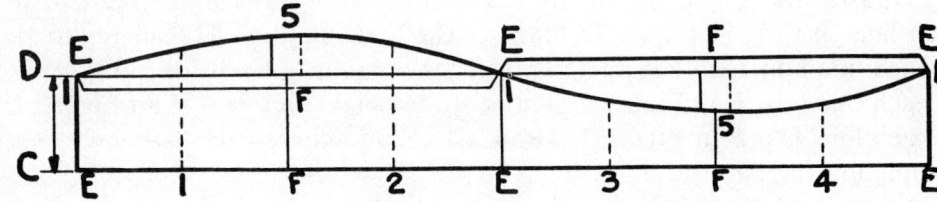

4 FULL PATTERN

PLATE 124 TAPERING-END ELBOW

It is not necessary to draw the full side view as in Figure 1. Draw only segment No. 1 equal in degrees according to the number of segments in the elbow.

To obtain the degree of angle for the first segment, multiply the number of segments by 2, then minus 2, and divide 90 deg. by the remainder.

EXAMPLE: $4 \times 2 = 8$; $8 - 2 = 6$; $90 \div 6 = 15$ deg., which is the angle of first segment.

The procedure for laying out the pattern for segment No. 4, in part, involves the method for laying out an equal taper joint as in Plate 20. Therefore, draw a one-quarter front view of the small end as in Figure 2. Divide the large circle into equal spaces, and draw a line from point 4 to the radius point, obtaining point 3.

Draw the true-length triangle as in Figure 3. The height A to B is equal to the height of the center line $A-B$ on segment No. 1 in Figure 1. Transfer the distances 1 to 2 and 2 to 3 in Figure 2 to the base line of the triangle in Figure 3.

To lay out the quarter pattern as in Figure 4, draw line 2 to 10 equal to one-quarter circumference of the large diameter in Figure 1, and then follow the same procedure as in Plate 123.

The distance 2 to 10 in Figure 5 is equal to one-quarter circumference of the large diameter in Figure 1, and represents the spaces 2 to 10 on the large curve in Figure 7.

The distance 1 to 9 in Figure 6 is equal to one-quarter circumference of the small diameter in Figure 1, and represents the spaces 1 to 9 on the small curve in Figure 7.

To lay out the pattern as in Figure 7, first develop the portion representing an equal taper joint by drawing line 1 to 2 equal to the slant true-length line 1–2 in Figure 3. Use the slant true-length line 2–3 in Figure 3 as a radius, and points 1 and 2 in Figure 7 as centers to strike arcs at points 3 and 4. Continue this procedure in the same manner as in Plate 20. Draw straight lines through points 3–4, 5–6, 7–8, 9–10. On one half of the pattern, extend these lines above points 4, 6, 8, and 10 to any length desired.

Transfer the height of line 10–C on the quarter pattern in Figure 4, to each line 10 to C in Figure 7. Transfer the height of line 8–D in Figure 4, to each line 8 to D in Figure 7. Transfer the height of line 6–E in Figure 4, to each line 6 to E in Figure 7. Transfer the height of line 4–F in Figure 4, to each line 4 to F in Figure 1. Draw a freehand curve from points $2'$ to $2'$ completing the pattern.

NOTE: To lay out the rivet holes, divide the freehand curved line from $2'$ to $2'$ into as many equal spaces as rivet holes may be desired.

To lay out the patterns as in Figure 8 by using the quarter pattern, follow the same procedure as in Plate 123.

PLATE 124

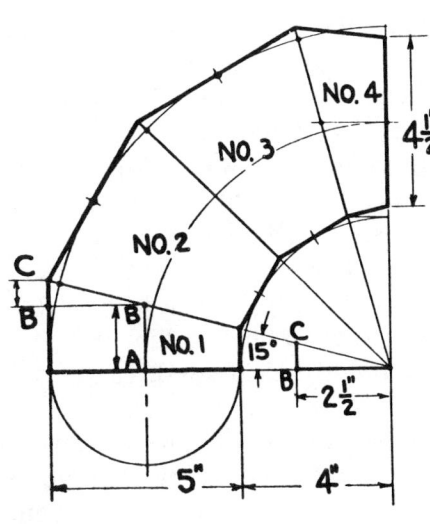

1 SIDE VIEW

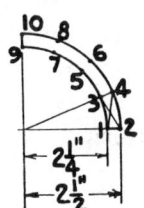

2 ¼ FRONT VIEW OF SMALL END

3 TRUE LENGTH LINES

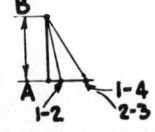

1/4 CIRCUMFERENCE

4 1/4 PATTERN

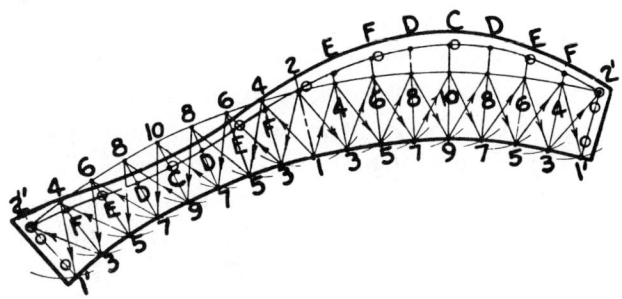

5 LARGE DIAMETER

6 SMALL DIAMETER

7 PATTERN 4

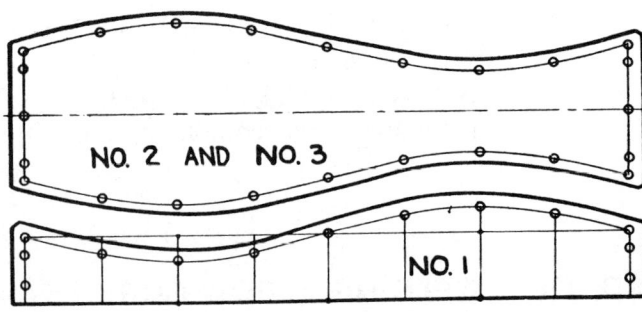

8 PATTERN 1, 2, & 3

237

PLATE 125 CENTER-TAPERED ELBOW

To lay out the side view as in Figure 1, draw the center line *A* to *B*, and divide it in 4 equal spaces (2 × 3 = 6, 6 − 2 = 4). Draw a line from the radius point through each end space to intersect lines 1′ and 2′, obtaining the height of the heel and throat on segments Nos. 1 and 3.

To lay out the pattern as in Figure 2, draw line 8 to 8 equal to the circumference of the small diameter in Figure 1. Transfer the lengths of the lines on segment No. 3 in Figure 1 to their respective lines in Figure 2.

To lay out the pattern as in Figure 3, draw line 7 to 7 equal to the circumference of the large diameter in Figure 1. Transfer the heights of the lines on segment No. 1 in Figure 1 to their respective lines in Figure 3.

NOTE: The quarter-pattern method may be used, if desired, to lay out segment patterns Nos. 1 and 3.

To obtain the true-length lines to lay out the pattern in Figure 4, transfer the lengths of lines 4, 6, 8, 10, and 12 from the half circle at segment No. 3 to their respective lines 3, 5, 7, 9, 11, and 13 on the half circle at segment No. 1. Transfer the slant lines on segment No. 2 to the base line, obtaining the slant true-length lines to lay out the pattern for segment No. 2 as in Figure 4. NOTE: The spaces 1′ to 13′ on the segment pattern No. 2 in Figure 4 are equal to spaces 1′ to 13′ on the freehand curved line on segment pattern No. 1 in Figure 3. The spaces 2′ to 14′ on the pattern in Figure 4 are equal to the spaces 2′ to 14′ on the freehand curved line on segment pattern No. 3 in Figure 2, thus completing the pattern as shown.

PLATE 126 WELDING ELBOW GORE SEAMS, METAL 18 GAUGE AND LIGHTER

This plate illustrates the forming of seam edges for welding elbows with smooth overlapping seams that will add strength to the seams.

The preparation and seam allowance for each gore section is illustrated in Plate 28. When the duct continues from the elbow and the seams welded, then the same $\frac{1}{8}$ inch allowance is made at the end of each end gore.

NOTE: When assembling and aligning of gore sections, the first tack weld must be made at the center of heel or back of elbow. This will facilitate the aligning of the throat to the required degree for that portion or segment of the elbow. Tack weld each side before completing weld.

PLATE 125

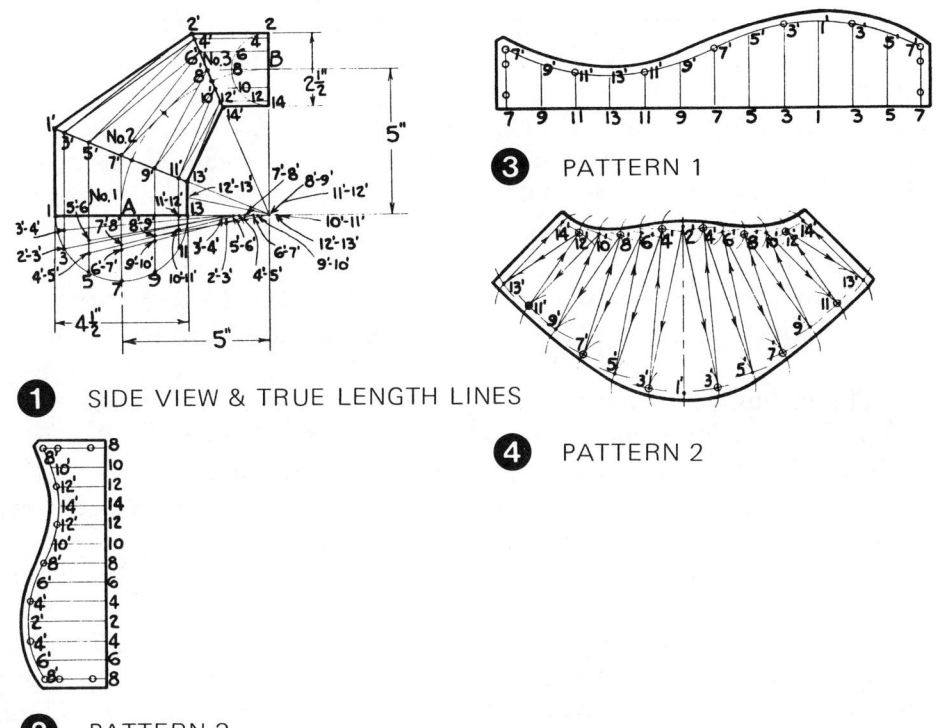

PLATE 126

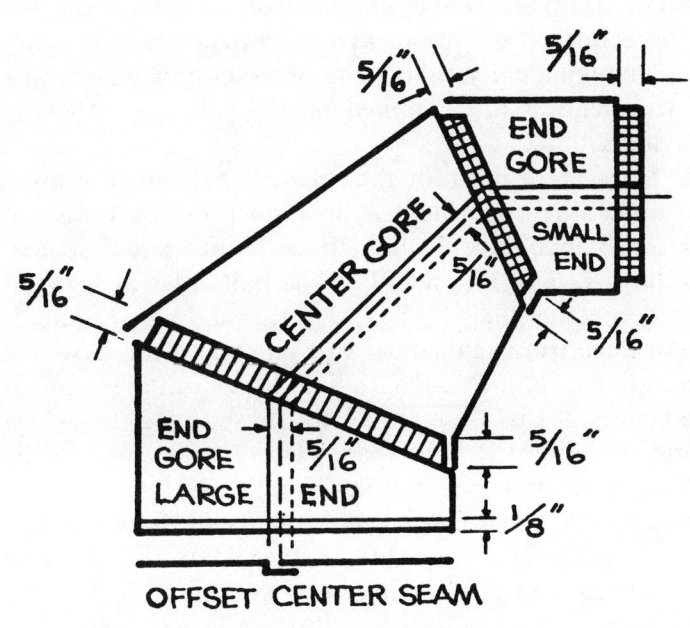

PLATE 127 EQUAL-TAPERING ELBOW WITH CENTER RADIUS

To lay out the side view as in Figure 1, draw the curved center line and divide it into 8 equal spaces (5 × 2 = 10; 10 − 2 = 8). Draw lines from the radius point through each end division point to intersect lines 1′ and 2′, obtaining the height of the heel and throat segments A and E. Again draw a line from the radius point to intersect lines 1 and 2, obtaining the height of the heel and throat on segments A and E. Again draw a line from the radius point to any length desired.

Draw a line from point 8′, Figure 1, tangent to the center arc to intersect the line draw from the radius point, obtaining point 7. Draw a line from point 7′ tangent to the center arc to intersect the line drawn from the radius point 8″.

To obtain the length for the miter lines 1–13 and 2″–14″ in Figure 1, draw line A' to E' in Figure 2 equal to the distance 2′ to 8′ on the slant miter line at segment A in Figure 1. Transfer the distance 1′ to 7′ on the slant miter line at segment E in Figure 1 to line A–E in Figure 2, from point E to D. Divide the distance A to D into as many equal spaces as the number of segments remaining (three: B, C, and D) in Figure 1, obtaining points B and C.

Transfer the distance from point E to B in Figure 2 to each side of point 7, Figure 1, obtaining points 1 and 13 representing the length of the miter at segment B in Figure 1. Transfer the distance from point E to C in Figure 2 to each side of point 8″, obtaining points 2″ and 14″ which represent the miter line at segment C.

Use point 7, Figure 1, as a center to draw the half circle 1 to 13. Use point 8″ as a center to draw the half circle 2″ to 14″. Divide each half circle into equal spaces, and draw a line from each division space to intersect their respective miter line.

Lay out the two end-segment patterns A and E in Figure 6 equal to the circumference of their respective diameters.

NOTE: The true-length triangles have been constructed away from the side view. This is not necessary; it is only to avoid confusion and simplify matters so that the reader may clearly follow the procedure. To save time in shop layout, these true lengths are obtained on the side view, thereby eliminating these extra steps.

To obtain the slant true-length lines for the segment pattern B, use the distance 1 to 7 on the miter line at segment B, in Figure 1, as a radius to draw the half circle 1 to 13 in Figure 3, and divide it into equal spaces. Transfer the heights of lines 4, 6, 8, 10, and 12 on the half circle at segment A in Figure 1, to their respective lines 3, 5, 7, 9, 11, and 13 in Figure 3. Transfer the slant length lines from segment B in Figure 1 to the base line B–13 in Figure 3, obtaining the slant true-length lines to lay out the segment pattern B. NOTE: The spaces 2′ to 14′ are equal to the spaces 2′ to 14′ on the freehand curved line on pattern A. The spaces 1 to 13 are equal to the spaces 1 to 13 on the half circle at segment B in Figure 1.

Again use the distance 1 to 7 in Figure 1 as a radius to draw the half circle 1 to 13 in Figure 4. Transfer the heights of lines 4″, 6″, 8″, 10″, and 12″ from the half circle on segment C to their respective lines 3, 5, 7, 9, 11,

PLATE 127

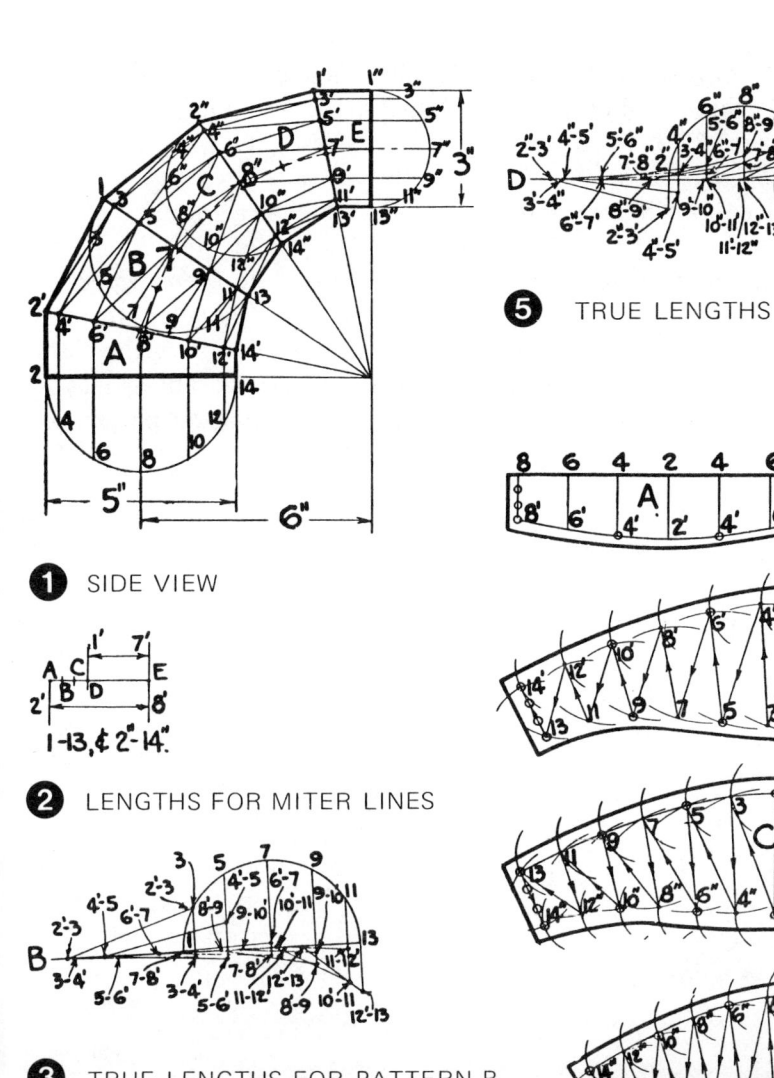

① SIDE VIEW
② LENGTHS FOR MITER LINES
③ TRUE LENGTHS FOR PATTERN B
④ TRUE LENGTHS FOR PATTERN C
⑤ TRUE LENGTHS FOR PATTERN D
⑥ FULL PATTERNS FOR A, B, C, D, & E

PLATE 127 EQUAL-TAPERING ELBOW WITH CENTER RADIUS (CONT.)

and 13 in Figure 4. Transfer the slant length lines on segment *C* in Figure 1 to line *C*–13 in Figure 4, obtaining the slant true-length lines to lay out pattern *C*. The spaces 2″ to 14″ are equal to the spaces 2″ to 14″ on the half circle at segment *C* in Figure 1.

Use the distance 2″ to 8″ at segment *C* in Figure 1 as a radius to draw the half circle 2″ to 14″ in Figure 5. Transfer the length of lines 3″, 5″, 7″, 9″, and 11″, from the half circle at segment *E* in Figure 1, to their respective lines 2″, 4″, 6″, 8″, 10″, and 12″ in Figure 5. Transfer the slant length lines from segment *D* in Figure 1 to line *D*–14″ in Figure 5, obtaining the slant true-length lines to lay out the pattern for segment *D*. NOTE: The spaces 1′ to 13′ are equal to the spaces 1′ to 13′ on the freehand curve on pattern *E*.

PLATE 127

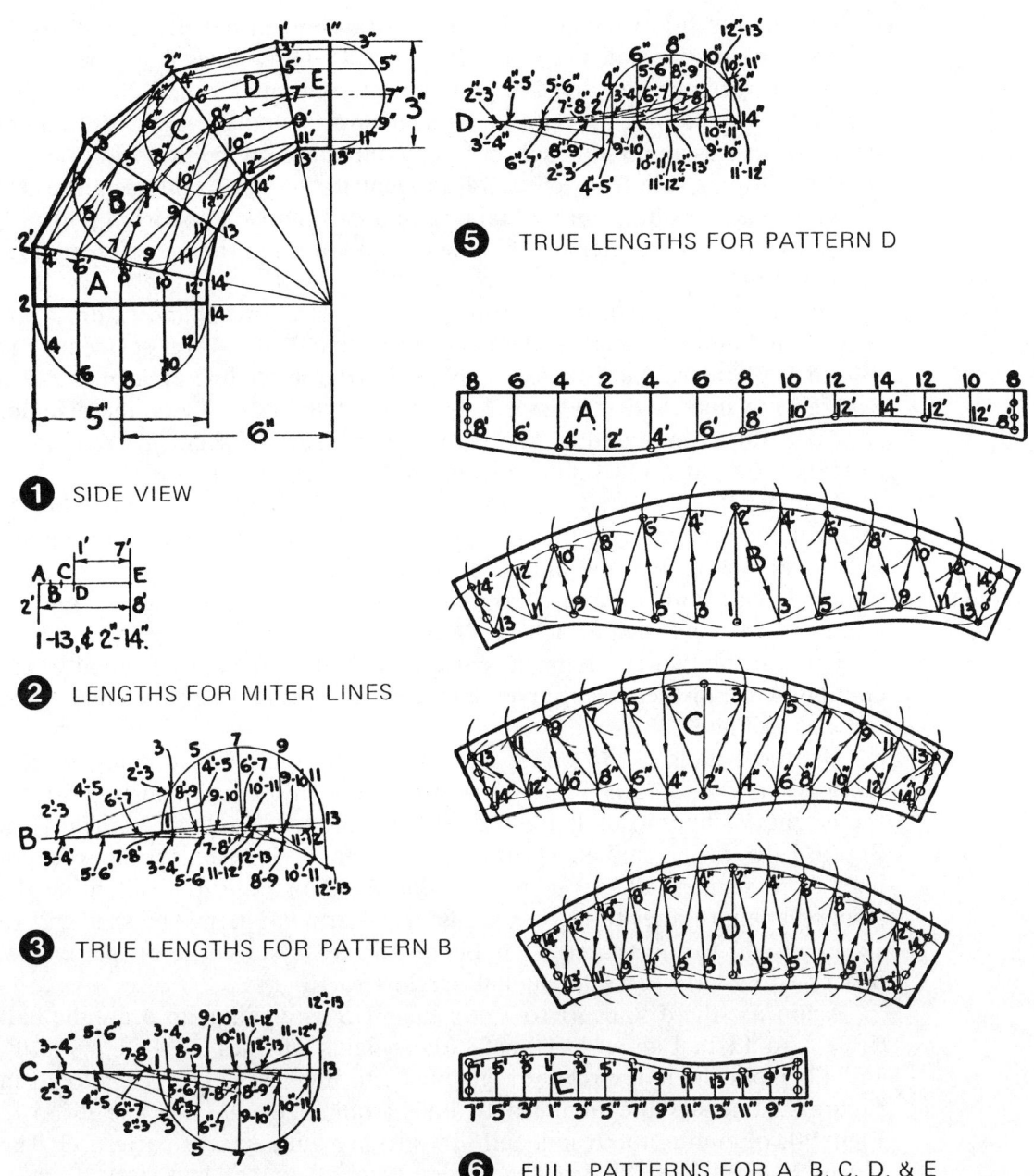

PLATE 128 TAPERING ELBOW WITH THROAT RADIUS

To lay out the side view as in Figure 1, draw the throat curve and divide it into 8 equal spaces (5 × 2 = 10; 10 − 2 = 8). Draw lines from the radius point through each end division point to intersect lines 1′ and 2′, obtaining the height of the heel and throat on segments A and E. Again draw lines from the radius point through every second division point to any length desired. Draw a line from point 14′ tangent to the arc to intersect the line drawn from the radius point, obtaining point 13. Draw a line from point 13′ tangent to the arc to intersect the line drawn from the radius point, obtaining point 14″.

To obtain the length for the miter lines for segments B and C, draw line A to E in Figure 2 equal to the slant miter line 2′ to 14′ on segment A in Figure 1. Transfer the distance 1′ to 13′ on the miter line at segment E in Figure 1, to line A–E in Figure 2, from point E to D. Divide the distance A to D into as many equal spaces as the number of segments remaining (three: B, C, and D) in Figure 1, obtaining points B and C.

Transfer the distance from point E to B in Figure 2 to the miter line at segment B, from point 13 to 1 in Figure 1. Transfer the distance from point E to C in Figure 2 to the miter line at segment C, from point 14″ to 2″ in Figure 1. Use points 7 and 8″ as centers to draw the half circles 1 and 13 and 2″ to 14″, and divide each into equal spaces.

Lay out the two end-segment patterns A and E in Figure 6 equal to the circumference of their respective diameters. The remaining procedure is the same as in Plate 127.

Use the distance 1 to 7 on the miter line B in Figure 1 as a radius to draw the half circle 1 to 13 in Figure 3, and divide it into equal spaces. Transfer the heights of lines 4, 6, 8, 10, and 12 from the half circle at segment A in Figure 1, to lines 3, 5, 7, 9, 11, and 13 in Figure 3. Transfer the slant length lines from segment B in Figure 1 to line B–13 in Figure 3, obtaining the slant true-length lines to lay out segment pattern B. NOTE: The spaces 2′ to 14′ are equal to the spaces 2′ to 14′ on the freehand curve on pattern A. Spaces 1 to 13 are equal to the half circle 1 to 13.

Again use the distance 1 to 7 at segment B as a radius to draw the half circle 1 to 13 in Figure 4. Transfer the heights of lines 2″, 4″, 6″, 8″, 10″, and 12″ from the half circle at segment C, to lines 3, 5, 7, 9, 11, and 13 in Figure 4. Transfer the slant length lines from segment C to line C–13 in Figure 4, obtaining the true-length lines to lay out segment pattern C. The spaces 2″ to 14″ are equal to the spaces 2″ to 14″ on the half circle.

Use the distance 2″ to 8″ at segment C as a radius to draw the half circle 2″ to 14″ in Figure 5. Transfer the lengths of lines 3″, 5″, 7″, 9″, and 11″ on the half circle at segment E, to lines 2″, 4″, 6″, 8″, 10″, and 12″ on the half circle in Figure 5. Transfer the slant length lines at segment D to the base line D–14″ in Figure 5, obtaining the slant true-length lines to lay out the segment pattern D. The spaces 1′ to 13′ are equal to the spaces 1′ to 13′ on the freehand curved line on pattern E.

PLATE 128

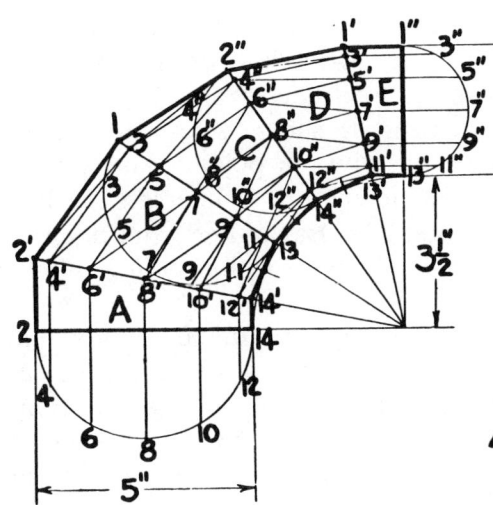
① SIDE VIEW

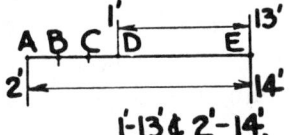
② LENGTHS OF MITER LINES 1'-13' & 2'-14'

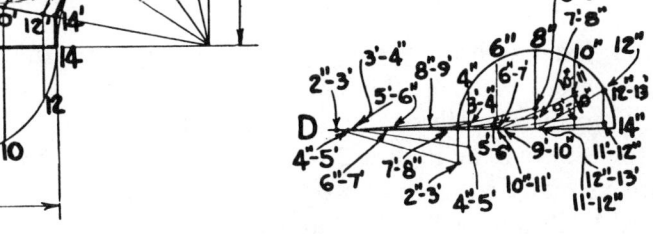
⑤ TRUE LENGTHS FOR D

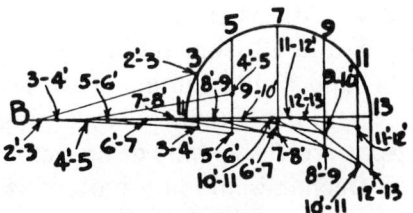
③ TRUE LENGTHS FOR B

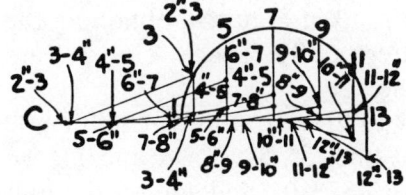
④ TRUE LENGTHS FOR C

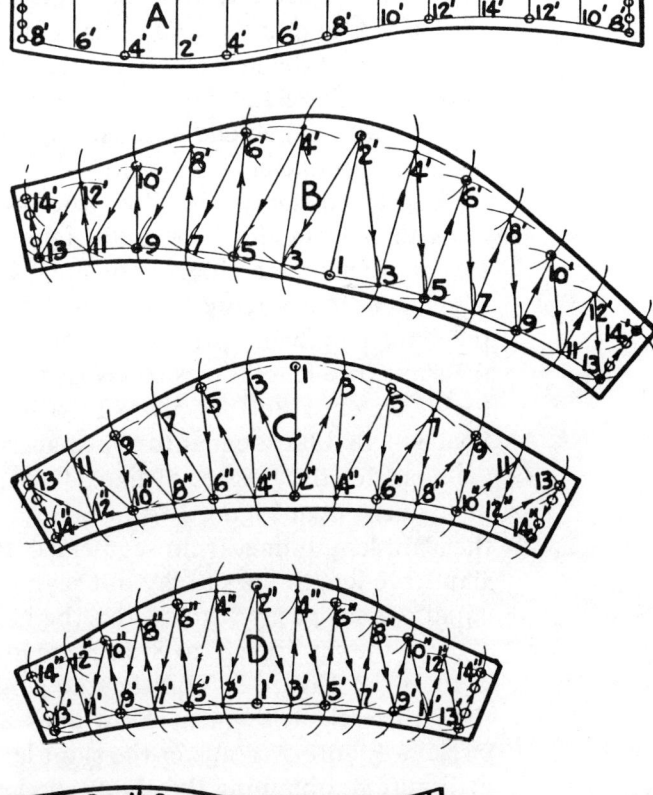
⑥ FULL PATTERNS FOR A, B, C, D, E

245

PLATE 129 TAPERING ELBOW WITH HEEL RADIUS

To lay out the side view as in Figure 1, draw the heel curve and divide it into 2 equal spaces (5 × 2 = 10; 10 − 2 = 8). Draw lines from the radius point through each end division point to intersect lines 1′ and 2′, obtaining the height of the heel and throat on segments A and E. Again draw lines from the radius point through every second division point to any length desired. Draw a line from point 2′ tangent to the arc to intersect the line drawn from the radius point, obtaining point 1. Draw a line from point 1′ tangent to the arc to intersect the line drawn from the radius point, obtaining point 2″.

To obtain the length for the miter lines for segments B and C, draw line A to E in Figure 2 equal to the slant miter line 2′ to 14′ on segment A in Figure 1. Transfer the distance 1′ to 13′ on the miter at segment E in Figure 1, to line A–E in Figure 2 from point E to D. Divide the distance A to D into as many equal spaces as segments remaining (three: B, C, and D) in Figure 1, obtaining points B and C.

Transfer the distance from point E to B in Figure 2 to the miter line at segment B, from point 1 to 13 in Figure 1. Transfer the distance from point E to C in Figure 2 to the miter line at segment C, from point 2″ to 14″ in Figure 1. Use points 7 and 8″ as centers to draw the half circles 1 to 13 and 2″ to 14″, and divide each into equal spaces. The remaining procedure is the same as in Plate 124.

Lay out the two end-segment patterns A and E in Figure 6 equal to their respective diameters.

Use the distance 1 to 7 on the miter line B in Figure 1 as a radius to draw the half circles 1 to 13 in Figures 3 and 4, and divide them into equal spaces. Transfer the heights of lines 4, 6, 8, 10, and 12 from the half circle at segment A in Figure 1, to lines 3, 5, 7, 9, 11, and 13 in Figure 3. Transfer the slant length lines from segment B to line B–13 in Figure 3, obtaining the slant true-length lines to lay out segment pattern B. NOTE: The spaces 2′ to 14′ are equal to the spaces 2′ to 14′ on the freehand curve on pattern A. Spaces 1 to 13 are equal to the spaces 1 to 13 on the half circle.

Transfer the heights of lines 4″, 6″, 8″, 10″, and 12″ from the half circle at segment C in Figure 1 to lines 3, 5, 7, 9, 11, and 13 in Figure 4. Transfer the slant length lines from segment C to line 1–13 in Figure 4, obtaining the slant true-length lines to lay out segment pattern C. The spaces 2″ to 14″ are equal to the spaces 2″ to 14″ on the half circle.

Use the distance 2″ to 8″ at segment C as a radius to draw the half circle 2′ to 14″ in Figure 5. Transfer the lengths of lines 3″, 5″, 7″, 9″, and 11″ on the half circle at segment E, to lines 2″, 4″, 6″, 8″, 10″, and 12″ on the half circle in Figure 5. Transfer the slant length lines on segment D to line 2″–14″ at Figure 5, obtaining the slant true-length lines to lay out segment pattern D. The spaces 1′ to 13′ are equal to the spaces 1′ to 13′ on the freehand curved line on pattern E.

PLATE 129

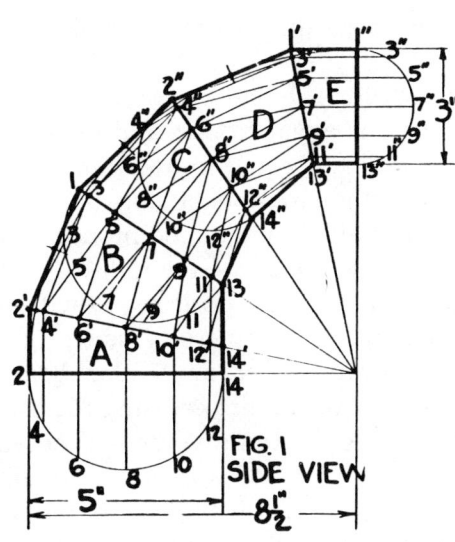

1 SIDE VIEW

2 LENGTH OF MITER LINES

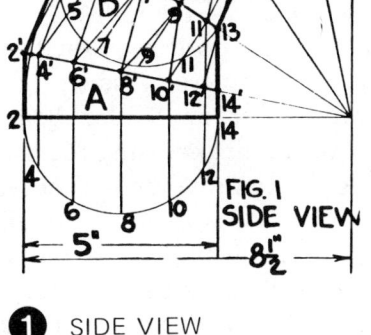

3 TRUE LENGTHS FOR B

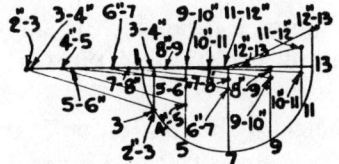

4 TRUE LENGTHS FOR C

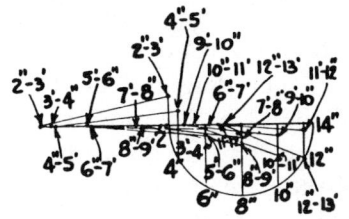

5 TRUE LENGTHS FOR D

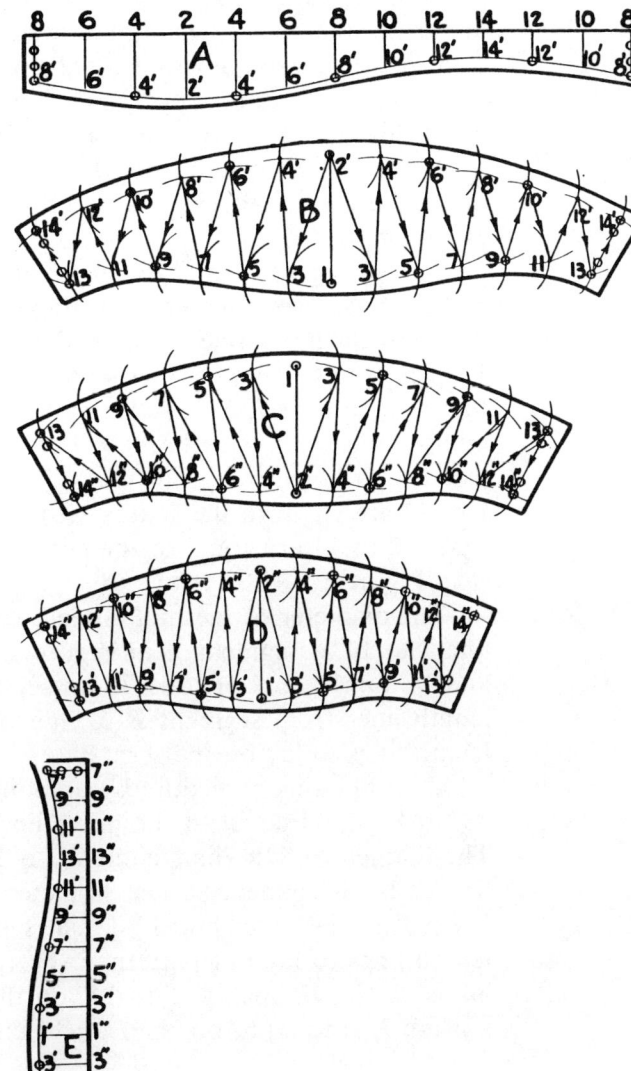

6 FULL PATTERNS FOR A, B, C, C, E

247

PLATE 130 SHIP'S VENTILATOR WITH ROUND MOUTH

The side view in Figure 1 must be drawn according to the formula for round-mouth ship's ventilator.

FORMULA: The diameter of the large mouth is equal to two times the diameter of the base. The radius of the heel is equal to one and one-half times the diameter of the base. The throat radius is equal to one fourth the diameter of the base. The horizontal angle of the mouth is 80 deg.

Draw the base diameter, the throat curve, the angle of the mouth and mark its diameter equal to the given dimensions. To obtain the radius point to draw the heel curve, and a line parallel to the straight portion of the heel to any height desired; the distance that this line is drawn from the straight portion of the heel must be equal to the heel radius. Set the dividers equal to the heel radius. Use point 1 as a center to strike an arc crossing the line drawn parallel to the heel, obtaining the radius point to draw the heel curve from point 1, striking the straight portion at the heel on a tangent. Divide the throat curve into equal spaces, and draw lines from the throat radius point through each division point to intersect the heel and obtain points 2, 1', 2', 1", and 2" at segments A, B, C, D, E, and F. Divide each miter line in half, draw a half circle at each segment, and divide into equal spaces as shown.

To obtain the slant true-length lines for the patterns, follow the same procedure as used for the round tapering elbow in Plate 127.

Draw the half circle at Figure 2 equal to the diameter of the large mouth. Transfer the heights of lines 4, 6, 8, 10, and 12 on the half circle 2–14 in Figure 1, to lines 3, 5, 7, 9, 11, and 13 in Figure 2. Transfer the slant length lines in segment A to the base line A–13 in Figure 2, obtaining the slant true-length lines to lay out the segment pattern A in Figure 6. The spaces 1 to 13 are equal to the spaces 1 to 13 on the half circle at the mouth. The spaces 2 to 14 are equal to the spaces on the half circle 2 to 14 at segment A in Figure 1.

The diameter of the half circle 2 to 14 in Figure 3 is equal to the miter line 2 to 14 at segment A in Figure 1. Transfer the heights of 3', 5', 7', 9', and 11' on the half circle 1'–13' at segment B in Figure 1. Transfer the slant length lines from segment B to line B–14 in Figure 3, obtaining the true-length lines to lay out pattern B.

Continue this procedure by drawing the diameter of the half circle 1' to 13' in Figure 4 equal to the miter line 1' to 13' at segment B in Figure 1. The diameter of the half circle 2' to 14' in Figure 5 is equal to the miter line 2'–14' at segment C. The diameter of the half circle 1" to 13" in Figure 7 is equal to the miter line 1"–13" at segment D, representing the slant true-length lines to lay out pattern E. NOTE: The spaces 2" to 14" on pattern E are equal to the spaces 2" to 14" on the freehand curved line on pattern F. Pattern F is equal to one half of the circumference of the base diameter.

PLATE 130

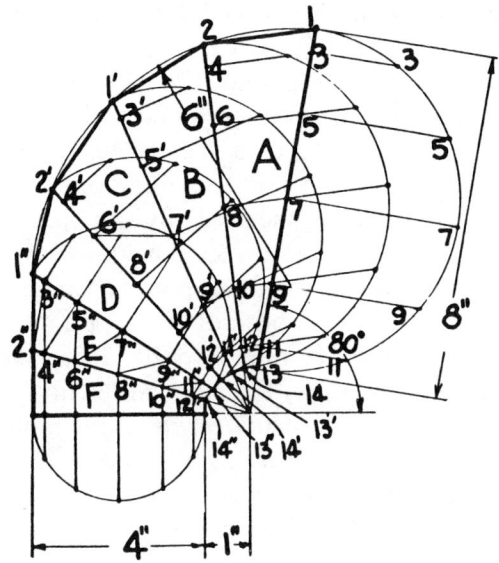

① SIDE VIEW

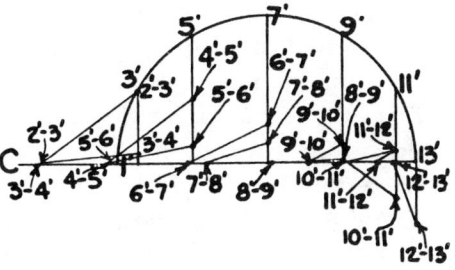

④ TRUE LENGTHS FOR C

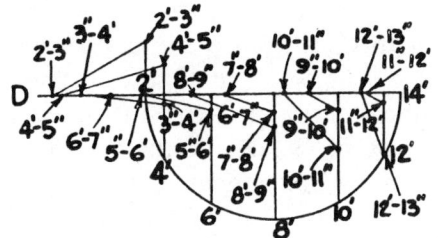

⑤ TRUE LENGTHS FOR D

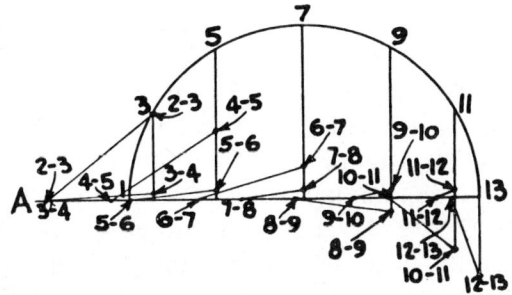

② TRUE LENGTHS FOR PATTERN A

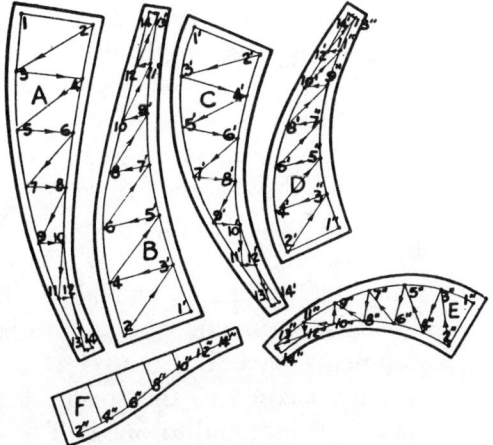

⑥ HALF PATTERNS

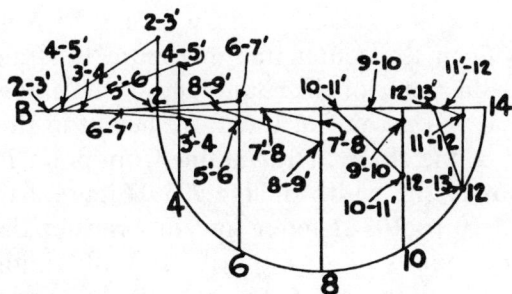

③ TRUE LENGTHS FOR B

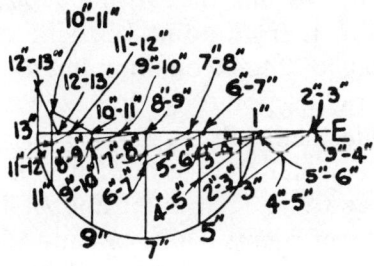

⑦

249

PLATE 131 RECTANGULAR-TO-ROUND TAPERING ELBOW

To lay out the side view as in Figure 1, draw the curved center line and divide it into 8 equal spaces (5-piece elbow; 5 × 2 = 10; 10 − 2 = 8) in the same manner as on the round tapering elbow in Plate 127. Draw lines from the radius point through each end division point to intersect the lines drawn up from the base obtaining points *A*, *B*, and *C*, and the lines drawn from point 1″, 7″, and 13″ obtaining points 1′, *F*, and *G* at the round opening. Draw a line from the radius point through every second division point. Draw a line from point *B* tangent to the arc to intersect the line drawn from the radius point obtaining point *D*. Draw a line from point *F* tangent to the arc to intersect the line drawn from the radius point, obtaining point *E*.

To obtain the length of the miter lines 1 to 19 and 2 to 20, draw line *B* to *C* in Figure 2 equal to the distance *B* to *C* on the slant miter line at segment No. 1 in Figure 1. Transfer the distance *F* to *G* on the slant miter line at segment No. 5 in Figure 1 to line *B–C* in Figure 2, from point *D–E* to point *G′*. Divide the distance *G′* to *C′* into as many equal spaces as the number of segments remaining (three: Nos. 2, 3, and 4) in Figure 1, obtaining points 1–19, and 2–20.

Transfer the distance from point *D–E* to 1–19 in Figure 2 to each side of point *D*, Figure 1, obtaining points 1 and 19, which represent the length of the miter line at segment No. 2 in Figure 1. Transfer the distance from point *D–E* to 2–20 in Figure 2 to each side of point *E*, obtaining points 2 and 20 and the length of the miter line at segment No. 3 in Figure 1.

To obtain the widths of the flats on the side at miter lines *D* and *E*, draw line *B* to *C* in Figure 3 equal to the distance *B* to *C* on the slant miter line at segment No. 1 in Figure 1. Divide this line into three equal spaces. Transfer the distance from point *D–E* to 9–11 in Figure 3 to each side of point *D*, obtaining points 9 and 11 on the miter line *D* in Figure 1. Transfer the distance from point *D–E* to 10–12 in Figure 3 to each side of point *E*, obtaining points 10 and 12 on the miter line *E* in Figure 1.

To obtain one half of the height or width of the heel and throat for segments Nos. 2, 3, and 4 at the miter line *D* and *E*, draw line *B* to *F* in Figure 4 equal to the spaces *B* to *F* on the center line in Figure 1. Draw line *F* to *F′* equal to one half of the diameter of the round end of the elbow in Figure 1. Draw line *B* to *B′* equal to one half of the (3-in. height in this plate) height at the rectangular end of the elbow. Draw a line from point *F′* to *B′* crossing lines *E* and *D*. Transfer the width of line *E* in Figure 4 to Figure 1, from point 12 to 12′ and 10 to 10′ at miter line *E*. Transfer the distance 12 to 20 on miter line *E*, from points 12′ to *J* and 10′ to *K*, obtaining the radius points to draw the quarter circles 12′ to 18 and 10′ to 4, and divide them into equal spaces.

Transfer the width of line *D* in Figure 4 to Figure 1, from point 11 to 11′ and 9 to 9′ at miter line *D*. Transfer the distance 11 to 19 on miter line *D*, from points 11′ to *M* and 9′ to *N*, obtaining the radius point to draw the quarter circles 11′ to 17 and 9′ to 3, and divide them into equal spaces.

NOTE: The spaces 1′ to 13′ on pattern No. 4 are equal to the spaces 1′ to 13′ on the freehand curved line on Pattern No. 5.

To erect the true-length triangles for segment pattern No. 4 as in Figure 5,

PLATE 131

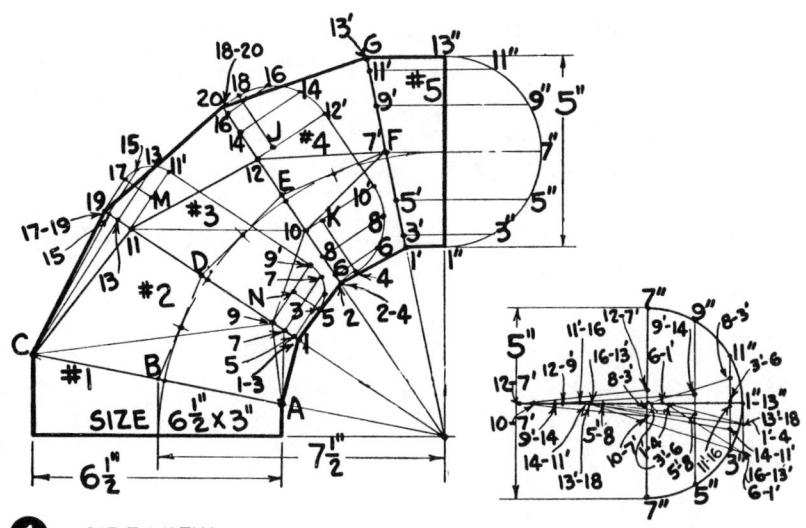

① SIDE VIEW

② LENGTH OF MITER LINES

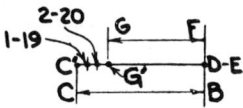

③ FLAT SIDE OF SEGMENTS

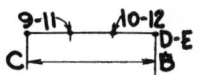

⑤ LENGTHS FOR 4

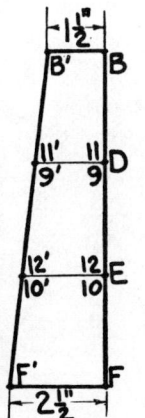

④ HALF WIDTHS OF HEEL & THROAT

⑥ LENGTHS FOR 3

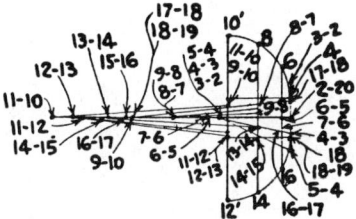

⑦ TRUE LENGTHS FOR 2

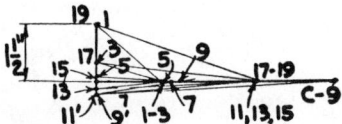

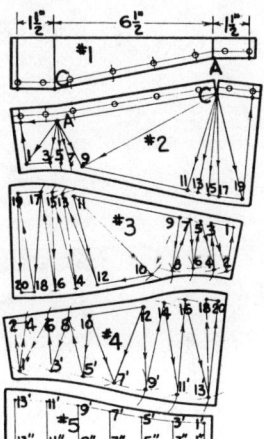

⑧ HALF PATTERNS FOR 1, 2, 3, 4, 5

251

PLATE 131   RECTANGULAR-TO-ROUND TAPERING ELBOW (CONT.)

draw the half circle equal to the diameter of the round end of the elbow.

To erect the true-length triangles for segment pattern No. 3 as in Figure 6, transfer the two quarter circles plus the straight 18 to 20 and 2 to 4 at miter line E in Figure 1, to each side of the center line in Figure 6.

Draw the height of the angle in Figure 7 equal to one half of the height of the rectangular end of the elbow.

The procedures for developing the patterns and obtaining the true-length lines are the same as for the round tapering elbows.

PLATE 131

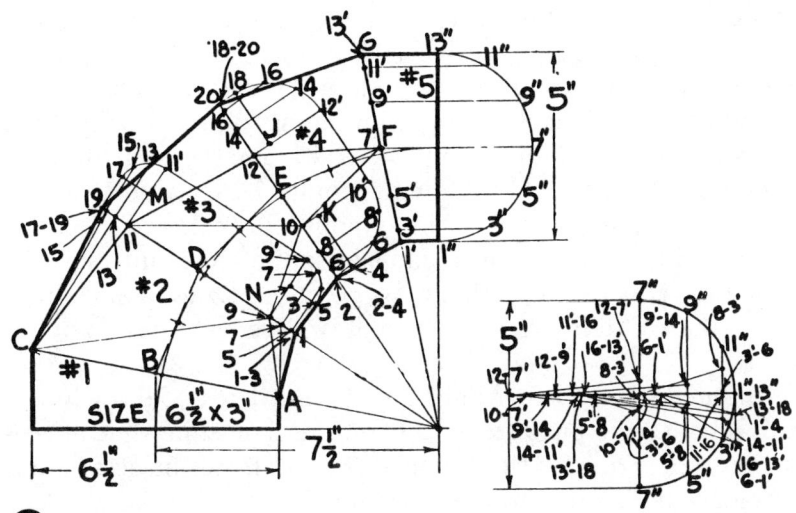

① SIDE VIEW

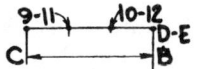

② LENGTH OF MITER LINES

③ FLAT SIDE OF SEGMENTS

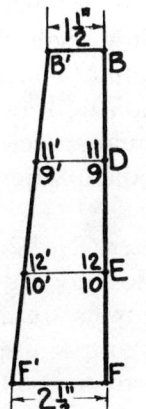

④ HALF WIDTHS OF HEEL & THROAT

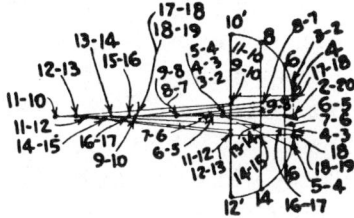

⑤ LENGTHS FOR 4

⑥ LENGTHS FOR 3

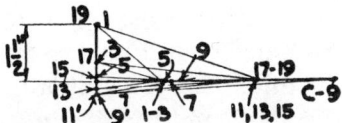

⑦ TRUE LENGTHS FOR 2

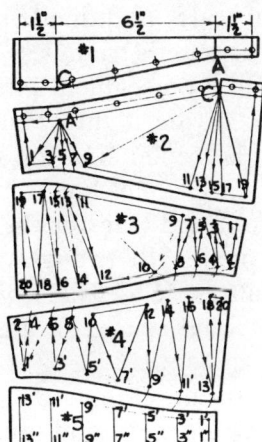

⑧ HALF PATTERNS FOR 1, 2, 3, 4, 5

PLATE 132 ROUND-TO-OBLONG TAPERING ELBOW

To lay out the side view as in Figure 1, draw the curved center line, and divide it into 8 equal spaces (5-piece elbow: 5 × 2 = 10; 10 − 2 = 8) in the same manner as on the round tapering elbow, Plate 127. Draw lines from the radius point through each end division point to intersect the lines drawn up from the base, obtaining points *A*, *B*, *C*, and *D*, and the lines drawn from points 2″, 8″, and 14″, obtaining points 2′, *G*, and *H* at the round opening. Draw a line from the radius point through every second division point. Draw a line from point *C* tangent to the arc to intersect the line drawn from the radius point, obtaining point *E*. Draw a line from point *G* tangent to the arc to intersect the line drawn from the radius point, obtaining point *F*.

To obtain the length of the miter lines 2 to 16 and 1 to 15, draw line *A* to *C* in Figure 2 equal to the distance *A* to *C* on the slant miter line at segment No. 1 in Figure 1. Transfer the distance *G* to *H* from the slant miter line at segment No. 5 in Figure 1 to line *A–C* in Figure 2, from point *E–F* to point *H′*. Divide the distance *A′* to *H′* into as many equal spaces as the number of segments remaining (three: Nos. 2, 3, and 4) in Figure 1, obtaining points 2–16 and 1–15. Transfer the distance from point *E–F* to 2–16 in Figure 2 to each side of point *E*, Figure 1, obtaining points 2 and 16. Transfer the distance from point *E–F* to 1–15 in Figure 2 to each side of point *F*, Figure 1, obtaining points 1 and 15. These represent the length of the miter lines at segment Nos. 2 and 3 in Figure 1.

To obtain the widths of the flats on the side at miter lines *E* and *F*, draw line *B* to *C*, Figure 3, equal to the distance *B* to *C* on the slant miter line at segment No. 1, Figure 1. Divide this line into three equal spaces. Transfer the distance from point *E–F* to 8–10 in Figure 3 to each side of point *E*, to get points 8 and 10 on the miter line *E* in Figure 1. Transfer the distance from point *E–F* to 7–9 in Figure 3 to each side of point *F*, obtaining points 7 and 9 on the miter line *F* in Figure 1. Use points 8 and 10 on miter line *E*, Figure 1, as centers, and the distances 8 to 2 and 10 to 16 as radii to draw the quarter circles 2 to 8 and 10 to 16, and divide each into equal spaces. Use points 7 and 9 on miter line *F*, Figure 1, as centers to draw the quarter circles 1 to 7 and 9 to 15; and divide each into equal spaces.

To erect the true-length triangles, Figure 4, for segment pattern No. 2, use the distance 2 to 8 on segment line *E*, Figure 1, as a radius to draw the half circle 8 to 10; divide it into equal spaces. Transfer the lengths from segment No. 2 to the center line for segment pattern No. 3, Figure 5, use the distance 1 to 7 on segment line *F* in Figure 1 as a radius to draw the half circle 7 to 9; divide it and transfer the spaces on segment No. 3 as before.

To erect the true-length triangles as in Figure 6 for segment pattern No. 4, draw the half circle 8″ to 8″ equal to the diameter of the round end, and divide it into equal spaces. Transfer the lengths from segment No. 4 to the center line.

The procedures for developing the patterns and obtaining the true-length lines are the same as for the round tapering elbows.

PLATE 132

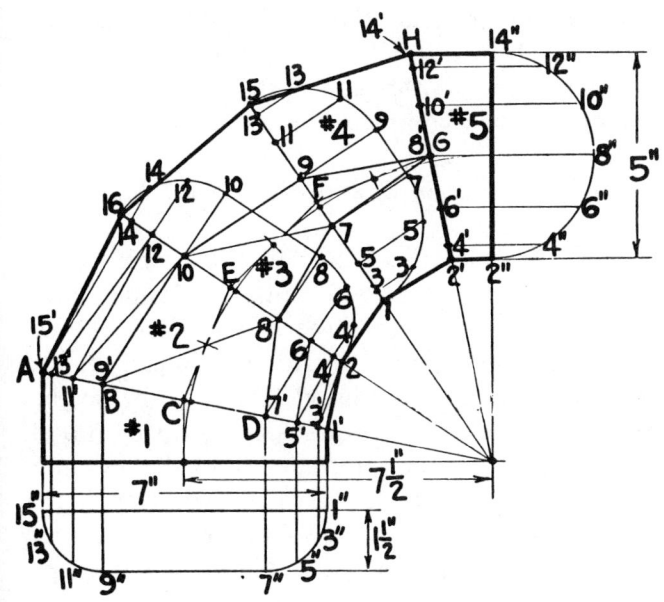

1 SIDE VIEW

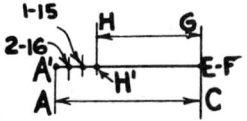

2 LENGTH OF MITER LINES

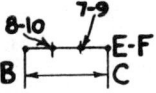

3 FLAT SIDE OF SEGMENTS

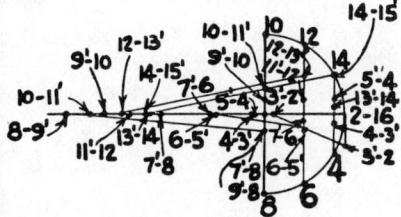

4 LENGTHS FOR 2

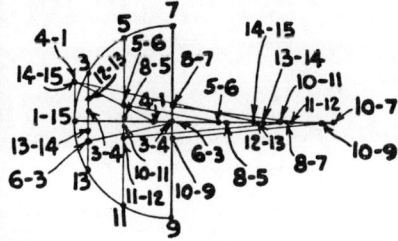

5 LENGTHS FOR 3

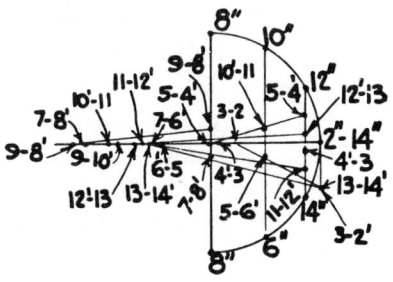

6 LENGTHS FOR 4

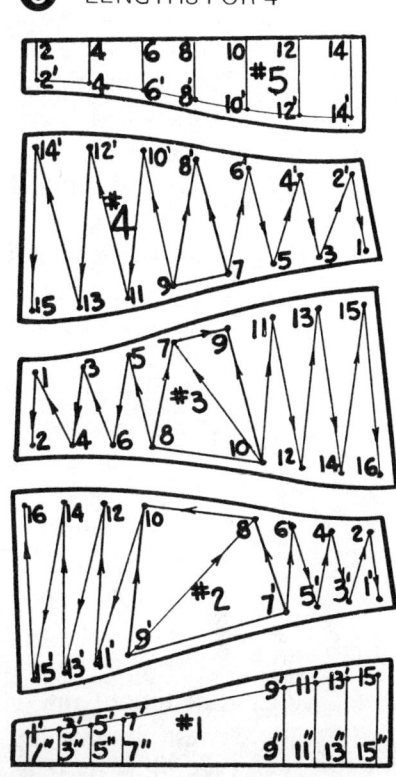

7 HALF PATTERNS FOR 1, 2, 3, 4, 5

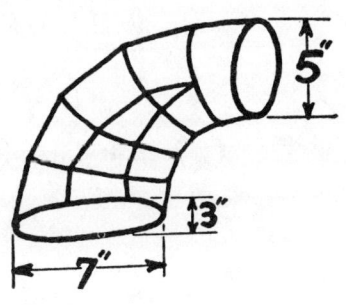

8 ISOMETRIC VIEW

255

PLATE 133   TWO-QUART MEASURE

To lay out the side view as in Figure 1, draw the height and the top and bottom diameters of the measure. Draw a line from point 6′ (on the center line of the half circle at the top diameter) through point 12′, to point 13, and through point 13′ to point 14; then complete the remaining side view of the hood as shown.

Draw the half bottom view as in Figure 2, and divide each half circle into equal spaces.

Transfer the heights of lines 6′, 8′, and 10′ on the half circle representing the top diameter of the measure to lines 7, 9, and 11 on the half circle representing the mouth opening of the hood. Transfer the heights of lines 8, 6, and 4 on the half circle at the spout opening on the hood to lines 7, 5, and 3 on the half circle at the mouth opening of the hood. Transfer the heights of lines 8, 10, and 12 on the half circle at the spout opening on the hood, to lines 6′, 9′, and 11′ on the half circle at the top diameter of the measure.

Transfer the slant lines on the hood to line 12′–13′ at the top edge of the measure, and to line 1–13 at the mouth of the hood, obtaining the required slant true-length lines to lay out the hood pattern as in Figure 5. The spaces 2 to 14 are equal to the spaces 2 to 14 on the half circle at the spout in Figure 1. The spaces 1 to 13 are equal to the spaces 1 to 13 on the half circle at the mouth opening of the hood. The spaces 12′ to 13′ are equal to the spaces 12′ to 13′ on the half circle representing the top diameter of the measure.

Lay out the handle pattern as in Figure 3 equal to the lengths C–D–13 in Figure 1.

To lay out the spout pattern as in Figure 4, use point A as a center to draw arcs from points B and 14 on the slant line on the spout in Figure 1 to Figure 4. The spaces 2 to 2 on the spout are equal to twice the number of spaces 2 to 14 on the half circle at the spout in Figure 1.

Transfer the slant lines from the half bottom view in Figure 2 to the base line of the true-length triangle E in Figure 1, obtaining the slant true-length lines to lay out the measure pattern as in Figure 6. Follow the same procedure as in Plate 24.

PLATE 133

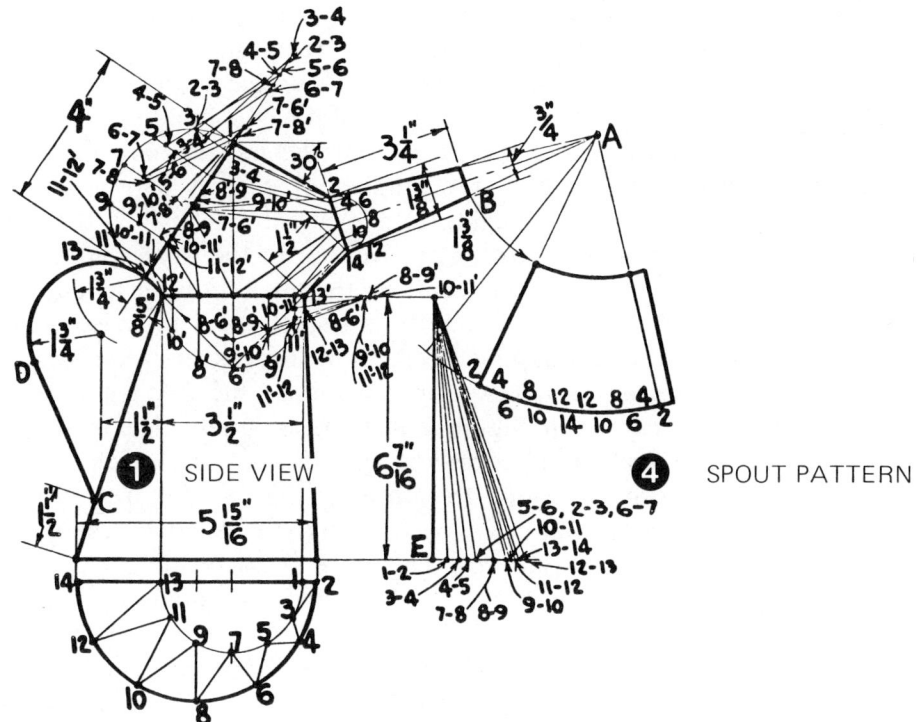

① SIDE VIEW
④ SPOUT PATTERN
② HALF BOTTOM VIEW
③ HANDLE PATTERN
⑤ HOOD PATTERN
⑥ MEASURE PATTERN

257

PLATE 134   360-DEG. SPIRAL CONVEYOR CHUTE

The patterns for a spiral conveyor chute are generally laid out in small segments, equal in proportion to the degrees of revolution and in accordance to the height of travel. When a conveyor chute has a spiral of about 360 deg., the height is equivalent to about one floor of a building. It is the general practice to design the conveyor so that the patterns are made in segments equal to one sixteenth of the number of degrees of revolution. However, when the degrees and height are considerably less, then the segments may be laid out in one eighth or one quarter of the total revolution.

To lay out the segment for one sixteenth of the revolution for a 360-deg. spiral that will be 16 in. high, draw only one quarter of a top view as in Figure 1, and divide the throat and heel curve into as many spaces as desired. It is not necessary to draw any portion of the side view as in Figure 2. The height A to B in Figure 2 is 1 in., or one sixteenth of the 16-in. height of the spiral, and represents the distance that each of the sixteen segments will drop or slope to complete the 360-deg. revolution, and may be obtained mathematically, by dividing the height by the number of segments (16 in. ÷ 16 = 1 in.). Therefore the 1-in. height A to B on the true-length triangles in Figures 3, 4, and 5 is equal to $A-B$ in Figure 2.

To complete the true-length triangle in Figure 3, draw the base line 2 to 3 equal to the slant line 2 to 3 in Figure 1. Complete the true-length triangle in Figure 4 by drawing the base line 1 to 3 equal to the distance 1 to 3 on the curved heel in Figure 1.

Complete the true-length triangle in Figure 5 by drawing the base line 2 to 4 equal to the distance 2 to 4 on the throat curve in Figure 1.

To lay out the pattern in Figure 6, draw line 1 to 2 equal to the given dimension. The distance 2 to 3' is equal to the distance 2 to 3' on the true-length triangle in Figure 3. The distance 1 to 3' is equal to the distance 1 to 3' on the true-length triangle in Figure 4.

The distance 2 to 4' is equal to the distance 2 to 4' in the true-length triangle in Figure 5. The distance 3' to 4' is equal to the given dimension.

NOTE: The second segment is laid out, only to establish three points from which to bisect to obtain an apex or center points M and N to draw the heel curve and the throat curve. The points 5' and 6' are obtained in the same manner as points 3' and 4'.

To form the pattern, use a hand-forming brake or a power-press brake to place several kinks between points 2 and 4', using point 3' as a center pivot. The severity of each kink depends on the height that each segment will slope.

To lay out the spiral guard equal to one-half revolution as in Figure 7, draw line J to K equal to one half of the circumference on the heel curve of the spiral as in Figure 1, and draw the line J to L equal to one half of the height of the spiral as A to I in Figure 2.

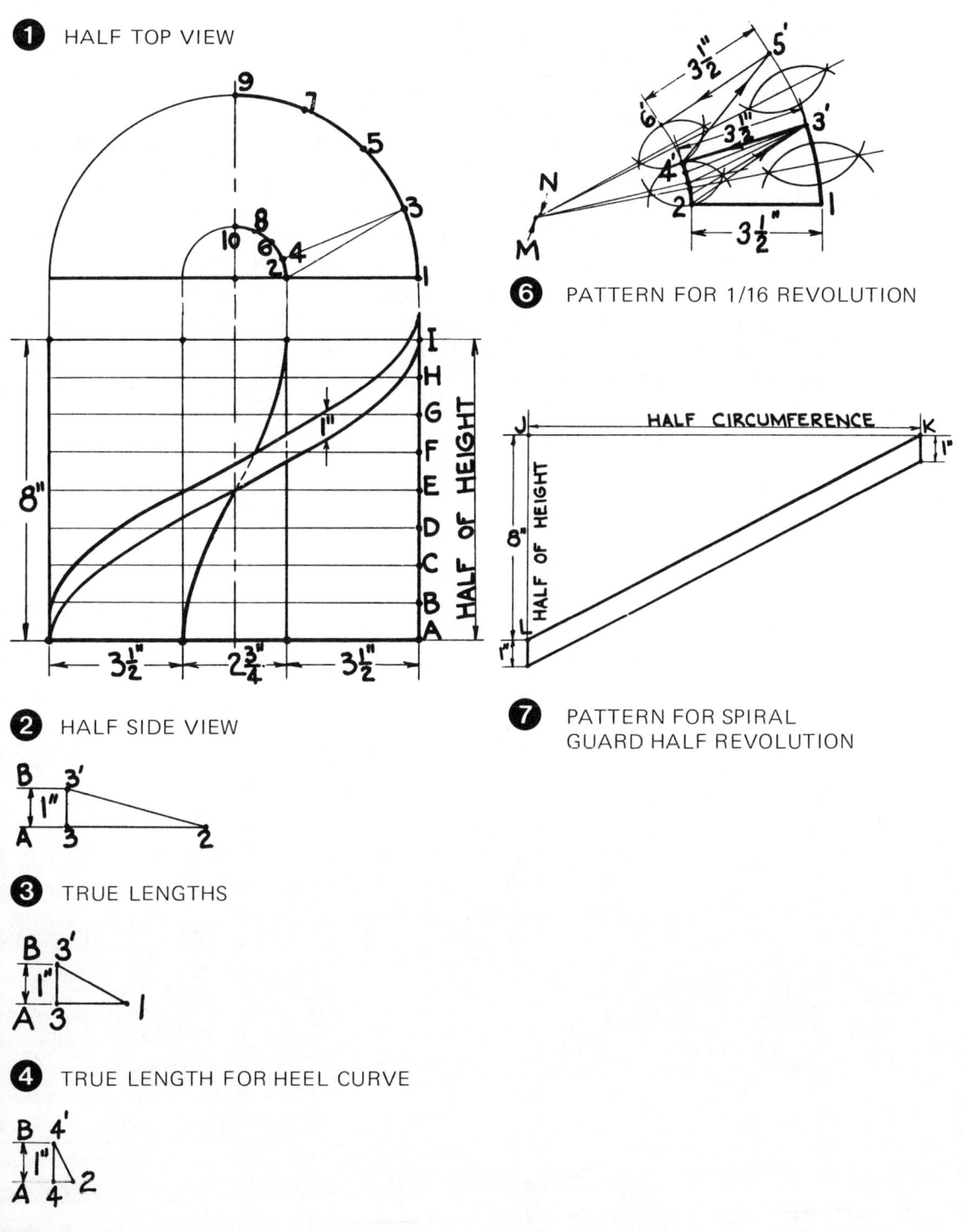

PLATE 134

PLATE 135 TOOLBOX WITH PITCHED COVER

Figures 1, 2, and 3 show how the cover on the box is pitched.

Figure 4 shows an end view of the open toolbox.

This type of box allows for a removable, shallow pan to be placed in the box, with the top edges of the pan resting on the top edges of the toolbox. Thereby the small, light tools can be separated from the larger and heavier tools. Also, the overalls may be rolled and placed in the cover when moving from job to job out in the construction field.

Figures 5 and 7 show how the patterns for the cover and the ends should be laid out and notched so that the edges may be double hemmed as shown, and the slant ends may be riveted in place.

Figures 6 and 8 show how the patterns for the box and the ends should be laid out so that top edges may be formed to a right angle as shown and the ends may be double seamed.

PLATE 135

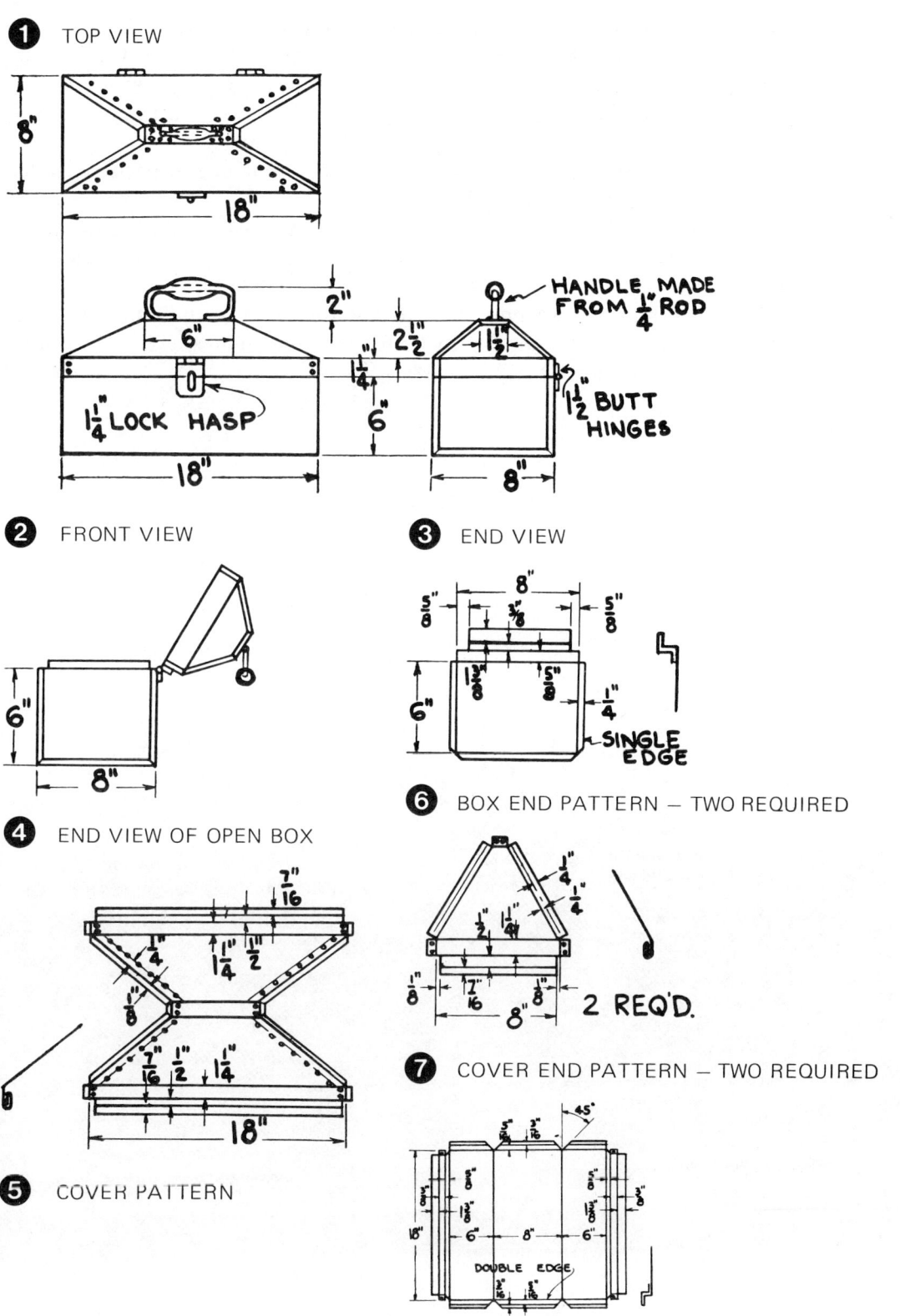

PLATE 136 FLAT-COVER, SUITCASE-STYLE TOOLBOX

Figure 1 shows an end view of the open toolbox. Note that the bottom of the box and the top of the cover are on the same level when the box has been opened. This eliminates the weight of the cover to bear on the hinges, and allows a portion of the tool to be placed in the cover without breaking the hinges.

This type of box is easy to carry, and is recommended especially for the men out in the construction field.

Figures 5 and 7 show how the patterns for the cover and ends should be notched, so that the top edges may be double hemmed, as shown, and the ends double seamed.

Figures 4 and 6 show how the patterns for the box and ends should be notched, so that the top edges may be formed to a right angle, and the ends may be double seamed.

PLATE 136

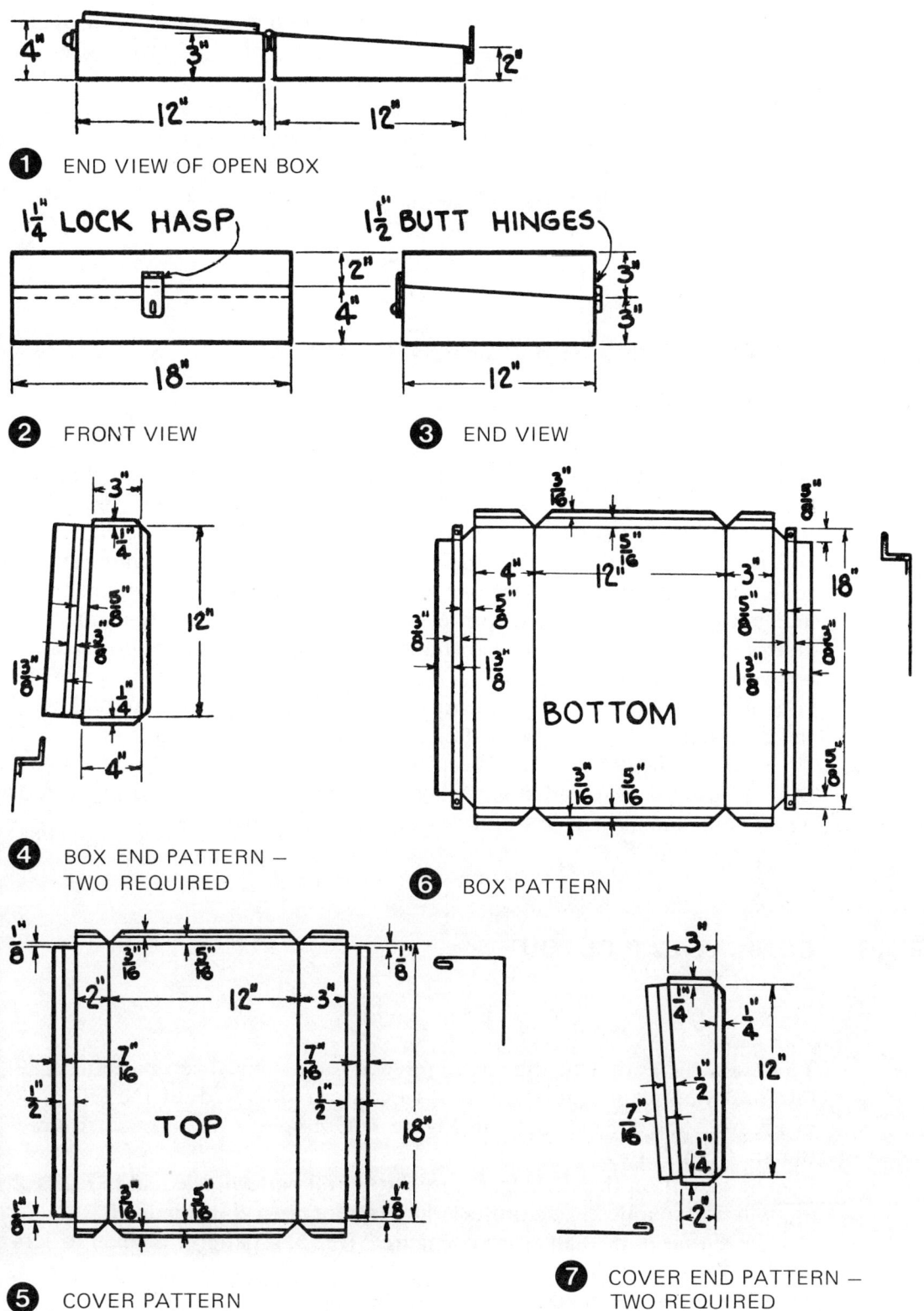

① END VIEW OF OPEN BOX

② FRONT VIEW

③ END VIEW

④ BOX END PATTERN — TWO REQUIRED

⑤ COVER PATTERN

⑥ BOX PATTERN

⑦ COVER END PATTERN — TWO REQUIRED

PLATE 137   ALIGNING ROUND PIPES WITH DRAW BAND WHEN ARC WELDING

This perspective view shows how a draw band with slots, or window openings cut in, may be used to align two round joints of pipes in preparation for arc welding.

    The draw band will keep the two joints of round pipes straight, and the electrode may be inserted through the various window openings, to allow the two joints to be tack-welded in as many places as there are windows cut in the band. This saves time and labor, and eliminates many tedious tasks, especially when several joints are to be welded together.

PLATE 138   FORMING CONE IN ROLLER WITH THE AID OF AN ANGLE IRON

A simple method for forming a cone pattern, especially when made of heavy-gauge metal is with the use of a short piece of angle iron, any size. Place the heel of the angle iron to rest between the top front and bottom front rollers, with one end of this angle iron resting against the frame of the roller at the handle side. Place one end of the cone pattern between the top and bottom front rollers, with the throat curve resting firmly against the end of the angle iron, and turn the handle. This will draw the remaining cone pattern through the rollers, thereby forcing the throat curve of the cone pattern to press firmly against the end of the angle iron, and allowing the cone pattern to form to its desired shape without any difficulty.

    NOTE: If the cone has not been formed to its desired shape at the first trial, adjust the rear and front rollers as may be needed; then proceed as above until the pattern has been formed to the desired cone shape.

PLATE 139   CONICAL-CAP CUTOUT

To obtain the cutout or the circumference for a conical-cap pattern, draw the front view as in Figure 1. Use A to C as a radius to draw the full circle in Figure 2 and the half circle in Figure 3. Use point C as a center to draw the arc from D to B.

    The cutout on the circle in Figure 2 (the full pattern) is equal to 6.28 (or $6\frac{1}{4}$ will be accurate enough) times the distance A to B in Figure 1.

    The cutout of the half circle in Figure 3 (the half pattern) is equal to 3.14 (or $3\frac{1}{8}$) times A to B in Figure 1.

    Allowances for seaming may be made as desired.

PLATE 137

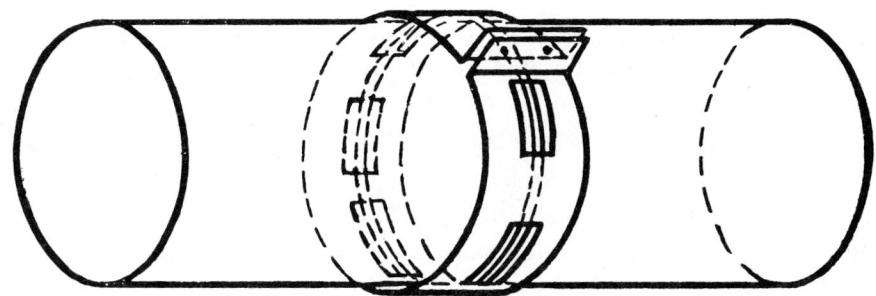

PLATE 138

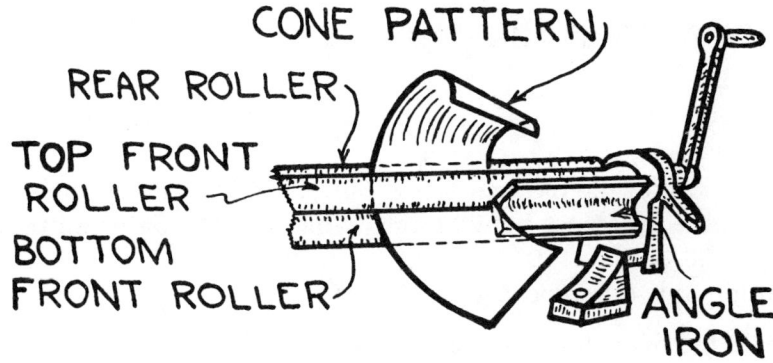

PLATE 139

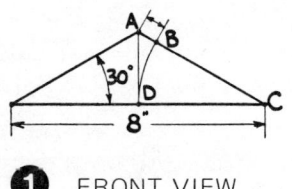

① FRONT VIEW

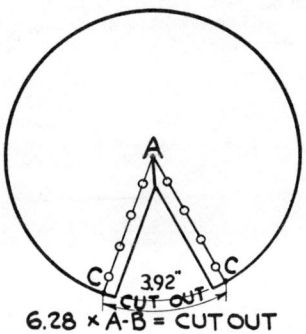

② FULL PATTERN
6.28 × A-B = CUT OUT

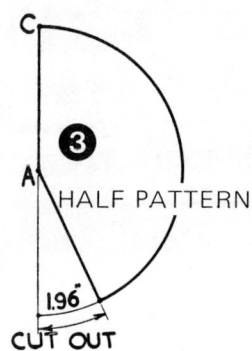

③ HALF PATTERN
CUT OUT
3.14 × A-B = CUTOUT
FOR HALF PATTERN

PLATE 140 FINDING THE DEGREE OF CUTOUT FOR STACK CAP

RULE: To find the number of degrees in the sector to be cut out of the flat pattern in Figure 2, find the difference between the diameters of the base and of the flat pattern.

Then multiply this difference by 360, and divide the product thus obtained by the diameter of the flat pattern.

SOLUTION: 36 in. is the diameter of the base shown in Figure 1.

$40\frac{1}{2}$ in. is the diameter of the flat pattern shown in Figure 2.
$40\frac{1}{4} - 36 = 4\frac{1}{4}$ in., or the difference.
$360 \times 4\frac{1}{4} = 1530$ in.
$\frac{1530}{40\frac{1}{4}}$ equals 38.01, or 38 deg.

The degrees may be laid off on the pattern with a protractor.

PLATE 141 FINDING CUTOUT FOR STACK CAP

RULE: To find the width of the piece to be cut out of the flat pattern shown in Figure 2, find the difference between the diameter of the base and the diameter of the flat pattern.

Then multiply this difference by 3.1416.

SOLUTION: 36 in. is the diameter of the base in Figure 1.

$2 \times 20\frac{1}{8} = 40\frac{1}{4}$ in., or the diameter of the flat pattern in Figure 2.
$40\frac{1}{4} - 36 = 4\frac{1}{4}$ in., or the difference.
$3.1416 \times 4\frac{1}{4} = 13.3518$ in. (or roughly, $13\frac{3}{8}$ in.), which is the width of the piece to be cut out.

This method can be made more efficient by using a handbook which contains the circumferences of circles. Then, by taking the circumference of a circle $4\frac{3}{8}$ in. in diameter, the dimension 13.3518 in. can be obtained very rapidly.

PLATE 140

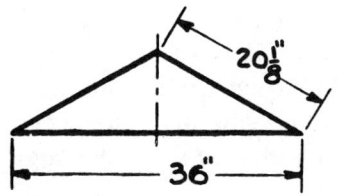

1 SIDE PROFILE

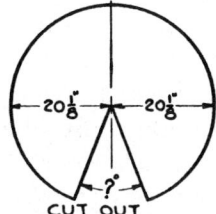

2 FLAT PATTERN

PLATE 141

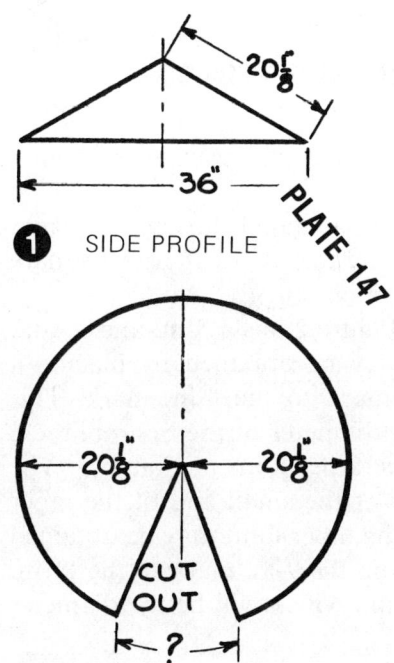

1 SIDE PROFILE

2 FLAT PATTERN

PLATE 142   OBTAINING RIGHT ANGLES

Figure 1 shows three combinations of whole numbers that may be used to form right angles without the use of a steel square. This the sheet-metal workers will encounter, especially out in the field, when a steel square is not available or when large right angles must be formed with accuracy where the small steel square may not be considered large enough to form the angle to a perfect 90 deg.

The combinations of numbers in Figure 1 may be used to represent inches or feet; this depends on the size of the right angle to be formed.

To form a right angle as in Figure 2, use one of the three combinations of numbers in Figure 1.

EXAMPLE: Draw a base line to equal 4 in. as in Figure 2. Set the divider to span 3 in., and use point A as a center to draw an arc at point C. Set the dividers to span 5 in., and use point B as a center to draw an arc to cross the arc at point C, obtaining the right angle. This combination is referred to as 3, 4, and 5. The remaining combination A to 6, A to 8, and 6 to 8 is referred to as 6, 8, and 10. The combination A to 12, A to 16, and 12 to 16 is referred to as 12, 16, and 20, meaning that the combination of each set of the three numbers will form a 90-deg. angle.

The combination of numbers 12, 16, and 20 is good for checking a steel square that has been used carelessly.

PLATE 143   OBTAINING THE DIAMETER OF THE MAIN BRANCH ON TWO- OR THREE-PRONGED FITTINGS

The theory for this plate is the same as for Plate 142; it illustrates combinations of numbers to form a right angle.

In Plate 143 the diameter of two prongs on a branch fitting are given; the third unknown diameter may be obtained by the use of the steel square as in Figure 1.

EXAMPLE: The two-pronged Y branch in Figure 2 has a 3-in. and a 4-in. branch. The diameter of the main branch may be obtained by placing a ruler across the steel square from the 3-in. mark to the 4-in. mark. This will show 5 in. on the ruler, which will be the diameter of the main branch.

The same principle applies to the T on the taper joint in Figure 3. The diameter of the T is 11 in., and the diameter of the small end of the taper joint is 9 in. The diameter of the large end of the taper joint may be obtained by placing the ruler across the steel square, from the 9-in. mark to the 11-in. mark. This will show on the ruler to be $14\frac{1}{4}$ in., which will be the diameter of the large end of the taper joint.

To obtain the diameter of a three-pronged branch fitting as in Figure 4, place the ruler across the square from the 12-in. mark (the diameter of branch 1) to the 16-in. mark (the diameter of branch 2). The dimension shown on the ruler will be 20 in. Again place the ruler across the square from the 20-in. mark just obtained to the 14-in. mark (the diameter of branch 3). The dimension shown on the ruler will be $24\frac{3}{8}$ in., which is the diameter of the main branch for the three-pronged branch fitting.

PLATE 142

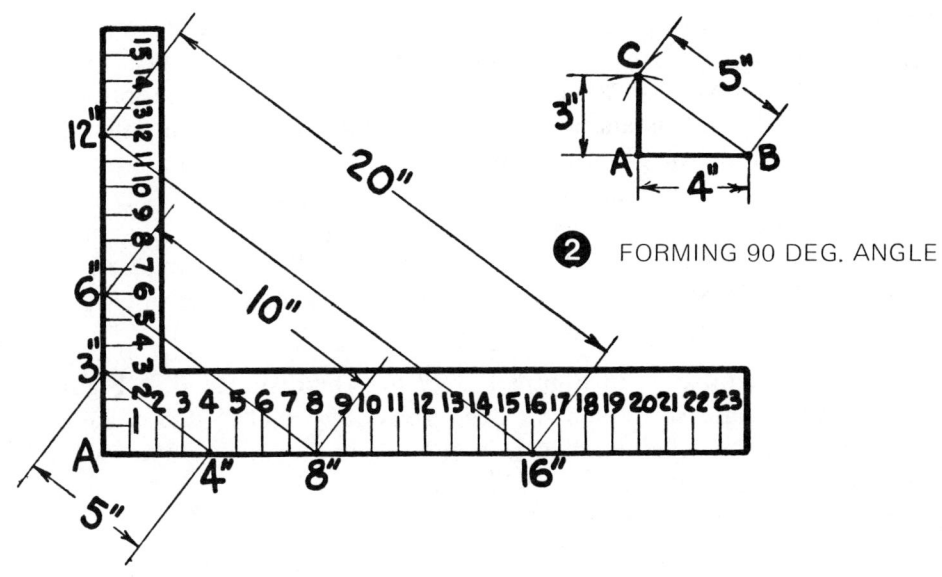

2 FORMING 90 DEG. ANGLE

1 COMBINATION OF NUMBERS

PLATE 143

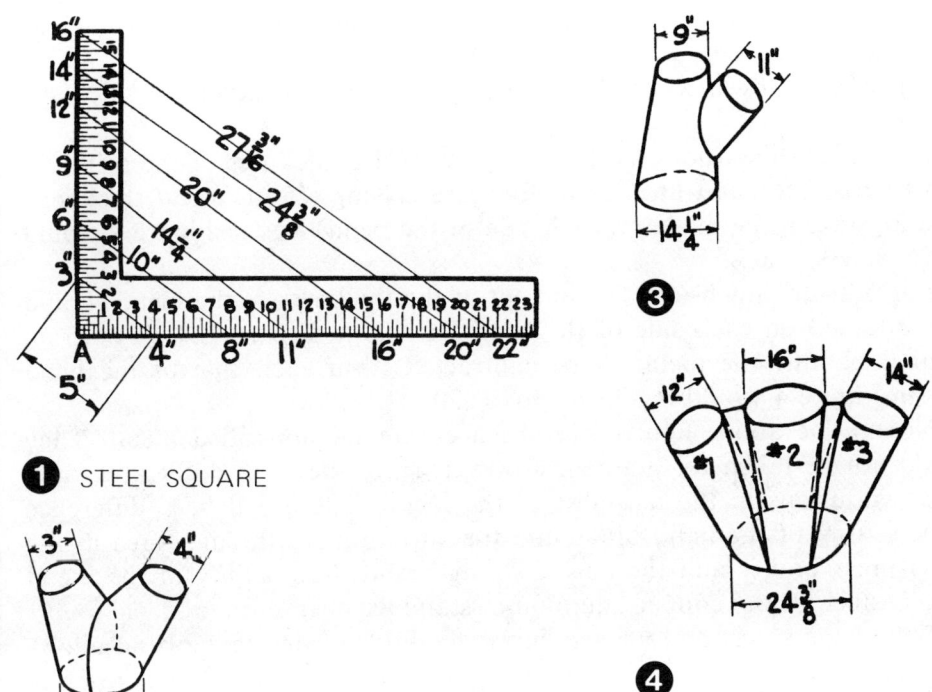

1 STEEL SQUARE

PLATE 144 FINDING THE LENGTH OF PIPE BETWEEN TWO ANGLES

Often a piece of pipe is used between two angles to represent an offset.

To find the length of the pipe between the two angles, it is necessary to calculate the distance between the two centers of the angles, as represented by C in Figure 1. This distance may be determined by multiplying the offset (A, Fig. 1) by the factor given according to the degrees of the angles. (For a 45-deg. angle, the factor is 1.414, for a 30-deg. angle, 2.00; for a 60-deg. angle, 1.154.) From the product obtained, subtract twice the distance of B, thus obtaining the length of the pipe as represented by D in Figure 1.

Distance B may be obtained by squaring in from the center of each side of one of the angles as shown in Figure 2.

EXAMPLE: The offset A in Figure 1 is 27 in.; the distance B in Figure 2 is $8\frac{1}{2}$ in.; the angles are each 45 deg., thus $A \times 1.414 - (B \times 2) = D$ (length of pipe). $27 \times 1.414 = 38.17$ in. $- (2 \times B$ in Fig. 2). $2 \times 8\frac{1}{2} = 17$ in.: $38.17 - 17 = 21.17$ or $21\frac{3}{16}$ in., which is the length of the pipe between the two 45-deg. angles, represented by D in Figure 1.

PLATE 145 BEND ALLOWANCE

This plate shows the approximate allowances for bends on inside dimensions as in Figure 1, and the subtractions for bends on outside dimensions as in Figure 3.

The metal allowances to be made for each bend depend on the thickness of metal.

For inside dimensions, add 20 per cent of the thickness of the metal on each side of the bend lines as in Figure 2. Using .25-in. metal thickness, the allowance to be made on each side of the bend lines in Figure 2 would be $.25 \times .20 = .05$.

For outside dimensions 80 per cent of the thickness of the metal should be subtracted on each side of the bend lines as in Figure 4. Using .25-in. metal thickness, the metal to be subtracted from each side of the bend lines in Figure 4 would be $.25 \times .80 = .20$.

NOTE: The above figures may be accurate on hot-rolled metal. They should not be taken for granted, however, since they will differ according to the variations in the hardness of the metal. There will be a difference in the setting of the hand brake and the variations in the dies used in the power-press brakes and the tensile strength and the ductility of the metal such as steel, brass, copper, aluminum, stainless steel, etc.

One of the surest and most accurate methods to determine bend allowances is to make a trial bend of the metal to be used, thereby determining the amount of the metal that is to be added or subtracted for each bend.

For a safe bend on aluminum and on some brass, the bend should have a radius equal to three times the thickness of the metal.

PLATE 144

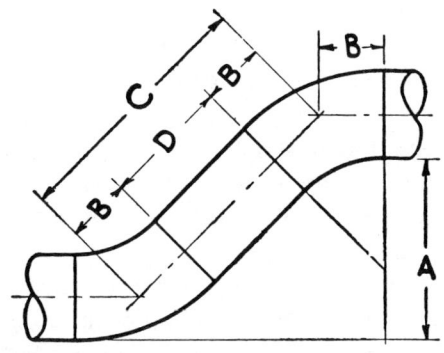

1 ASSEMBLED ANGLES

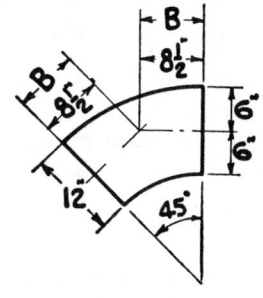

2 45° ANGLE

45° ANGLES, C = A x 1.414
30° ANGLES, C = A x 2.00
60° ANGLES, C = A x 1.154

PLATE 145

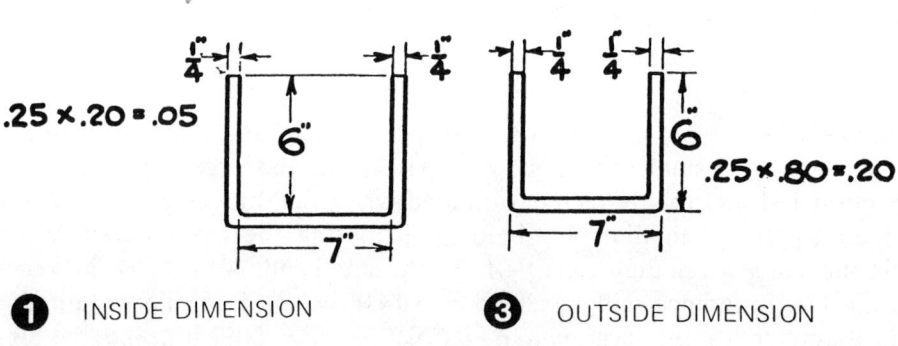

.25 x .20 = .05

1 INSIDE DIMENSION

.25 x .80 = .20

3 OUTSIDE DIMENSION

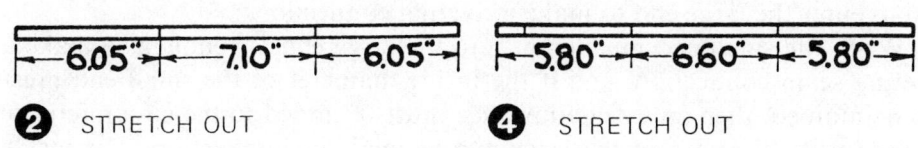

2 STRETCH OUT

4 STRETCH OUT

PLATE 146 METAL LOST IN ROLLING

When metal or band iron is rolled to form a round hoop or cylinder, a shrinkage takes place in the course of rolling. This reduces the inside diameter equal to one thickness of the metal, but increases the outside diameter equal to one thickness of the metal.

This plate shows the different diameters that may be obtained when no allowance or reduction is made in the circumference (to compensate for the thickness of the metal) before the pattern is rolled into a cylinder or hoop.

When no allowance or reduction has been made in the circumference in accordance to the thickness of the metal, after the pattern has been rolled into a cylinder, the diameter to which the circumference was figured will appear as the mean diameter. The inside diameter will be reduced equal to one thickness of the metal, and the outside diameter will be increased equal to one thickness of the metal as shown. Here ¼-inch metal has been rolled to a 20-in. diameter cylinder.

When the inside diameter is desired, add to the given diameter one thickness of the metal. In this project $20 + \frac{1}{4} = 20\frac{1}{4}$ in.; thus the circumference will be $20.25 \times 3.1416 = 63.61$ in. When the outside diameter is desired, subtract from the given diameter one thickness of the metal. In this project 20 minus ¼ equals 19¾ in., and the circumference will be $19.75 \times 3.1416 = 62.05$ in.

PLATE 147 SMALL AND LARGE ENDS

When patterns are laid out for round stacks, a small and a large end must be made so that the small end of one joint will fit into the large end of another.

Section 1 shows the diameters obtained when the thickness of the metal is added to the given diameter before calculating the circumference. To obtain the inside given diameter $A-A$, for the small end, $20 + \frac{1}{4} = 20\frac{1}{4}$ in.; thus the circumference will be $20.25 \times 3.1416 = 63.61$ in. To obtain the inside diameter for the large end $B-B$, $20\frac{1}{2} + \frac{1}{4} = 20\frac{3}{4}$ in., and the circumference will be 65.18 in.

The above method is true and accurate, but it will not allow the small end to enter the large end to make a riveting connection.

When the small end must enter into the large end far enough to make a riveting seam connection, and if the inside diameter of the small end must be maintained, then an extra allowance must be added to the circumference of the inside diameter of the large end to facilitate connection. The inside diameter of the small end is 20 in.; then $20 + \frac{1}{4} = 20\frac{1}{4}$ in., and the circumference will be 63.61 in. The circumference for the inside diameter of the large end will be equal to the circumference for the small end plus seven times the thickness of the metal. In this case $7 \times \frac{1}{4} = 1\frac{3}{4}$ or $1.75 + 63.61 = 65.36$ in. This allows the seam to be loose enough to make an easy connection. If the looser connection is desired, then eight or nine times the thickness of the metal may be added.

PLATE 146

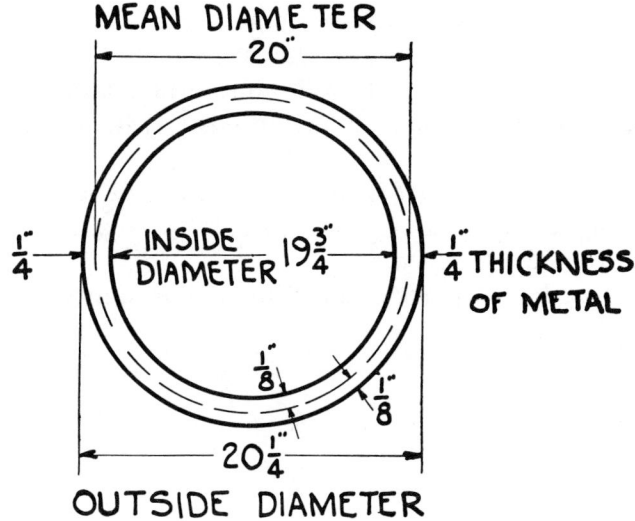

PLATE 147

Section 2 shows the diameter of the large end $C-C$ and the mean diameter $D-D$ when the thickness of the metal is not added to the given diameter.

When the inside diameter of the small end does not have to be maintained, then the circumference for the large end is obtained, which will be $20 \times 3.1416 = 62.83$ in. The circumference for the small end is obtained by subtracting seven times the thickness of the metal from the circumference of the large end. In this case $7 \times \frac{1}{4} = 1\frac{3}{4}$ or 1.75, then $62.83 - 1.75 = 61.08$ in. This allows the small end to be inserted into the large end without any difficulty.

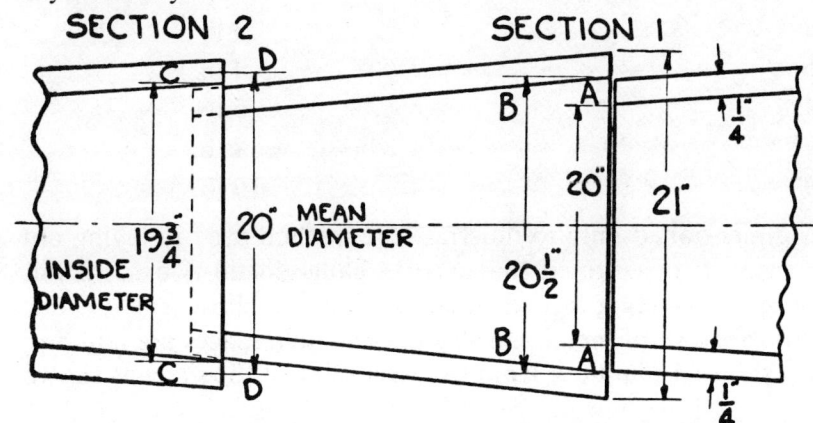

PLATE 148 AREA OF ELLIPSE—CIRCUMFERENCE OF ELLIPSE

The ellipse is closely related to the circle. If a round pipe is cut on a slant, the sectional view is a true ellipse. For this reason the sheet-metal worker often has to consider the properties of an ellipse in laying out certain patterns.

The area of an ellipse is equal to one half the major axis (A) times one half the minor axis (B), times 3.1416.

EXAMPLE: One half of the major axis (A) is equal to 20 in.; one half of the minor axis (B) is equal to 12 in. Thus, $A \times B \times 3.1416$ = area, $20 \times 12 \times 3.1416 = 240 \times 3.1416 = 753.98$ sq. in.

The method for finding the circumference of an ellipse differs somewhat from the method for finding the area.

PLATE 149 RIVETED LAP SEAM

When seaming metals of No. 18 gauge and heavier, the riveting lap seam is used. While No. 18-gauge metal may be groove seamed, it is not recommended unless a special grade of metal is used or there is a definite reason why a grooved seam should be used.

Metals lighter than No. 18 gauge also may be riveted, because there are cases where it is absolutely necessary that riveting be used. The amount of lap to be allowed and the distance the holes are to be spaced apart will vary in accordance with the gauge of metal that is used.

The following table shows the lap allowance, the rivet spacing, and the size rivets that are to be used for the different gauges of metal:

Gauge	Lap Allowance in Inches	Rivet Space in Inches	Size Rivets
28	1/4	2 1/2	12 oz.
26	1/4	2 1/2	1 lb.
24	3/8	2 1/2	1 1/2 lb.
22	3/8	3	1 1/2 lb.
20	3/8	3	2 1/2 lb.
18	1/2	3	3 lb.
16	1/2	3 1/2	4 lb.
14	1/2	3 1/2	5 lb.
12	5/8	4	8 lb.
10	3/4	4	10 lb.

Figure 1 is dimensioned only to illustrate the method used in laying out a riveting lap seam. It is assumed that No. 18-gauge metal is used in this project. To lay out a riveting lap seam as in Figure 1, draw a 1/2-in. line on either side of the piece of metal. On this line the hole spaces are laid out. If possible, they should be laid out from the center line of the sheet as shown in Figure 1.

PLATE 148

The circumference for an ellipse is equal to one half the major axis squared, plus one half the minor axis squared, times two; then extract the square root from the product times 3.1416.

EXAMPLE: One half of major axis (A) is equal to 20 in., one half the minor axis (B) equals 12 in.

Thus, $3.1416\sqrt{2(A^2 + B^2)}$ = circumference

$3.1416\sqrt{2(20^2 + 12^2)} = 3.1416\sqrt{2(400 + 144)}$

$3.1416\sqrt{2 \times 544} = 3.1416\sqrt{1088}$

$3.1416 \times 33 = 103.67$ in. circumference

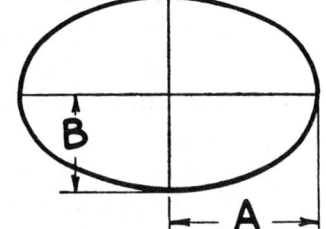

PLATE 149

Find the center and mark the hole spaces on each side of it as shown, by 3-in. spacings. This method is referred to as *marking the holes from center*.

Sometimes this method cannot be applied because there will be too much space between the end hole and the end of the sheet of iron. In that case, another method may be used, whereby half of the desired hole space is marked on each side of the center line. From these points the remaining holes then are spaced. This method is known as *marking the holes off center*. Either method is accurate and practical and makes the patterns foolproof so that either piece may be reversed and the holes will match at all times.

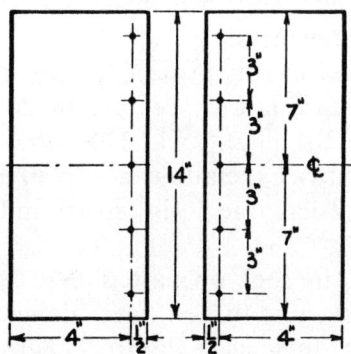

❶ METHOD OF LAYING OUT HOLES

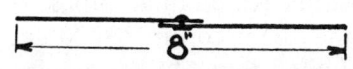

❸ END VIEW OF RIVETED SEAM

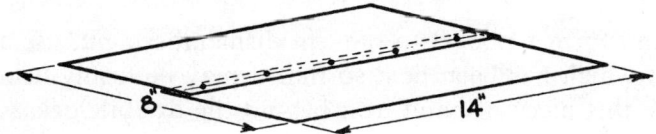

❷ PERSPECTIVE VIEW OF RIVETED LAP SEAM

PLATE 150 GROOVED LOCK SEAM

This type of seam is used when joining together two or more pieces of metal and when making round or rectangular pipes of metal No. 20 gauge and lighter.

Figure 1 shows how seam allowance is made when two pieces of metal are to be seamed together, using a $\frac{3}{8}$-in. grooved lock seam. The allowance to be made on each piece of metal equals $1\frac{1}{2}$ times the size of seam desired.

This project calls for a $\frac{3}{8}$-in. grooved lock seam; therefore a $\frac{9}{16}$-in. allowance is made on each piece. A $\frac{3}{8}$-in. line is marked on one side of each piece of metal, and that same amount is turned up, one hooked to the inside and one to the outside, as shown in Figure 2.

To assemble the pieces of metal, they are locked together as shown in Figure 3. Then, with the aid of a hand grooving tool, the locks are flattened and the top piece of metal is brought down to form a pocket as in Figure 4. This will prevent the locks from dismembering.

Figure 5 shows a perspective view of two pieces of metal that are grooved lock seamed together.

PLATE 151 HAMMER-GROOVED LOCK SEAM

This type of seam is used for joining long sheets or small pieces of metal and for locking rectangular pipes. It is an efficient seam which is easy to form, and may be used on No. 20-gauge and lighter metal.

It is assumed that No. 24-gauge metal is used in this project. Figure 1 shows how the patterns are laid out and the seam allowances are made. On pattern No. 1, a $\frac{3}{8}$-in. allowance is made for the single lock. The allowance for the double lock on pattern No. 2 equals twice the allowance for the single lock plus $\frac{1}{16}$ in. Therefore, pattern No. 2 has one $\frac{3}{8}$-in. space and one $\frac{7}{16}$-in. space.

To form the locks on the patterns, turn a $\frac{3}{8}$-in. lock on pattern No. 2, as in Figure 2. Reverse the sheet so that the lock, just turned, faces down. Now, bend up the double lock to about a 60-deg. angle as in Figure 4. Turn the single lock on pattern No. 1 to about a 60-deg. angle as in Figure 3. Insert the single lock on pattern No. 1, in the double lock on pattern No. 2 as in Figure 5.

To flatten the lock and form a pocket to prevent dismemberment, use a short piece of band iron which has been bent so that it may be firmly held with the left hand. Place this piece of band iron behind the double lock as shown in Figure 6. Keep the band next to the seam, and use a mallet to strike the lock seam and the band with each blow. This flattens the lock seam, and the band iron offsets the metal and forms a pocket as in Figure 7, preventing the lock from coming apart.

PLATE 150

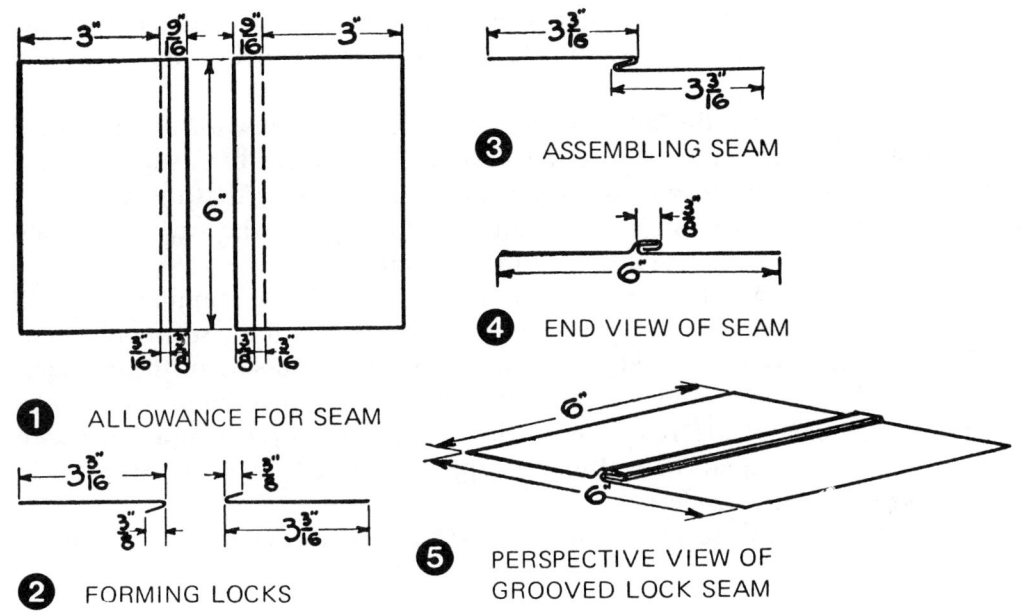

① ALLOWANCE FOR SEAM
② FORMING LOCKS
③ ASSEMBLING SEAM
④ END VIEW OF SEAM
⑤ PERSPECTIVE VIEW OF GROOVED LOCK SEAM

PLATE 151

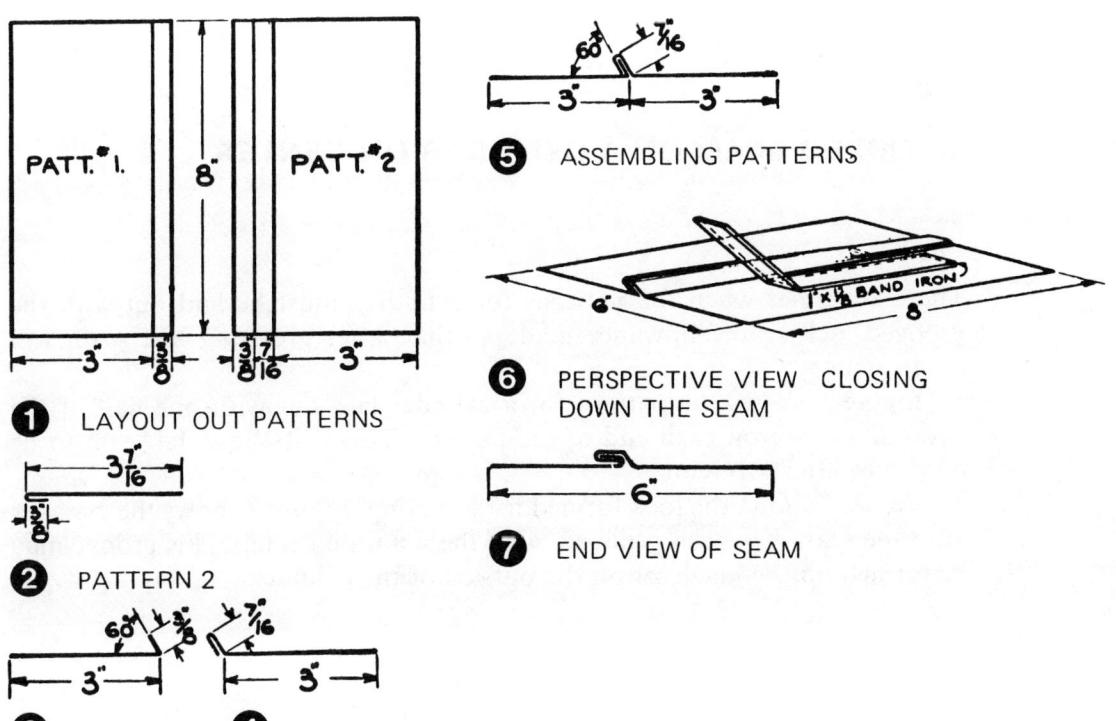

① LAYOUT OUT PATTERNS
② PATTERN 2
③ PATTERN 1
④ PATTERN 2
⑤ ASSEMBLING PATTERNS
⑥ PERSPECTIVE VIEW CLOSING DOWN THE SEAM
⑦ END VIEW OF SEAM

PLATE 152 GROOVED SEAM ON A CYLINDER OFF CENTER

There are times when the patterns for cylinders must be laid out with the grooved lock seam allowance made so that, after grooving, the seam will be off center.

Figure 1 shows the pattern for a cylinder laid out with a single-edge allowance on one end and a double edge allowed on the opposite end. This will allow the seam to be off center after grooving. Note that the allowance for a grooved lock seam must be equal to three times the size of the lock to be turned, and must be added to the circumference.

Figure 2 shows how the locks are formed for groove seaming. Figure 3 shows the cylinder after the seam has been grooved with the seam off center. This groove may be formed on the inside or on the outside of the cylinder. If a hand grooving tool is used to form the groove, then the groove will be formed on the outside as shown in Figure 3. If a rail or a stake is used with a groove cut to equal the width and depth of the lock, then the groove will be formed on the inside.

Both methods are correct and practical, but grooving with the rail is the faster method.

PLATE 153 GROOVED SEAM ON A CYLINDER ON CENTER

There are times when the patterns for cylinders must be laid out with the grooved lock seam allowance made so that, after grooving, the seam will be on center.

Figure 1 shows the pattern for a cylinder laid out with one half of the seam allowance on each end of the pattern. This will allow the seam to be on center after grooving.

Figure 2 shows the lock formed for grooving. Figure 3 shows the cylinder after the seam has been grooved, with the seam on center. This groove may be formed on the inside or on the outside of the cylinder.

PLATE 152

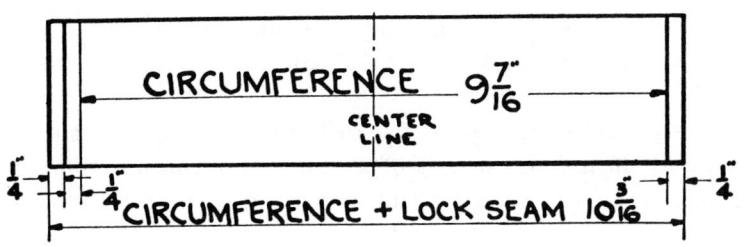

❶ PIPE PATTERN

❷ LOCK SEAM BENT

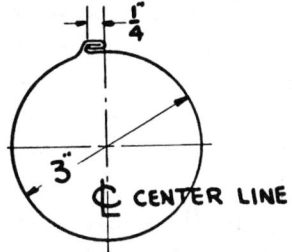

❸ GROOVED LOCK SEAM OFF CENTER

PLATE 153

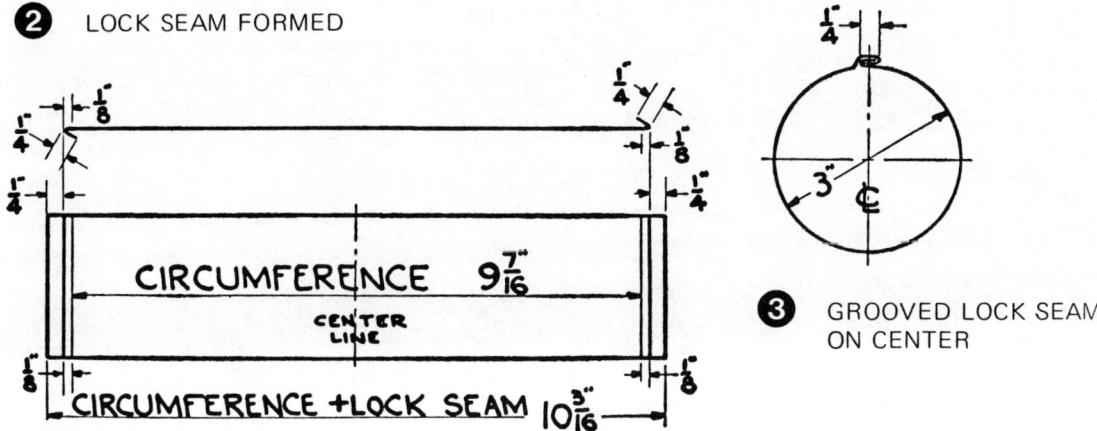

❷ LOCK SEAM FORMED

❸ GROOVED LOCK SEAM ON CENTER

❶ PIPE PATTERN

279

PLATE 154 FORMING WIRE EDGE

To reinforce an edge by enclosing wire or rod with metal, the metal to be allowed is equal to $2\frac{1}{2}$ times the thickness of the wire or rod to be enclosed as shown in Figure 2.

To form the wire-edge allowance by the use of the brake, two bends must be made. The first bend should be equal to the minimum amount that the brake will bend (about $\frac{3}{32}$ in. as in Fig. 3) and be bent up to about a 45-deg. angle.

The second bend will be the remainder of the allowance, and should be bent up to less than 90 deg. on the outside bend or equal to about 97 deg. on the inside, as shown in Figure 4.

With the aid of a mallet, the metal may be bent over the wire, and the front edge may be dressed up, if necessary, with a square-faced hammer, enclosing the wire as in Figure 1.

When reinforcing the top of a cylinder with a wire edge, the metal should be rolled, then straightened out again, the wire edge formed, and the locks turned for seaming. Before the wire is enclosed, it should be allowed to extend out of the wiring edge at the end, with the seaming lock turned to the inside as at point *A* in Figure 5. At the opposite end the wire will be pulled in from the end of the wiring edge as at point *B*. A short piece of wire is inserted at point *B* to form that portion of the wire edge and to prevent from flattening when the cylinder is formed in the rolls.

When using heavy metal to enclose a round rod, the thickness of the metal must be considered. The allowance of metal for a wire edge should be twice the wire thickness plus five times the metal thickness.

PLATE 154

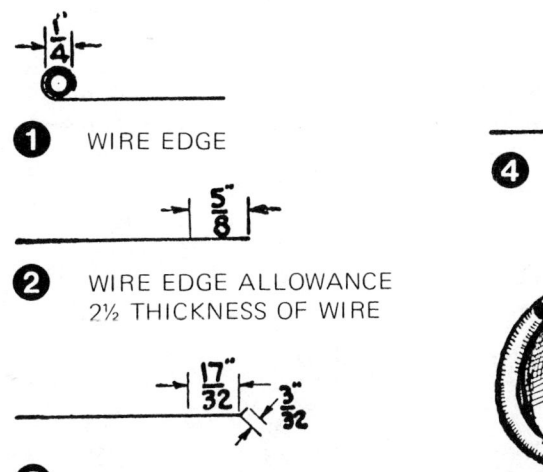
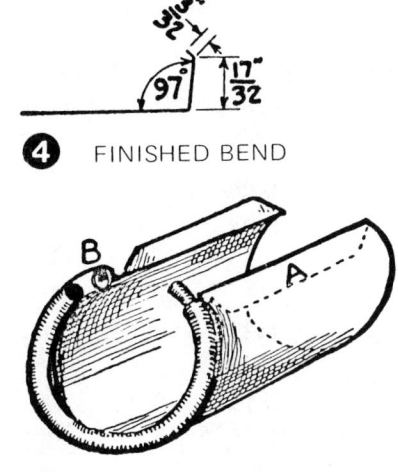

① WIRE EDGE

② WIRE EDGE ALLOWANCE 2½ THICKNESS OF WIRE

③ FIRST BEND

④ FINISHED BEND

⑤ CYLINDER WIRED EDGE

THICKNESS OF GALVANIZED IRON

No. of Gauge	Thickness Decimal	Thickness Approximate Fraction	Pounds per Sq. Ft.	Approximate Wire Thickness
7–0's	.5	1/2	20.00	—
6–0's	.46875	15/32	18.75	—
5–0's	.4375	7/16	17.50	—
0000	.40625	13/32	16.25	.460
000	.375	3/8	15.	.409
00	.34375	11/32	13.75	.365
0	.3125	5/16	12.5–	.325
1	.28125	9/32	11.25	.289
2	.265625	17/64	10.625	.258
3	.2391	1/4	10.	.229
4	.2242	15/64	9.375	.204
5	.2092	7/32	8.75	.182
6	.1943	13/64	8.125	.162
7	.1793	3/16	7.5	.144
8	.1644	11/64	6.875	.128
9	.1495	5/32	6.25	.114
10	.1345	9/64	5.625	.102
11	.1196	1/8	5.	.091
12	.1046	7/64	4.375	.081
13	.0897	3/32	3.75	.072
14	.0747	5/64	3.125	.064
15	.0673	1/16	2.8125	.057
16	.0598	1/16	2.5	.051
17	.0538	3/64	2.25	.045
18	.0478	3/64	2.	.040
19	.0418	3/64	1.75	.036
20	.0359	1/32	1.50	.032
21	.0329	1/32	1.375	.028
22	.0299	1/32	1.25	.025
23	.0269	1/32	1.125	.022
24	.0239	1/32	1.	.020
25	.0209	1/64	.875	.018
26	.0179	1/64	.75	.016
27	.0164	1/64	.6875	.014
28	.0149	1/64	.625	.013
29	.0135	1/64	.5625	.011
30	.0120	1/64	.5	.010
31	.01094	1/64	.4375	.009
32	.01016	1/64	.40625	.008

COMPARISON OF THICKNESSES OF SHEET IRON AND COPPER

Copper Thickness			Iron or Steel Thickness		
Gauge by Wt. in Oz.	Decimal Thickness	Pounds per Sq. Ft.	Gauge U.S.S.	Decimal Thickness	Pounds per Sq. Ft.
8	.0108	1/2	31	.0109	.4375
9	.0120	9/16	30	.0120	.50
10	.0135	5/8	29	.0135	.5625
11	.0146	11/16	28	.0149	.6875
12	.0162	3/4	27	.0164	.6875
13	.0173	13/16	26	.0179	.750
14	.0189	7/8	—	—	—
15	.0202	15/16	25	.0209	.875
16	.0216	1	—	—	—
18	.0243	1 1/8	24	.0239	1.
20	.0270	1 1/4	23	.0269	1.125
24	.0324	1 1/2	21	.0329	1.375
28	.0378	1 3/4	20	.0359	1.50
32	.0432	2	19	.0418	1.75
36	.0486	2 1/4	18	.0478	2.
40	.0540	2 1/2	17	.0538	2.25
44	.0596	2 3/4	16	.0598	2.5
48	.0648	3	15	.0673	2.8125
56	.0756	3 1/2	14	.0747	3.125
64	.0864	4	13	.0897	3.75
72	.0972	4 1/2	—	—	—
80	.1080	5	12	.1046	4.375
88	.1188	5 1/2	11	.1196	5.
—	.1296	6	10	.1345	5.625
—	.1404	6 1/2	9	.1495	6.25
—	.1512	7	9	.1495	6.25
—	.1620	7 1/2	8	.1644	6.875
—	.1728	8	7	.1793	7.5

NONFERROUS METAL GAUGE

Brown and Sharpe gauge is used for measuring aluminum, brass, and all other nonferrous metals.

Gauge	Thickness	Pounds per Sq. Ft.	Gauge	Thickness	Pounds per Sq. Ft.
10	.1019	1.44	26	.0159	.225
	.1000	1.41	27	.0142	.200
	.0938	1.32	28	.0126	.178
11	.0907	1.28	29	.0113	.159
	.0900	1.27	30	.0100	.141
12	.0808	1.14			
	.0800	1.13			
	.0781	1.10			
13	.0720	1.01			
	.0703	.990		Tin Plate	
	.0700	.986			
14	.0641	.903	Tin Plate		Pounds per
	.0625	.880	Number	Thickness	Sq. Ft.
	.0600	.845			
15	.0571	.804	1 C	.0125	.491
	.0563	.793	1 X	.0156	.620
	.0550	.775	2 X	.0189	.711
16	.0508	.716	3 X	.0203	.804
	.0500	.704	4 X	.0150	.638
17	.0453	.638			
	.0450	.634			
	.438	.617			
18	.0403	.568			
	.0400	.563	D C	.0150	.638
	.0375	.528	D X	.0203	.827
	.0360	.507	D 2 X	.0250	.964
19	.0359	.506	D 3 X	.0281	1.102
20	.0320	.450	D 4 X	.0313	1.231
21	.0285	.401			
22	.0253	.357			
23	.0226	.318	NOTE: The above is thickness of black sheet iron before tinning.		
24	.0201	.283			
25	.0179	.252			

METAL GAUGE AND RIVET SIZES

GAUGE OF METAL USED FOR ROUND PIPES OR RECTANGULAR DUCTS

Size in Inches	1 to 12	13 to 30	31 to 42	43 to 60	61 and up
Gauge	26	24	22	20	18

FLATHEAD RIVETS

Size Weight per 1000 Rivets Oz. and Lb.	Diameter	Length
4 oz.	.070	$1/8$
6	.080	$9/64$
8	.090	$5/32$
10	.094	$11/64$
12	.101	$3/16$
14	.109	$3/16$
1 lb.	.115	$13/64$
$1\,1/4$	.120	$7/32$
$1\,1/2$	.125	$15/64$
$1\,3/4$	.133	$1/4$
2	.140	$17/64$
$2\,1/2$	.147	$9/32$
3	.160	$5/16$
$3\,1/2$	.163	$21/64$
4	.173	$11/32$
5	.185	$3/8$
6	.200	$25/64$
7	.215	$13/32$
8	.225	$7/16$
9	.230	$29/64$
10	.233	$15/32$
12	.253	$1/2$
14	.275	$33/64$
16	.293	$17/32$

MEASURES

Linear Measure

12 inches = 1 foot
3 feet   = 1 yard

The word inch or inches may be indicated by the abbreviation *in.*, or by the symbol ("). A measurement of 3 inches is expressed as 3 in. or 3". The abbreviation for foot, or its plural, feet, is *ft.*; the symbol is ('). The measurement of 4 feet would be expressed as 4 ft. or 4'.

Board Measures

A board foot (bd. ft.) is equal to a piece of lumber one foot long, one foot wide, and one inch thick: 12" × 12" × 1".

Square Measure

A square foot is equal to the length times the width.

(12" × 12") 144 square inches (sq. in.) = 1 square foot (sq. ft.)
(3' × 3')   9 square feet   = 1 square yard (sq. yd.)
            100 square feet   = 1 square

Cubic Measure

Cubic measure deals with three dimensions: length, width, and thickness. The abbreviation for cubic inches is (cu. in.). In square measures we find that a square foot is equal to length × width (12" × 12" = 144 sq. in.). In cubic inches we must first obtain the square inches. The cubic inches are obtained by multiplying the product of the base or one side by the height or thickness, such as length times width times the height. The cubic inches in a 12-inch cube are equal to 12 × 12 × 12 or 1728 cubic inches (or 12 × 12 = 144 sq. in. × 12 = 1728 cu. in.).

1728 cubic inches (cu. in.) = 1 cubic foot (cu. ft.)
27 cubic feet   = 1 cubic yard (cu. yd.)

Avoirdupois or Commercial Weight

16 ounces (oz.) = 1 pound (lb.)
100 pounds = 1 hundredweight (cwt.)
20 hundredweight = 1 long ton (T) = 2240 pounds

Liquid Measure

16 fluid ounces = 1 pint
2 pints   = 1 quart
4 quarts   = 1 gallon
$3\frac{1}{2}$ gallons   = 1 barrel

A gallon of water at 62 deg. F. weighs 8.3356 pounds.
The U.S. gallon contains 231 cubic inches.

Dry Measure

2 pints or 57.75 cu. in. = 1 quart
4 quarts or 231 cu. in. = 1 gallon
2 gallons or 8 quarts = 1 peck
4 pecks or 2150.42 cu. in. = 1 bushel
1 gallon = 231 cu. in. or 1.34 cu. ft.
7.48 gallons = 1 cu. ft.
1 liquid quart = $57\frac{3}{4}$ cu. in.
1 gallon water = $8\frac{1}{3}$ lb.
1 gallon gasoline = 5.84 lb.
1 gallon linseed oil = 7.84 lb.
1 cu. ft. of water weighs $62\frac{1}{2}$ lb.
1 cu. ft. of ice weighs 57.25 lb.

Angles are measured in degrees, minutes, and seconds.

1 degree = 60 minutes
1 minute = 60 seconds
1 circle = 360 degrees

Symbols are commonly used as follows:

Degrees = °
Minutes = ′
Seconds = ″

For example, 35 degrees, 23 minutes, and 41 seconds is written 35° 23′ 41″.

DECIMALS TO FRACTIONS

In calculating sheet-metal work, the results often are in decimals. When laying out a job with decimals involved, they must be changed to fractions so they can be read on a rule. This may be done mentally or mathematically.

To determine the fractional equivalents, the results often need to be carried only to the nearest $\frac{1}{16}$ of an inch. If, therefore, the results obtained in an answer have decimals, reduce them to the nearest $\frac{1}{16}$ of an inch.

To do this, multiply the decimal portion of the answer by 16. Thus, if the answer is 4.82, multiply the .82 by 16. $.82 \times 16 = 13.12$ sixteenths, say $\frac{13}{16}$ in. The answer, then, is given the value of $4\frac{13}{16}$.

To change a decimal to 8ths, multiply the decimal by 8. To change a decimal to 32nds, multiply the decimal by 32.

These conversions can be made more readily if the following table is committed to memory:

.875 in. = $\frac{7}{8}$ in.   .75 in. = $\frac{3}{4}$ in.   .5 in. = $\frac{1}{2}$ in.
.250 in. = $\frac{1}{4}$ in.   .125 in. = $\frac{1}{8}$ in.   .0625 in. = $\frac{1}{16}$ in.
.03125 in. = $\frac{1}{32}$ in.

DECIMAL EQUIVALENTS OF ONE INCH

$1/64$ = .015625	$11/32$ = .34375	$11/16$ = .6875
$1/32$ = .03125	$23/64$ = .359375	$45/64$ = .703125
$3/64$ = .046875	$3/8$ = .375	$23/32$ = .71875
$1/16$ = .0625	$25/64$ = .390625	$47/64$ = .734375
$5/64$ = .078125	$13/32$ = .40625	$3/4$ = .75
$3/32$ = .09375	$27/64$ = .421875	$49/64$ = .765625
$7/64$ = .109375	$7/16$ = .4375	$25/32$ = .78125
$1/8$ = .125	$29/64$ = .453125	$51/64$ = .796875
$9/64$ = .140265	$15/32$ = .46875	$13/16$ = .8125
$5/32$ = .15625	$31/64$ = .484375	$53/64$ = .828125
$11/64$ = .171875	$1/2$ = .5	$27/32$ = .84375
$3/16$ = .1875	$33/64$ = .515625	$55/64$ = .859375
$13/64$ = .203125	$17/32$ = .53125	$7/8$ = .875
$7/32$ = .21875	$35/64$ = .546875	$57/64$ = .890625
$15/64$ = .234375	$9/16$ = .5625	$29/32$ = .90625
$1/4$ = .25	$37/64$ = .578125	$59/64$ = .921875
$17/64$ = .265625	$19/32$ = .59375	$15/16$ = .9375
$9/32$ = .28125	$39/64$ = .609375	$61/64$ = .953125
$19/64$ = .296875	$5/8$ = .625	$31/32$ = .96875
$5/16$ = .3125	$41/64$ = .640625	$63/64$ = .984375
$21/64$ = .328125	$21/32$ = .65625	1 = 1.
	$43/64$ = .671875	

CIRCUMFERENCES AND AREAS OF CIRCLES

	Of One Inch			Of Inches or Feet		
Fract.	Dec.	Circ.	Area	Dia	Circ.	Area
1/64	.015625	.04909	.00019	1	3.1416	.7854
1/32	.03125	.09818	.00077	2	6.2832	3.1416
3/64	.046875	.14726	.00173	3	9.4248	7.0686
1/16	.0625	.19635	.00307	4	12.5664	12.5664
5/64	.078125	.24545	.00479	5	15.708	19.635
3/32	.09375	.29452	.00690	6	18.850	28.274
7/64	.109375	.34363	.00939	7	21.991	38.485
1/8	.125	.39270	.01227	8	25.133	50.265
9/64	.140625	.44181	.01553	9	28.274	63.617
5/32	.15625	.49087	.01917	10	31.416	78.540
11/64	.171875	.53999	.02320	11	34.558	95.033
3/16	.1875	.58905	.02761	12	37.699	113.10
13/64	.203125	.63817	.03241	13	40.841	132.73
7/32	.21875	.68722	.03758	14	43.982	153.94
15/64	.234375	.73635	.04314	15	47.124	176.71
1/4	.25	.78540	.04909	16	50.265	201.06
17/64	.265625	.83453	.05542	17	53.407	226.98
9/32	.28125	.88357	.06213	18	56.549	254.47
19/64	.296875	.93271	.06922	19	59.690	283.53
5/16	.3125	.98175	.07670	20	62.832	314.16
21/64	.328125	1.0309	.08456	21	65.973	346.36
11/32	.34375	1.0799	.09281	22	69.115	380.13
23/64	.359375	1.1291	.10144	23	72.257	415.48
3/8	.375	1.1781	.11045	24	75.398	452.39
25/64	.390625	1.2273	.11984	25	78.540	490.87
13/32	.40625	1.2763	.12962	26	81.681	530.93
27/64	.421875	1.3254	.13979	27	84.823	572.56
7/16	.4375	1.3744	.15033	28	87.965	615.75
29/64	.453125	1.4236	.16126	29	91.106	660.52
15/32	.46875	1.4726	.17257	30	94.248	706.86
31/64	.484375	1.5218	.18427	31	97.389	754.77
1/2	.5	1.5708	.19635	32	100.53	804.25
33/64	.515625	1.6199	.20880	33	103.67	855.30
17/32	.53125	1.6690	.22166	34	106.81	907.92
35/64	.546875	1.7181	.23489	35	109.96	962.11
9/16	.5625	1.7671	.24850	36	113.10	1017.88
37/64	.578125	1.8163	.26248	37	116.24	1075.21
19/32	.59375	1.8653	.27688	38	119.38	1134.11
39/64	.609375	1.9145	.29164	39	122.52	1194.59
5/8	.625	1.9635	.30680	40	125.66	1256.64
41/64	.640625	2.0127	.32232	41	128.81	1320.25

CIRCUMFERENCES AND AREAS OF CIRCLES

	Of One Inch				Of Inches or Feet	
Fract.	Dec.	Circ.	Area	Dia.	Circ.	Area
21/32	.65625	2.0617	.33824	42	131.95	1385.44
43/64	.671875	2.1108	.35453	43	135.09	1452.20
11/16	.6875	2.1598	.37122	44	138.23	1520.53
45/64	.703125	2.2090	.38828	45	141.37	1590.43
23/32	.71875	2.2580	.40574	46	144.51	1661.90
47/64	.734375	2.3072	.42356	47	147.65	1734.94
3/4	.75	2.3562	.44179	48	150.80	1809.56
49/64	.765625	2.4054	.45253	49	153.94	1885.74
25/32	.78125	2.4544	.47937	50	157.08	1963.50
51/64	.796875	2.5036	.49872	51	160.22	2042.82
13/16	.8125	2.5525	.51849	52	163.36	2123.72
53/64	.828125	2.6017	.53862	53	166.50	2206.18
27/32	.84375	2.6507	.55914	54	169.65	2290.22
55/64	.859375	2.6999	.58003	55	172.79	2375.83
7/8	.875	2.7489	.60132	56	175.93	2463.01
57/64	.890625	2.7981	.62298	57	179.07	2551.76
29/32	.90625	2.8471	.64504	58	182.21	2642.08
59/64	.921875	2.8963	.66746	59	185.35	2733.97
15/16	.9375	2.9452	.69029	60	188.50	2827.43
61/64	.953125	2.9945	.71349	61	191.64	2922.47
31/32	.96875	3.0434	.73708	62	194.78	3019.07
63/64	.984375	3.0928	.76097	63	197.92	3117.25

Of Inches or Feet			Of Inches or Feet		
Dia.	Circ.	Area	Dia.	Circ.	Area
64	201.06	3216.99	96	301.59	7238.23
65	204.20	3318.31	97	304.73	7389.81
66	207.34	3421.19	98	307.88	7542.96
67	210.49	3525.65	99	311.02	7697.69
68	213.63	3631.68	100	314.16	7853.98
69	216.77	3739.28	101	317.30	8011.85
70	219.91	3848.45	102	320.44	8171.28
71	223.05	3959.19	103	323.58	8332.29
72	226.19	4071.50	104	326.73	8494.87
73	229.34	4185.39	105	329.87	8659.01
74	232.48	4300.84	106	333.01	8824.73
75	235.62	4417.86	107	336.15	8992.02
76	238.76	4536.46	108	339.29	9160.88
77	241.90	4656.63	109	342.43	9331.32
78	245.04	4778.36	110	345.58	9503.32
79	248.19	4901.67	111	348.72	9676.89

CIRCUMFERENCES AND AREAS OF CIRCLES (CONT.)

Of Inches or Feet			*Of Inches or Feet*		
Dia.	Circ.	Area	Dia.	Circ.	Area
80	251.33	5026.55	112	351.86	9852.03
81	254.47	5153.00	113	355.00	10028.75
82	257.61	5281.02	114	358.14	10207.03
83	260.75	5410.61	115	361.28	10386.89
84	263.89	5541.77	116	364.42	10568.32
85	267.04	5674.50	117	367.57	10751.32
86	270.18	5808.80	118	370.71	10935.88
87	273.32	5944.68	119	373.85	11122.02
88	276.46	6082.12	120	376.99	11309.73
89	279.60	6221.14	121	380.13	11499.01
90	282.74	6361.73	122	383.27	11689.87
91	285.88	6503.88	123	386.42	11882.29
92	289.03	6647.61	124	389.56	12076.28
93	292.17	6792.91	125	392.70	12271.85
94	295.31	6939.78	126	395.84	12468.98
95	298.45	7088.22			

INDEX

B

Bend allowances, 270
Bullhead Y branch, 138

C

Circumferences and areas of circles, 290
Cone, forming in roller, 264
Conical-cap cutout, 264, 266
Conveyor chute, spiral, 258
Copper and sheet iron, comparison of thickness, 283

D

Decimal equivalents of one inch, 289
Decimals to fractions, 288
Diameter of main on two- or three-pronged fitting, obtaining, 268
Dividing line into equal spaces, 2; line into four, eight, or sixteen equal spaces, 4; line into six, twelve, or twenty-four equal spaces, 4; quarter circles, 2
Double-offset rectangular-to-round, 66
Double-offset transformer in one piece, 112

E

Elbow, center segment tapered, 238; equal-tapering, rectangular-to-round, 250; equal-tapering round-to-oblong, 254; equal-tapering with center radius, 240; equal-tapering, with heel radius, 246; equal-tapering with throat radius, 244; irregular-shaped, 96; oblong-to-rectangular, three-piece, 78; round, quarter pattern, 234; tapering one end, 236; three-piece, oblong-to-rectangular, 80; three-piece, rectangular-to-round, 74; three-piece, rectangular-to-round, one side straight, 76; transition, curved heel and throat patterns, 100; transition, double-flare, 94; transition, irregular-to-rectangular, 98; transition, rectangular-to-rectangular, 84; transition, rectangular-to-rectangular, equal taper, 86; transition, with slant throat, 88, 90, 92
Elbow boot, center taper, 50; offset, 54; on center, slant throat, 56, 58; one side straight, 52; rectangular-to-round, with curved back, 82; straight on one side, slant throat, 60; two-piece, 72
Elbows, two, entering a tapering base, 232
Ellipse area, 274
Ellipse, drawn by compass, 6; drawn by use of string, 6
Elliptical-to-round taper joint, 42; one side straight, 44

F

Flat-crotch Y branch, 176; one side straight, 190
Four-pronged Y branch, 204

G

Gable-pitched rectangular-to-round, 22
Galvanized iron, thickness of, 282

I

Irregular-shaped elbow, 96
Irregular-to-rectangular transition, elbow, 98

L

Line, dividing into equal spaces, 2; dividing into four, eight, or sixteen equal spaces, 4; dividing into six, twelve, or twenty-four equal spaces, 4

M

Measures, avoirdupois or commercial weight, board, cubic, linear, liquid, square, 286; dry, 287
Metal gauge, and rivet sizes, 285; nonferrous, 284
Metal lost in rolling, 272

N

Nonferrous metal gauge, 284

O

Oblong-to-oblong transformer, 46, off center, 48
Oblong-to-rectangular elbow, three-piece, 80
Oblong-to-round offset, 130, 132
Offset, double, rectangular-to-round, 28, 136; double, tapering, 134; oblong-to-round, 130, 132; rectangular-to-round, 26; round-tapering, 128; twisted rectangular, 120
Offset elbow boot, 54

P

Pipe length between two angles, finding, 270

Q

Quarter circles, dividing, 2

R

Rectangular T, 45-deg., on cylinder, 212
Rectangular-to-irregular-shaped transition, 102, 106
Rectangular-to-rectangular elbow transition, 84; equal taper, 86
Rectangular-to-round, 14, 16; gable-pitched, 22; one side straight, 24; pitch at the top, 62; straight on one side, pitch at the top, 64; with pitch at base, 30
Rectangular-to-round double offset, 28, 136
Rectangular-to-round elbow, three-piece, 74
Rectangular-to-round elbow boot with curved back, 82

INDEX

Rectangular-to-round 50-deg. double offset, 66
Rectangular-to-round 50-deg. twisted double offset, 68
Rectangular-to-round offset, 26
Rectangular-to-round three-piece elbow, one side straight, 76
Rectangular-to-triangular transition, 104
Right angles, obtaining, 268
Round elbows, quarter pattern, 234
Round equal taper joint, 34
Round pipes, aligning when welding, 264
Round tapering offset, 128
Round taper joint, one side straight, 38; two sides straight, 40
Round taper with base mitered, 126
Round-to-oblong Y branch, 150, 152, 154, 156
Round-to-rectangular Y branch, 168, 170, 172, 174
Round Y branch, one side flat, 180, 182, 184, 186, 188; with center splitter, 178

S

Seam, grooved, on cylinder off center, 278; grooved, on cylinder on center, 278; grooved-lock, 276; lock, hammer-grooved, 276; riveted-lap, 274
Sheet iron and copper, comparison of thickness, 283
Ship's ventilator with round mouth, 248
Small and large ends, 272
Square-to-intersecting hip on roof, 70
Square-to-round, 10, 12, 18; four-piece, 20; with fan bracket, 32
Streamlined Y branch, 208; three-way, 210; two-way, 206

T

T, 45 deg. on taper joint, 228; 45 deg. on taper joint, straight on one side, 220, 230; rectangular, 45 deg. on cylinder, 212; tapering, 40 deg. on taper joint, one side flat, 198; tapering, 45 deg. flat on one side, 226; tapering, 45 deg. on cylinder, 222; tapering, 45 deg. on cylinder off center, 224; tapering, one side straight, on cylinder, 214; tapering 30 deg. on a taper joint, 196; tapering offset, on cylinder, 216; tapering offset, straight on one side on cylinder, 218
Taper, round, with base mitered, 124, 126; small and large ends, 36
Tapering double offset, 134
Tapering-end elbow, 236
Taper joint, elliptical-to-round, 42; elliptical-to-round, one side straight, 44; round, equal, 34; round, one side straight, 38; round, two sides straight, 40
Taper-joint pattern, 36
Three-pronged Y branch, flat-crotched, 192; 30 and 45 deg., 194; two at 35 deg., 200
Three-way Y branch, 202
Toolbox, flat-cover, 262; pitched-cover, 260
Transformer, double-offset, in one piece, 112; in one piece, 108, 110, 112, 114; oblong-to-oblong, 46; oblong-to-oblong, off center, 48; quarter round-to-rectangular, 122; twisted, 118
Transition, rectangular-to-irregular-shaped, 102, 106; rectangular-to-triangular, 104
Transition elbow, double-flare, 94; with curved heel and throat patterns, 100; with slant throat, 88, 90, 92
True-length triangles, illustrating, 8
Twisted double-offset rectangular-to-round, 68
Twisted rectangular offset, 120
Twisted transformer, 118
Two-quart measure, 256

W

Wire edge, forming, 280

Y

Y branch, bullhead, 138; different diameters, 146, 148; different spreads, 144; equal spread, 140, 142, 160; flat crotch, 176; four-pronged, 204; horizontal and vertical branches, 164, 166; horizontal branches, 162; horizontal branches, flat on one side, 184; one side straight, flat-crotch, 190; round, one side flat, 180; round-to-oblong, 150, 152, 154, 156; round-to-rectangular, 168, 170, 172, 174; round-to-rectangular, flat on one side, 186, 188; round with center splitter, 178; streamlined, 208; streamlined, three-way, 210; streamlined, two-way, 206; 30-deg., 158; 30-deg., flat on one side, 182; three-pronged, flat-crotch, 192; three-pronged, 30 and 45 deg., 194; three-pronged, two at 35 deg., 200; three-way, 202